The Genesis Grid

P.M. Woolford

authorHOUSE®

AuthorHouse™ UK Ltd.
500 Avebury Boulevard
Central Milton Keynes, MK9 2BE
www.authorhouse.co.uk
Phone: 08001974150

First published by AuthorHouse 4/8/2010.

ISBN: 978-1-4490-8117-1 (sc)

This book is printed on acid-free paper.

THE GENESIS GRID by P.M. Woolford

PART ONE: THE DISCOVERY

Chapter 1:	The Meteor	7
Chapter 2:	The little Horn	15
Chapter 3:	The Mystery of 11	21
Chapter 4:	Heavenly Bodies	25
Chapter 5:	Unearthing the number 28	31
Chapter 6:	Prime Numbers of the Bible	35
Chapter 7:	A first Grid	37
Chapter 8:	A Picture Pattern	41
Chapter 9:	A 19 year Middle East Cycle	45
Chapter 10:	Tongues of Gold	51
Chapter 11:	An Abstract Machine	55
Chapter 12:	The Deity String	63
Chapter 13:	Mesmeric Eyes	69
Chapter 14:	The Signal	73
Chapter 15:	A self-similar Gap	79
Chapter 16:	Three defining numbers of the Grid in pi	83
Chapter 17:	Segments in Genesis	87
Chapter 18:	The Great Breach	91
Chapter 19:	Genesis: Game, Set & Match	97
Chapter 20:	The Two Mounts	103
Chapter 21:	John confirms the Grid	107
Chapter 22:	Commonplace Patterns	115

PART TWO: WHO AND WHY?

Chapter 23:	Tides that encode the Grid	119
Chapter 24:	Grid vibrations	123
Chapter 25:	Numbers: Figment or Fact	129
Chapter 26:	The Ovum & Spermatozoon	139

Chapter 27: And then there were Two   147
Chapter 28: Never before seen, the number 28 155
Chapter 29: Revelation encodes the Grid   163
Chapter 30: The Pattern of a Throne   167

PART THREE: OUTCOMES

Chapter 31: A brief history of History   175
Chapter 32: Decline and fall   189
Chapter 33: The Prophetic Grid   203
Chapter 34: A strange 2003   209
Chapter 35: The Crisis   215
Chapter 36: The Lady   221
Chapter 37: Germani   233
Chapter 38: Samaria on Thames   239
Chapter 39: Kelvedon Hatch   253
Chapter 40: Rome in the Crosshairs   269
Chapter 41: The Sickness   275
Chapter 42: Two Americans   299
Chapter 43: The God Brand   305
Chapter 44: The Mark   321

PART FOUR: INSIGHTS

Chapter 45: Subterranean numbers   331
Chapter 46: A Secret Architecture   335
Chapter 47: The Rainbow   339
Chapter 48: That they may be taken   345
Chapter 49: The 1711171 system   355
Chapter 50: The Deity Prime Sequence   365
Chapter 51: The proof of Eleven   369
Chapter 52: The key pattern 888   373
Chapter 53: The Ultimate Plan   379
Chapter 54: The Alien   383
Chapter 55: Bullinger's Question   387

APPENDICES		RELATING TO CHAPTER

I	Claimed Bible codes	4: *Heavenly Bodies*
II	28 attacks on teachers	5: *Unearthing the number 28*
III	About the number 19	9: *A 19 year Middle East Cycle*
IV	107 occurrences of the Father	21: *John confirms the Grid*
V	Four further Grid verifications	22: *Commonplace Patterns*
VI	Further patterns in music	24: *Grid Vibrations*
VII	Three to the power of four law	25: *Numbers: Figment or Fact*
VIII	Mathematical findings	25: *Numbers: Figment or Fact*
IX	More on the number 44	26: *The Ovum & Spermatozoon*
X	Mysteries of the Godhead	27: *And then there were Two*
XI	The Grid in Revelation	29: *Revelation encodes the Grid*
XII	The Snake on a Pole	36: *The Lady*
XIII	Migrations to the British Isles	38: *Samaria on Thames*
XIV	Global Warming	42: *Two Americans*
XV	The Great White Throne	49: *The 1711171 System*
XVI	The 888 name pattern	52: *The Key pattern 888*

ABOUT THE AUTHOR

"He still does the minimum possible work and seems unable to turn more than a fraction of his mind to it even at his most attentive moments. The remainders of his intellectual powers seem to be as inaccessible to him as they are to me."
J.R.B., February 1968, W.B.G.S. Mathematics school report.

INTRODUCTION

Why do some people experience strange coincidences or feel inexplicable hunches? Maybe they have premonitions, telepathic experiences or witness psychic events; some predict the manner of their own death.

People have found the answer to a vexing scientific question on an impulse, visiting a library to search a corner they've never been in before. Picking up the nearest book they find that it contains detailed information they needed.

Some people take a book on holiday. Imagine a small ramshackle caravan, the roof held up with a wooden pole, on the shores of Scotland's Loch Lomond. Fleeing from the clattering rain you crouch over your book with some lukewarm coffee, trying to ignore the biting midges. Soon you find something highly irregular. Fifteen pages into the nineteenth century book *Number in Scripture* a phenomenal and intriguing discovery is shown: one which baffled its author. He had identified a definite but camouflaged numerical tag for the two leading personalities of the Bible. Such a compelling finding demands an explanation, yet none was offered. Then in a flash you yourself conceive the beginnings of an answer.

Back at home you begin to research. It soon becomes an obsession. You discover things undreamt of, notions so bizarre they would seem to be at the outer boundaries of science fiction. Is it conceivable that the first chapter of the book of Genesis in the Bible could be built on the first 31 digits of pi? Does the human race carry a cryptic message in its reproductive cells? What of a link between the Great Recession of 2007-09, the Vatican and a signal from another world?

One could expect any mathematician, physicist, economist or biologist to debunk such outlandish ideas, but PhD qualified scholars representing all of those professions have reviewed the Grid discovery. They have found no fault.

Some of the discoveries involve events in ancient history: others reveal frightful disasters yet to come. They are encapsulated in six simple numbers and one rather complex one, the number *pi* - the most memorised number in the world today.

At the turn of the sixteenth century the Dutch mathematician Ludolf van Ceulen first calculated pi to thirty-five places. For fully four-hundred years, a message encoded in the lichen encrusted decimals carved into his gravestone has been concealed from the world …until now.

PART ONE: THE DISCOVERY

1

THE METEOR

"Coincidence is God's way of remaining anonymous." Albert Einstein

A chill wind assailed the silent pre-dawn assemblies of ordinary looking folk. There had not been a happening like it in living memory. To the fanciful, the scene had something of a science-fiction feel to it, but it was not to be a story like *The Day of the Triffids,* or *The Andromeda Strain.* This was no lobotomised army massing to unleash unspeakable horror, but citizens fully possessed of their senses.

Yet it will be conclusively proved that what had struck terror across Britain was indeed due to something *out there.*

Huddled in jeans, sweaters and anoraks this crowd of English people faced in the same direction as they waited for the rising Sun. At the onset of the greatest financial crisis in history they were waiting for *their money.* The Northern Rock panic of September 2007 was the first run on a British bank in over one hundred and forty years. It was symbolic of - and a causative factor in - the subsequent global financial holocaust and collapse in confidence. Tremors had been heard a few weeks before in certain European institutions, but the worldwide alarm that greeted the seismic collapse of Northern Rock was unique in modern times. Its impact on the international scene was meteoric.

It was as if the world had been struck by a disease carrying meteor, releasing financial contagion around the globe. From that time on, week by interminable week, successive hammer blows rained down on families, businesses, banks and governments the world over. Every dire warning and prediction from economists was exceeded by events. Soon the international community faced the greatest economic catastrophe ever known, outstripping in its rapacious speed even the onset of the Great Depression. It was as if a giant cosmic screwdriver had been jabbed into the heart of the entire global system, smashing its mechanism.

The timing of the Northern Rock debacle was particularly strange: it came *exactly* four years after another astonishing - and rather sinister - event on planet earth. ROME'S DARKEST DAY, the front page of London's *Daily Mail* had proclaimed. On that occasion the entire Italian peninsula had been plunged into darkness, disrupting the ailing Pope John Paul's ordination ceremony for 31 new Cardinals. It had been the final blow in a sequence of eight huge and inexplicable power cuts across the globe. These eight power cuts had unfolded over a period of 44 days. The eighth struck Rome in the night. But the amazing thing was this: the worldwide financial calamity arrived *the very hour* four years later! Four years after the Pope's humiliation: that is, according to the calendar of the Bible, as will be later explained.

In a meeting with the newly elected President Obama, the global crisis that followed was described by the UK Prime Minister:

> *"It's like a power cut that went through the global banking system." Gordon Brown, March 2009*

Yet within a few weeks of that meeting the crisis had mysteriously abated. A mere nineteen months after the Northern Rock disaster, the sense of fear and foreboding melted away like the morning mist.

Back in 2003 the series of eight electrical power cuts had started in the great economic power centre of Manhattan, where Times Square is 'the crossroads of the world.' That first power cut, considered at the time to be the largest in history, spread into other major US cities as far away as Detroit and even into Canada. It was only the first of a set of eight. Within a few days the Asian republic of Georgia was stricken, to be followed by Helsinki, London, Kuala Lumpur and Cancun, most of Sweden and Denmark and finally Rome: eight vast power cuts had struck huge regions across the planet in rapid succession. The world's *two great power centres*, both temporal and spiritual - New York and Rome - had been humbled first and last.

Although they made front page news, nobody seemed to notice that in total *eight* power cuts occurred over 44 days. Perhaps the strangest thing of all was that the eight *locations* described a giant figure of eight across planet earth, with Rome sitting at the centre cross. Incredibly, following that pattern, the total of the positional numbers at successive points equaled 44! If this was a 'sign from someplace' what could it possibly mean?

These strange power cuts occurred in 2003, a year that has proved to be an historic turning point for the world. March had seen the Iraq invasion, at exactly the same time that an obscure Iranian professor of civil engineering was elected Mayor of Tehran. He was soon to become President Mahmoud Ahmadi-Nejad of Iran. All efforts by the international community to curb the nuclear ambitions of this maverick leader have failed.

Ahmadi-Nejad's election (and subsequent re-election) was a consequence of the Iraq invasion. In July 2007 the Iraqi Prime Minister's security advisor predicted *"There is going to be a hurricane in Iraq..."* Iran stands to gain territorially from such. At the close of 2006 that nation hosted an international conference, *Global Vision*, for the express purpose of denying the Jewish Holocaust. Describing the State of Israel as a *"tumour"* and a *"stinking corpse"* Iran's bombastic leader has stated his intention to *"wipe Israel off the face of the Earth."* Such vitriol has not been spouted by any elected politician since the 1930's. Harvard law Professor Alan Dershowitz has branded Ahmadi-Nejad *"The Hitler of the 21st Century."*

Soon after the Global Vision fiasco the UN imposed sanctions forbidding the export of nuclear materials to Iran. Ahmadi-Nejad's response was to predict that Britain was *"doomed to disappear"* along with Israel and America. In his view the three countries are to *"vanish like the Pharaohs"* of Egypt as they have *"moved far away from the teachings of God."*

By the time Gordon Brown became British Prime Minister the US presidency had reached its nadir. Even the unlamented ex-President Jimmy Carter could brand that second Bush administration *"...the worst in history...we now have endorsed the concept of pre-emptive war...even though our own security is not directly threatened."*

After squandering international good will and credibility in Iraq the Anglo-Saxon nations were powerless to deal with Iran. The emerging consensus was that a nuclear device would be required to halt the Iranian uranium enrichment program. The front cover of *The New Statesman* of 16th April 2007 headlined: *"We are being led into the most serious crisis in modern history."* Influential former speaker of the US House of Representatives Newt Gingrich said

"We are in the early stages of World War III."

But the world soon grows accustomed to firebrand leaders. The Credit Crunch was to bounce Ahmadi-Nejad from the front pages of the newspapers. Disquiet over Iran was all too easy to forget in the wake of Northern Rock, Bear Sterns, Lehman's, and the collapse of giant US mortgage corporations Fannie Mae and Freddie Mack. These events were followed by the failure of the Icelandic banking system, and the hasty launch of several emergency rescue packages cobbled together by governments of major nations. As the largest dominoes of the world's banking system began to topple the crisis over Iran was sidelined. Central banks pumped trillions into a teetering banking system as giants 'too big to fail' fell anyway. Leading economists invoked such terms as *scary; panic; toxic; a perfect storm; a vicious chain reaction, a devastating domino effect and a financial tsunami.* The deputy governor of the Bank of England, Charles Bean, described it as

"The greatest financial crisis in human history."

Then, almost unnoticed, there came a pivotal moment in the 21st century: for the first time, the eurozone countries overtook the USA in Gross Domestic Product. That event prompts two vital questions: firstly, how much longer can a discredited America maintain world leadership? US retail tycoon Howard Davidowitz put it this way:

"As a country we are out of control, we are in a death spiral."

Social scientist and New York Times best-selling author Martin L. Gross considers that the U.S. is heading for *"national suicide"* due to overspending and high taxes. Gross has testified before six Congressional committees on government inefficiency.

Secondly, how long can the weakened dollar hold its own as the world's reserve currency? That position faces almost weekly challenges from global political figures and central bankers. But while the frenzied thoughts of Western leaders focus on economic survival, might cooler minds be engineering the stealthy emergence of a Middle Eastern Shiite kingdom? Although Arab unity remains a dream, could the Persians of Iran succeed in using the sinew of religion - beginning in Iraq - to bind together a new fundamentalist empire?

Islam is consolidating its power and influence in every Western nation by building mosques, particularly in Germany. In 1997 Turkish Prime Minister Recep Tayyip Erdogan stated: *"The minarets are our lances, the domes our helmets, the believers our army."* Meanwhile the institutions of the EU are themselves on a mission. Relentlessly, EU bureaucrats seek an ever greater centralisation of power while Pope Benedict XVI labels the capitalism and liberalism of the West 'fascism.' Following the July 2007 Treaty negotiated by Tony Blair, Britain was told it was *"...part of a new European empire by the Brussels bureaucrat who would be emperor"* - a reference to the much derided Spanish Commission President José Manuel Barroso, considered a puppet of the German and French governments and re-elected in September 2009.

Europe has the critical mass to engineer an eventual new world trading system based on the Euro, the new Deutschmark in all but name. Meanwhile China and Russia have advanced the idea of an IMF based currency. The status quo was challenged openly in September 2009 when China purchased $50 billion in IMF bonds. Evidently the terror of what might have happened to the world's financial system has prompted some leaders toward radical policies. The Governor of the Bank of England Mervyn King, speaking to bankers at the city of London's Mansion House, described:

11

"An extraordinary, sudden, severe and simultaneous downturn of activity and trade in every corner of the world economy."

No event, no crisis, comparable to this 'meteor' has ever been recorded. Unlikely though it may sound, proof is given here that the breadth, depth and timing of this staggering world-changing panic was pre-determined by powers outside of our solar system, alien to this world. The crisis had been timed by them to *the very day and the very hour*, as will be amply demonstrated. The period of panic spanned nineteen months from mid-September 2007 to mid-April 2009. Then suddenly, in April, President Ahmadi-Nejad was back on the front pages. An Israeli defence official was quoted as saying:

"Iran will have the bomb within two years...once they have the bomb it will be too late, and Israel will have no choice but to strike - with or without America."

That month Ahmadi-Nejad was true to form at a UN conference where he branded the 'Zionist' Israeli government as racist. British foreign secretary David Miliband responded by saying *"Such hate-filled rhetoric is an intolerable abuse of free speech."*

The appearance of a leader such as Ahmadi-Nejad is not a random event. Iran is now by far the most powerful Middle East nation. Its regime represents the return of an ancient empire with a theocratic vision incompatible with Western and European values. In September 2009 Tehran erupted as tens of thousands of opposition supporters poured onto the streets. From the time of the disputed June elections the authorities waged a brutal war on protestors, with tear gas, batons, sniper fire on the streets and torture in the jails. The *Times* reported one elderly woman at a rally saying:

"They have raped, murdered and tortured our youth after stealing the election. May God's wrath come down on them."

A young female student said:

"The cheating, the raping, the killing and the torture drive you mad. I've come to express my hatred for Ahmadi-Nejad."

Tehran University was cordoned off as the President gave an oration to bussed-in supporters. He delivered another verbal onslaught against Israel, claiming the Holocaust was the West's pretext for establishing a Jewish state and *"a lie based on a mythical and unprovable claim."*

This statement attracted further opprobrium from around the world. Britain's Foreign Secretary David Miliband described them as *"abhorrent"* while the White House called them *"baseless, ignorant and hateful"* adding that Iran must come to the negotiating table, or face isolation over its assumed nuclear weapons program.

Clearly many people in Iran do not want war with Israel; neither would they welcome a pre-emptive military strike that might spray nuclear contamination over a wide area. But this fanatical leader will never relinquish his grip on power, nor will he cease his grandstanding at the UN. He is taking the country ever closer to confrontation. Ahmadi-Nejad wants an eventual war and the US is impotent to do anything about him.

There is a sub-theme to this story of impending, potentially nuclear, war. History offers some fascinating parallels. Just as a pact between Russia and Germany was the prelude to Hitler's first territorial snatch, there are today signs of an alliance with the aggressor state. According to the Moscow-based commentator Dmitry Sidorov:

> *"By blocking sanctions, Moscow is trying to deprive the international community of any leverage against Tehran."*

Sidorov considers this to be the Kremlin's adaptation of Stalin's approach before World War Two, namely, to maneuver opponents into a fight and then move in afterwards and pick up the pieces.

Now that the USA as global policeman is visibly fading away, a massive confrontation between Iran and the EU is inevitable. But we do not face the coming calamity alone. Since 1960 the University of Berkeley has run a program called S.E.T.I. – the *Search for Extra-Terrestrial Intelligence*. Even utilising the Hubble space telescope, the project has proved completely fruitless. An even more powerful instrument, the Kepler Telescope, was launched in March 2009 with the remit to discover planets friendly to life. One scientist described it as *"Trying to find E.T.'s home."* But no meaningful signals have ever been detected from amongst the myriad stars sprinkled across the inky blackness of space.

If there really exists any other race of sentient beings in our galaxy, individuals to whom a human life may seem as fleeting as that of a butterfly to us, why have they remained silent for so long? If such eyes have witnessed our empires galvanizing their masses in military conquest - only to sedate them with religion and sport - to be superseded by yet another civilisation, why didn't these great ones make themselves known to us? If the rhythm of tribes and nations springing up and decaying back like the waves of the seashore has been watched by a species more ancient than ourselves, why have they not contacted us until now?

The assumptions behind the S.E.T.I. program have never been tested. But now there is proof: the Earth recently received a signal from entities more powerful than man. Will these powers intervene to save us from ourselves?

Their fingerprint is 44 and their calling card is pi.

2

THE LITTLE HORN

"Old men are dangerous: it doesn't matter to them what is going to happen to the world." George Bernard Shaw

It had been the largest gathering of heads of state in history. No Roman emperor could have equalled it. Presidents, prime ministers and dignitaries from nearly every nation were crowded together, cheek by jowl, to pay their respects to a colossus of the twentieth century. The funeral in April 2005 of John Paul II, a man now described by his successor as *The Great,* was a truly global event. With four million in attendance it eclipsed in scale the 1965 state funeral of Britain's wartime saviour Winston Churchill.

By the start of the 21st century the prestige of the papacy had reached its zenith in modern times. In John Paul II, the world lost a power broker of unrivalled stature. This was the pope who had defied the Soviet Union with his threat to mobilise the Polish people against invasion, a stance widely considered to have sounded the death knell for the Russian empire. And it had been John Paul, the first non-Italian pope in almost five hundred years, who had cried out for Europe to 'rediscover its roots,' invoking the spirit of Charlemagne.

How did this ancient office and modern power house of the papacy come to exist? Concerning its origin there is here offered an explanation which, if shown to be true, would rock the world of religion to its foundations. Here is a scenario that, once seen to be irrefutable could - in the age of the internet - rapidly shipwreck the faith of millions, if not billions, of people.

Down through the ages, has there been a dark secret withheld from the masses? If so, the unmasking of it could be imminent. Is the unique institution of the papacy in all its power, wealth and grandeur really 'inspired by God' - or is there an alternative explanation?

One clue may be found in a legend concerning the first pope. According to certain ancient records, that pope or 'father' was a magician imbued with the power of *levitation.* The personality of Simon Magus appears just once in the Bible. The book of Acts relates his attempt to buy a position of authority in the primitive Church. The second century Christian apologist Justin Martyr gives what is accepted as a reliable account of the man, revered by his followers as "that great power of God." Early Church history is of course shrouded in mystery, but it is a fact that the religion bearing a Christian label that was crystallised at the Council of Nicaea under Constantine in 325 AD differed radically from the original. Apart from that Christian label it was largely unrecognisable, its teachings and practices having morphed into something *alien* to the holy book it purported to represent.

There have been deviations from holy writings in Islam, but nothing compares to the chicanery of Catholicism and its offshoots. Much blood was spilt in the enforcing of the fourth century Catholic paradigm. But one might reasonably ask how those changes to a corpus of religious belief, determined many centuries ago, could affect today's world. Yet they have affected it drastically.

Before Benedict XVI assumed office in 2005 he had, as Cardinal Ratzinger, already earned himself the nickname of 'the Pope's Rottweiler.' As prefect of the Congregation for the doctrine of the Faith, known prior to 1965 as the Holy office of the Inquisition, he was the successor to many Inquisitors. As Pope he rapidly stirred controversy in many areas by provoking Muslims, debating Evolution, rehabilitating holocaust deniers and denouncing the use of condoms in Africa.

What manner of man is Ratzinger? In choosing the name Benedict he was inspired, at least in part, by the life and works of Pope Benedict XV whose seven year pontificate was overshadowed by the First World War. That pope had been an ardent mariologist, a believer in Mary as 'co-redemptrix with Christ.' He had been devoted to the veneration of Mary through prayer, art, music and architecture. He also authorised a festival: the Feast of Mary

Mediator of all Graces. An emphasis on the theme of Mary veneration had also been maintained by John Paul II. He had himself lauded Mary as the Mediatrix between Christ and man, piling myth upon mystery. How paradoxical it is that, in a supposedly rational age of scientific enquiry, the world can be so powerfully swayed by a belief system so implausible and self-contradictory. Can there be any continuum between ideas many regard as absurd superstition and a modern technological society? Yes, there is a link: since the dark ages it has always been the bid for *power*.

Speaking of power, Pope John Paul II had been struck by the biggest power cut in history, the eighth in a series of huge inexplicable blackouts. Four years later to 'the day and the hour' the Northern Rock experienced the first run on a British bank in over 140 years. This triggered a slide into worldwide panic. The crisis in the world's financial system has brought calls for a new globally coordinated approach. This amounts to a challenge to US economic, and therefore ultimately military, supremacy. It is a call for a new world order.

At the G20 summit meeting in March 2009 the voice of Russia's President Dmitry Medvedev, calling for a new supra-national currency, resulted in a radical new development:

> *"The G20 leaders have activated the IMF's power to create money and begin global quantative easing. In doing so they are putting a de facto world currency into play. It is outside of the control of any sovereign body." The Telegraph, London*

Some kind of one-world entity would seem to be the only way to eliminate the massive global financial imbalances and restore confidence. The influential journal *Foreign Affairs* of the Council on Foreign Relations ran a 2007 article *The End of National Currency*:

> *"National currencies and global markets simply do not mix...countries should abandon monetary nationalism...it is simply incompatible with globalisation...economic development outside the process of globalisation is no longer possible."*

Nations are increasingly leaning towards regional currencies, and the panic induced in the corridors of power can only accelerate the process. Leaders and opinion makers are largely agreed on the severity of the crisis. US President Barack Obama observed that:

"The whole world has been touched by this devastating downturn."

Paul Volker, former Federal Reserve chairman, and chairman of President Obama's economic recovery board said:

"I don't remember anytime, maybe even in the Great Depression, when things went down quite so fast, quite so uniformly around the world."

An editorial in London's *Times* newspaper of April 2009 observed:

"In raw numbers, the global recession of 2009 is as severe as the Depression. Industrial output is collapsing...a sustained fall in prices...would be catastrophic...the Great Depression offers a terrible precedent...hardship, hunger, penury and bankruptcy destroyed lives...so much wealth has been destroyed in the asset-price meltdown that the banking system is unable to extend credit..."

Is a new global system, purged of strangulating debt and institutional distrust, even possible? Which nation or block of nations has the clout to assume the leadership of such a new world-girdling system? On the other hand, how can a wounded USA with unemployment rocketing, a bankrupt car and banking industry, exploding government debt and a cruelly divided body-politic put the world to rights? It no longer has the capacity to act as the world's economic locomotive.

The world may soon be looking for a greater kind of leadership. Can any dignitary on the world scene today move men to put aside narrow self-interest for what is perceived as the greater good? To embrace a great cause, even a compelling myth? A myth can be created from virtually nothing. The myths of Jewish guilt and German omnipotence once galvanised a modern industrial nation. What might be the next great myth to capture the minds of men?

On October 7th 2008 Pope Benedict remarked that the world's financial system was *"built on sand,"* a comment hardly calculated to promote confidence in the existing framework, in which the US dollar acts as the reserve currency. Is another world economic system secretly being prepared? And what schemes might be secretly afoot to wrench world leadership from the USA? It is certainly possible and even likely that the Vatican, the world's greatest perpetrator of myths, could galvanise the hearts and minds of men with some striking initiative or bold proclamation. That a pope might do so is perfectly logical: his perceived moral authority could provide both a diagnosis of economic woe and an inspired path by which nations could see a return to prosperity.

The roots of the present world crisis stretch back through successive post-war recessions. Short sighted crowd-pleasing policies have kept warding off the evil day of a correction. It should be obvious that greed, avarice and deception, as exemplified by certain larger than life personalities on Wall Street, have characterised many of the countless billions of transactions leading up to the greatest economic crisis in history. The Anglo-Saxon policy of unbridled capitalism and unprecedented personal debt, excessive leverage and the deceptive 'slicing and dicing' of repackaged mortgages sold to banks around the globe must take a large part of the blame.

The moral dimension to this debacle has enjoyed very little discussion thus far; that might change as recriminations grow in the years ahead. The time may come when a great religious leader, seemingly above the common fray, could coordinate a 'rescue operation' bringing in a new international order. To do this he would need an accomplice in the political sphere, a close confidant. Is there a leader today who could provide the rhetorical inspiration for a brave new Babel? Hitler was able to stir the German people to unity of purpose, though that was just one nation.

A power vacuum is developing and many leaders are searching for that thing, that great cliché which is rapidly becoming a necessity: a new world order. In the words of Japan's former finance minister Eisuke Sakakibara *"The American Age is over."* Speculation about the future of the dollar is mounting. A Middle Eastern currency basket backed by gold for the Gulf and Arab countries was again under discussion in March 2009. The Arab nations' peg to the dollar, and with it their link to the weak US economy, is slated to be broken in 2010 with the establishment of a new central bank for the Gulf. Both China and Russia have called for a new world currency.

All power is ultimately economic. Has Europe - headed in reality by Germany, still the world's biggest exporter in 2009 - the wealth and the pedigree to catch this falling sword of world power? Writing in the London *Financial Times* May 2009 Wolfgang Münchau, anticipating a decade of recession, commented on the German banking scene:

> *"Governments are not coming clean on the scale of the crisis...write offs would be more than €800 billion...the entire capital and reserves of its monetary and financial institutions were only €441.5 billion...if the leaked number is true, it would mean the German financial system is broke."*

But the American power vacuum cannot last; somebody must step in and fill it. In the *Financial Times* star economist Andy Xie made a sobering prediction that same month for the dollar:

> *"If global stagflation takes hold, as I expect it to, it will force China to accelerate its reforms to float its currency and create a single, independent and market-based financial system. When that happens, the dollar will collapse."*

By August 2009, as Germany began leading Europe out of recession, there was a growing lack of belief in the long term ability of the dollar either to sustain world trade or provide a reliable store of wealth. For a reserve currency, such a perception must eventually alter reality. The dollar will one day be superceded. Could world economic leadership then shift to the continent that has the strongest currency - the new Deutschmark in all but name?

Perhaps this German Pope, like his ancient progenitor, will be a magician too.

3

THE MYSTERY OF 11

"Why is there something, rather than nothing?" Gottfried Leibniz

To make sense of the strange power cuts of 2003 and the world economic power cut that struck in September 2007, it is necessary to review a discovery that has lain dormant for more than a century.

Born 1837 in Canterbury, Kent, the theologian Ethelbert W. Bullinger lived at a time when the privileged classes on both sides of the Atlantic experienced a time of rising optimism, travel, exploration and scientific progress. It was the heyday of a global British Empire and human know how was expanding at an unprecedented pace.

Such was the world in which this nineteenth century theologian began work on a new project involving the numbers of the Bible. Bullinger found that foreign lands were not the only places where groundbreaking knowledge was to be uncovered. Early in his investigations he made a monumental discovery which is the key to this book, *The Genesis Grid*. Bullinger himself should have gone on to solve the questions he raised, but something held him back.

Bullinger's interest in the numbers of the Bible had been sparked by an earlier theologian, Professor M. Mahan, author of *Palmoni - or, the Numerals of Scripture* first published in 1863. That book was heavy reading and the subject of numbers of the Bible had, thought Bullinger, been unduly neglected. Therefore Bullinger embarked on one of the larger projects of his career to investigate, quantify and explain the numbers of the Bible. His book entitled *Number in Scripture* was first published in 1894 and surveyed the Bible to determine, amongst other things, the frequencies with which key words appear.

21

Bullinger uncovered patterns in the regular recurrence of certain names, one in particular. He was astonished to discover that the number 11 was linked to the most important Bible personality of all. This did not accord with his idea of a supreme being. The appearance of the number 11 in connection with the Bible's supreme 'Father' became something for which Bullinger was unable, or unwilling, to attempt an explanation. Could Bullinger have been afraid of the implications? His findings certainly did not sit comfortably with the Trinitarian ideas of mainstream religion.

This present book takes up Bullinger's discovery of Bible number patterns where he left off. Claims have been made before about patterns in the Bible, supposedly revealed by shuffling and reshuffling vast streams of Hebrew letters into thousands of permutations, after which various unrelated names and objects are cherry-picked to show a 'match.' But there can be no mistaking the veracity of the present discovery for which Bullinger pointed the way. The non-mathematician can draw his own conclusions without the help of an expert. This he should do quite capably, because the facts are simple, narrow and specific and invoke only the one basic theme. There can be no arbitrary sifting or shuffling of the sparse information assembled. The jigsaw fits in only one way.

Bullinger was on a search for Bible numbers and the meaning of them. He was searching below the surface, delving down for hidden messages. Counting the occurrences of various words, he discovered that the names Jesus Christ and 'the Father' occurred in multiples of 7 and 11. He made this statement on page 25 of *Number in Scripture*:

> *"Why should it be these two numbers seven and eleven? Why not any other two numbers? Or why two at all? Why not three? We may or may not be able to explain why, but we cannot close our eyes to the fact."*

It is unusual for experts to admit defeat in this way. But one question Bullinger did not ask was: why *any* numbers at all? Yes, why should there be any clearly identifiable multiples attached to the names of these Bible personalities *at all*, and how did such patterns get to be there? Bullinger died not knowing - and maybe not wanting to know - the true answer.

As a theologian he hadn't bargained for *two* numbers connected with the God of the Bible; he wanted and expected three. Finding two numbers was a clash, cutting across all convention. It is this religious 'trinitarian' presumption of three, not two, that has been a boulder concealing the entrance to an Aladdin's cave of discovery. Yet Bullinger did not suppress the truth about his finding concerning the numbers 7 and 11, awkward to him though it was. His insight concerning the number 7 in connection with the name of Christ, quoting from *Number in Scripture,* is shown below. Bullinger used the King James Version (KJV) of the Bible, the version on which Strong's Exhaustive Concordance is also based, and it is the convention adopted in the present work. Comments in brackets [ ] have been added for clarification. Bullinger uses the * sign for multiplication:

> *Number in Scripture*
> Page 26-32, words in the Old Testament
> [excerpts from these pages]
> *The "wave offering"* [ref. to Christ] 28 times   *(7 * 4)*
> *The "heave offering"* [ref. to Christ] 28 times   *(7 * 4)*
> [from the New Testament]
> "The Lamb" as used of Christ 28 times   *(7 * 4)*
> *The Apocalypse* [meaning: the book of Revelation]
> "Jesus" occurs 14 times   *(7 * 2)*
> "Lord" occurs 21 times   *(7 * 3)*

The number seven[1] in respect of Jesus Christ (as 'Lord of the Sabbath') should not seem so strange. But what comes into view in Bullinger's next discovery is quite remarkable:

> *Number in Scripture*
> Page 28: words in the New Testament

"The Father" occurs in Matthew	44 times	*(4 * 11)*
Mark	22	*(2 * 11)*
Luke	16	*(4 * 4)*
John	121	*(11 *11)*
The rest of the N.T.	77	*(7 * 11)*

23

In all the above sections but one, multiples of the number 11 are linked to 'the Father' (The exception of the 4's in Luke may offer a secondary, or supportive, number. This is explained later). The facts so far, which one could treat as no more than powerful clues, are sufficient to justify the adoption as a working hypothesis the following idea: that the numbers 7 and 11 stand for the Bible personalities Jesus Christ and the Father[2]. Would it be such a remarkable thing that the Bible assigns, if indeed it does, such numbers to its greatest personalities? After all, the less significant (but well known) 'beast' of Revelation is marked with a number. So why should not the two most senior personalities be assigned numbers, albeit cryptically?

The numbers 7 and 11 are of course prime numbers: the series beginning 2, 3, 5, 7, 11, 13, 17, 19… Prime numbers can only be divided by themselves or the number 1. This observation of Bullinger's concerning 7 and 11 will have been perused by thousands of scholars. Yet it has been passed over, lying dormant for over a century. Now its meaning will be revealed.

1. Regarding this first discovery of a pattern concerning the number 7, there need be little mystery as to why Jesus Christ *"The Lord of the Sabbath"* (Matthew 12:8) is attached to it. The Sabbath is the seventh day and Christ observed it 'as his custom was.'

2. There is an association between the number 11 and 'the fathers' in the Old Testament, as in the phrase *"of the patriarchs"* which occurs 11 times in the book of Genesis. Noah, who is in Bible terms the father of the present world, is 11th in the genealogy given in the book of Luke. Various biblical judgments handed down are linked to the number 11. Judgment on Egypt (11 judgments over 7 chapters of Exodus), Tyrus (Ezekiel 30), Samson and the 1100 pieces of silver, Solomon to have the kingdom rent from him in I Kings 11:11, Pharaoh whose arm was broken in the eleventh year, Christ who had 11 apostles living at the time of the crucifixion, the judgment on mankind via the two witnesses of Revelation 11 and Psalm 119 which contains 22 judgments. The nation of Israel wandered 40 years after 11 witnesses caused them to turn from the promised land. They were judged in Kadesh-Barnea 11 days from Horeb and the kings Jehoiakim and Zedekiah brought down dire punishment: both ruled 11 years.

24

4

HEAVENLY BODIES

"No great discovery was ever made without a bold guess." Isaac Newton

The numbers 7 and 11 form a secret number code in respect of the two leading personalities of the Bible. What else can these numbers be related to? The number eleven will remind astronomers of the Sun and its strange eleven year sunspot cycle. But there is something else very odd about the Sun and its nocturnal partner, the Moon.

Why does the Moon appear to be the same width as the Sun? As a disc it appears - to the naked eye observer - to possess the same diameter. Living in a world where there are 'two great lights' in the sky of *identical* diameter, as viewed from Earth, could seem very strange if one stopped to dwell on the fact. This phenomenon, unique in the solar system, can be observed because although the Sun is 400 times wider than the Moon it is also exactly 400 times further away from the Earth, see figure 1.

Chemist and science fiction author Isaac Asimov called the Sun/Moon 400:1 ratio of width to distance *"...the most unlikely coincidence imaginable."*

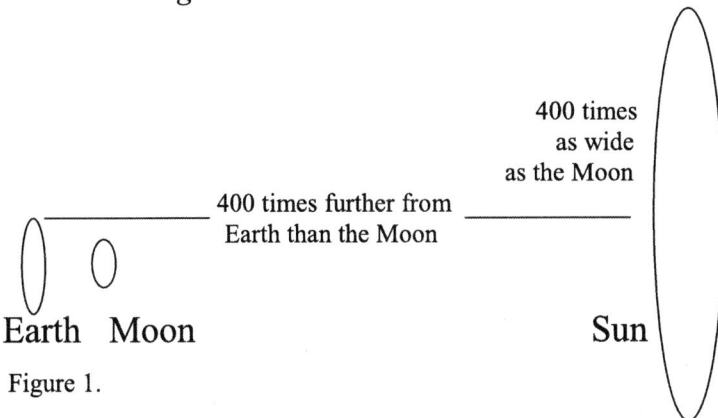

400 times
as wide
as the Moon

400 times further from
Earth than the Moon

Earth   Moon   Sun

Figure 1.

Many mysteries are hidden in the Moon. Science has had huge difficulty accounting for the proximity of the Moon to the Earth in view of that satellite's almost circular orbit. Had the Moon joined our planet as a captured body it would possess a far flatter elliptical motion. There are no viable theories to explain the Earth-Moon System. Scientists have half wondered if the Moon wasn't actually manœuvered into position[1].

It has recently come to light that the dimensions, speeds of rotation and relative distances of the three spheres of the Earth-Moon-Sun system are unmistakably interrelated in a way unique in our solar system[2]. The Moon's width is 27.31% of the Earth's. This number corresponds closely to the time the Earth takes to circle the Sun, 366 days, because the reciprocal 1/366 gives 0.0027322, the same number pattern as the value for the sidereal lunar orbit of 27.322 days. So width and orbit carry the same number pattern. It is also remarkable that the Sun is just over 109.2 times wider than the Earth, and that 109 Earth diameters would fit into the void between these two bodies at its maximum. Therefore the two bodies are linked in respected of diameter and their separating void by the number 109. Furthermore, the rate of rotation of the Earth relates to that of the Moon by a factor of exactly 1:100. (These discoveries are described by Christopher Knight and Alan Butler in their book *Who Built the Moon?*)

All this is evidence of an intelligent fine-tuning to the system encompassing the three bodies. It was shown in the previous chapter that the numbers 7 and 11 are hidden number tags for the two leading Bible personalities. But there may be other ways in which these two personalities are represented. The Bible is a highly symbolic book, and the very first symbols mentioned are also a pair, the Sun and the Moon:

> " And God said, let there be lights in the firmament of the heaven to divide the day from the night; and let them be for **signs**, and for seasons, and for days and for years...v16 And God made two great lights..." Genesis 1:14

This passage is an imperfect translation. The original Hebrew word rendered as the English word 'signs' is *owth* and it means: *"a signal (literally or figuratively), flag, beacon, monument, omen, prodigy, evidence"* (*Strong's Exhaustive Concordance*). Thus the word 'sign' falls short in conveying the full import and uniqueness of these objects. Throughout the book of Genesis the Sun and the Moon are the only *signs* given.

These two heavenly bodies serve as foundational *monuments* or symbols. As already mentioned, one of the most striking things about the Sun is its well known 11 year cycle of spots. These were first observed in modern times by a German father and son team, David and Johannes Fabricius, a year before Galileo observed them in 1612. If one were looking for a simple number associated with the Sun, the most obvious would be the number 11. The Sun's spots appear as dark earth-sized blotches on its surface. These are capable of emitting huge flares hundreds of thousands of miles in length. What is less commonly known is that the polarity (direction of magnetic field) of the spots reverses in each subsequent 11 year cycle, creating a 22 year cycle once that additional feature is taken into account. A more accurate figure for the basic cycle is 11.1 years, although some variations have been observed. Sunspot activity fell to almost zero during a seventy year span from 1645 to 1715, a period called the Maunder Minimum[3].

If the Sun is strongly associated with the number 11 due to its sunspot cycle, what of the Moon? Both bodies have light as their common factor. The Moon acts as a giant mirror reflecting the light of the Sun, but in this it is less efficient than one might expect. It reflects a mere 7% of the light falling on it and is therefore said to possess an albedo of seven!

An overriding theme of Genesis chapter one is that of light, and a creative sequence is described beginning in verse 3 with the statement *"Let there be light."* The appearance of light is the first act of the creative process. Therefore the all important first number for the Moon is 7, because as the other 'Great Light' it reflects 7% of all sunlight reaching it.

In this way the Sun and the Moon are marked by the numbers 11 and 7, light being the common factor.

The two great lights of Genesis are associated with one another by virtue of the number 400, as shown in figure 1. The number 400 makes its first biblical appearance in Genesis15:13 in relation to the number of years of the Israelites' slavery in Egypt. The human writers of *Genesis* could have had no accurate knowledge of cosmological distances, yet the number appears in that book alone of the Torah (the first five books in the Old Testament, authored by Moses) as if by coincidence.

If there is an emerging picture, growing in the manner of a jigsaw puzzle, it can be summarised as follows:

MOON	*SUN*
Jesus Christ	The Father
7	**11**

Suggesting the Sun and Moon as symbols to match 7 and 11 goes radically further than Bullinger, but it does not answer his question *"Why these two numbers?"* Bullinger, like most theologians, held to the idea of a 'trinity' - a triune Godhead. Though that idea is based on tradition, it cannot be denied that the 'Holy Ghost' (as translators have sometimes rendered it) is a hugely important factor connected in some way to these two leading personalities of the Bible. It is mentioned over three hundred times throughout the Bible. Therefore the above explanation concerning 7 and 11 begs certain questions such as:

1. Is there any symbol for the Bible's 'Holy Ghost'?
2. If so, is there any number for the 'Holy Ghost'?
3. And if so could he, she or it be expected to be represented by a prime or non-prime number?
4. What might any number for the 'Holy Ghost' indicate as regards its nature; whether it is described as possessing personhood, or other characteristics?

A toyshop will sell a jigsaw puzzle in a box with a picture on the front. The parts to be assembled are inside, of a known number and the purchaser knows that, if he perseveres, he is assured of a meaningful picture. But in this case there is no picture at the start and no certainty that one even exists.

Assuming that the discovered pattern so far constitutes the tip of an iceberg, the nature or quantity of undiscovered parts cannot yet be seen. To find them one must survey the landscape of Bible knowledge, scrutinizing it for clues to the hidden message which has lain dormant since Moses wrote the book of Genesis.

Before finding a third or fourth number or factor in the hoped for picture, a vital piece of supportive background information must be explained. Because in addition to 7 and 11 there is another number that though not a part of the eventual Grid, is key to figuring it out. This new number, explained in the next chapter, is a completely unknown and never before noticed secret factor in the Bible. Its discovery is so important that it would merit, for the amusement of Bible aficionados if nothing else, a book all of its own. It is, in and of itself, a world first.

The hidden entities that have struck this planet in 2007 with the greatest economic calamity in history are intent on teaching us something. They have a teaching number: the number 28.

1. Regarding the origin of the Moon, the best explanation is that the moon is an illusion; so said Irwin Shapiro in a 1970's university astronomy class, as cited by J.J. Lissauer. Several theories have been offered to explain the origin of the Moon, the principal ones being The Fission Theory, The Capture Theory, The Condensation Theory and The Impact Theory, but it is admitted that all but one have insoluble problems. Even the current favourite, the *Impact Theory*, would require an Earth-impact of an object twice as massive as Mars – a most unlikely event, the legacy of which would have been excess angular momentum the loss of which cannot be explained. [*Nature* 389 (6649):353-357, September 25, 1997 – *"Lunar accretion from an impact generated disc"*]. J.Sarfati PhD wrote in *Creation* 20 (40) September 1998 "But even if the Moon had started receding from being in contact with the Earth, it would have taken only 1.37 billion years to reach its present distance."

2. In the 2005 book *Who made the Moon?* Authors Christopher Knight and Alan Butler assign the number 366 as the "pin number" of the Earth, on the basis that this is the number of rotations on its own axis in one revolution around the Sun. But $100/366 = 0.27322$; 27.322 being the sidereal days in a *lunar* orbit. The anomalistic month differs from the synodic month which is based only on what we see of the Moon from the Earth's surface, i.e. the lunar phases from a crescent to a full Moon. Because this "lunation" is a phenomenon due to sunlight it is dependent on the Earth-Sun line which shifts as the Earth moves around the Sun. This has the effect of continually stretching the cycle, hence the synodic period is 29.53058885 days; the anomalistic period is 27.55454988. However the lunar sidereal period of 27.322 days is considered the true period of revolution for the Moon around the Earth (see www.iap.ac.cn). It is the period of revolution as measured using a star as a datum point (see www.scienceworld.wolfram.com). They also claim that the number of Earth diameters compared to that of the Sun is 109.245, and that this matches the number of Sun diameters contained within Earth's aphelion (its greatest distance to the Sun) which is 109.267 Sun diameters. This in turn matches the rate of rotation of the Moon which has an equatorial circumference of 10,920.8 km and rotates at a rate of 1 kilometre per second exactly. They also discovered that the product of the circumferences of the Moon and the Earth is 100 times the circumference of the Sun.

3. The Maunder Minimum (thus named by astrophysicist Jack Eddy) was a seventy year period from 1645 – 1715 when sunspot activity fell to about $1/1000^{th}$ of the normal level as recorded on French tracings of the Sun and as evidenced in tree rings, as well as the regular freezing over of the River Thames during the period.

5

UNEARTHING THE NUMBER 28
"Great magicians are also great liars." Adolf Hitler

About three thousand years ago a self-styled prophet set out to perform a most unusual task. In those days even dignitaries traveled by means of the donkey.

According to the story a king had engaged this prophet to pronounce a curse on a neighbouring nation, one with which he had been competing for natural resources. The prophet's means of travel, an ass, had been considered quite ordinary until this particular journey: there was certainly no reason to suppose that the animal was endowed with psychic powers. But on the journey to pronounce this curse the prophet's ass saw an apparition that he himself could not see. The animal slumped to the ground terrified, and despite the beatings administered by its owner it refused to get up. That was when it raised its head and, according to the Bible's account, spoke! The ass said *"What have I done to you that you should have beaten me these three times?" Numbers 22*

Wolves and dogs are used in the Bible as similes for certain types of men, but the ass is the only animal man is directly likened to by his birth[1]. The ass is mentioned 28 times, beginning and ending with reference to this particular animal, whose fame was rekindled in the epistle of 2Peter 2:16 *"...the dumb ass speaking with man's voice forbade the madness of the prophet."* This is the twenty-eighth mention made of asses in general, at which point one is brought back by Peter to the same original ass. It is as if one is returning full-circle via the number 28. In this the ass provides a clue.

The discoveries of the present book are due, in great measure, to the observation that words and names in the Bible occur in exact numbers, or multiples of numbers. Of vital importance to the discoveries show in this book is the fact that 28 excoriating criticisms of religious teachers[2] are contained within the 'set' of the four gospels: Matthew, Mark, Luke and John (Jesus Christ passed almost all of these comments). That result harmonises with another completely hitherto unknown Bible factor: the existence of 280 key teaching statements within those same four gospels. But then there are more and more instances of 28 in connection with teachers, particularly the two great luminaries. Christ's alternative title 'The Lamb' also occurs 28 times in the New Testament, synchronising with the leading Bible personality 'the Father' who is mentioned 280 times[3] within that same section.

This shows two powerful results for an interchange between the numbers 28 and 280 involving 'teachers.' It would have been in Bible terms the leading personage - the Father - who handed down the teaching program[4] to the teacher. Therefore the number 28 is a number associated with the theme of teaching. Findings for this are summarised more fully in the later chapter *Never before seen, the number 28.*

Let the reader realise that this is a world first discovery concerning the Bible. The number 28 is a secret key, like a sliding stone in the wall of an ancient pyramid. The Bible has been constructed rather like a building: it displays order and symmetry and this will of necessity involve numbers. Many of these numbers are obvious such as twelve in respect of twelve apostles or the twelve tribes of Israel, but others have been craftily hidden. The number 28 is what could be termed a 'submerged' number. It is as if one were handed the technical drawings for a huge building in which some of the measurements are explicitly stated, but certain other critical measurements have been concealed.

In this investigation the number 28, it will be shown, is *the teaching number* of the Bible. No one has ever detected this fact yet its veracity will be demonstrated beyond all doubt. No other number suggests itself in this way as the teaching number. The preliminary findings for occurrences of the number 28 are summarised here:

THE NUMBER TWENTY-EIGHT:

The Father, in the New Testament	280
Total of teaching statements in the four gospels	280
Denunciations of teachers in the four gospels	28
The wave-sheaf symbol (Christ "offered") Old Testament	28
The heave offering (prophecy of Christ) Old Testament	28
The Lamb (Christ)	28
The Passover (four gospels)	28
"As it is written" in the New Testament	28

Encoded in the first verse of the Bible:
Genesis 1:1 has 28 Hebrew letters   28

Key verse number:
Genesis 1 verse 28, the first instructions given to man   28

Next it is necessary, before explaining the discovery in full, to demonstrate how prime numbers are found in the Bible. Prime numbers are frequently used for encoding, as in credit card security. But the human race is not alone in hiding secrets in prime numbers.

For the twenty-eight denunciations of teachers see Appendix II.

1. The comparison between the birth of men and donkeys is found in Job 11:12 *"For vain man would be wise, though man be born like a wild ass's colt."*

2. A list of the 28 attacks on teachers in the four gospels is shown in Appendix II.

3. As shown in *Number in Scripture (chapter 3, p28)* Bullinger gives the tally of occurrences of "the Father" in the New Testament:

"The Father" occurs in Matthew	*44 times*	*(4 * 11)*
Mark	*22*	*(2 * 11)*
Luke	*16*	*(4 * 4)*
John	*121*	*(11 *11)*
The rest of the N.T.	*77*	*(7 * 11)*

Bullinger never commented on the total of 280. He never realised that 28 was the teaching number of the Bible.

4. In this instance, Christ the teacher stated in John 8:28 *"I do nothing of myself: but as the Father has taught me, so I speak these things."*

6

PRIME NUMBERS OF THE BIBLE
"Nature, even the entire universe, is coded in prime numbers." Peter Plichta

The first two symbols of the Bible, the 'Two Great Lights' of the book of Genesis, are linked to the numbers 7 and 11. These two are classified as prime numbers.

Prime numbers are special in as much as they cannot be factorised into whole numbers. For example, 16 isn't a prime but 17 is. So a prime number can only be divided by itself and the number 1. The series of prime numbers begins as follows:

2, 3, 5, 7, 11, 13, 17, 19, 23, 29, 31, 37, 41, 43, 47, 53, 59, 61, 67…

Modern data encryption methods utilise prime numbers and so does the Bible. There is a very important set of primes, perhaps the only such set in the Bible. It relates to the first five books, the Torah. These five foundational books are well known by scholars to correspond with the Psalms in the following way:

Genesis	Psalms 1 – 41,	41 chapters; 41 is the 13th prime
Exodus	Psalms 42 – 72,	31 chapters; 31 is the 11th prime
Leviticus	Psalms 73 – 89,	17 chapters; 17 is the 7th prime
Numbers	Psalms 90 – 106,	17 chapters; 17 is the 7th prime
Deuteronomy	Psalms 107 – 150,	44 chapters; **44 is not a prime**

What may not have been noticed before is that four of these sets have totals that are prime numbers: 41, 31, 17 and 17 as shown above. This is the case for each set of Psalms except the last, corresponding to Deuteronomy. That last set totals the non-prime number 44. *As the exception to the rule, attention is drawn to it.* This number takes on an almost mystical importance later in this investigation. It was highlighted during the strange events of 2003 and its meaning will shortly be revealed.

It is commonly claimed that the Bible is full of patterns. The truth is that there are few known patterns, other than sevens, that can be shown to convey any meaning. However, prime numbers feature in the working hypothesis below:

MOON	SUN
Jesus Christ	The Father
7	**11**

Again, how can one answer Bullinger's question *"why these two numbers?"* Firstly, one might hope to find additional numbers for the two heavenly bodies beyond 7 and 11; more pieces in a jigsaw as it were. Could the periods of motion of heavenly bodies offer clues? A further number for the Moon is the 173 day cycle of lunar eclipses. These occur when the Earth is in alignment between the Moon and the Sun. The number 173 looks rather unhopeful until its position in the series of prime numbers is considered:

2, 3, 5, 7, 11, 13, 17, 19, 23, 29, 31, 37, 41, 43, 47, 53, 59, 61, 67, 71, 73, 79, 83, 89, 97, 101, 103, 107, 109, 113, 127, 131, 137, 139, 149, 151, 157, 163, 167, 173

Suddenly there is a fascinating possibility, because 173 is the 40th prime number! This looks to be a very meaningful result, because the number 40 is strongly associated with the activities of Christ in the Bible. He fasted for 40 days (Matthew 4:2), and was seen (according to the account in Acts chapter one) by his disciples for 40 days following the resurrection. Furthermore, in I Corinthians 10:1-4 the apostle Paul claims that during the 40 years of wandering in the wilderness the Israelites were followed by a 'Rock' and *" that Rock was Christ."* And King Solomon was a type or forerunner of Christ: he reigned 40 years.

Is it possible that the number 40 is the next piece in a mysterious Bible jigsaw puzzle hidden for millennia?

7

A FIRST GRID

"Logic will get you from A to B.
Imagination will take you everywhere." Albert Einstein

Is it conceivable that a hidden message encoded by the first symbols mentioned in the Bible, the Sun and the Moon, could overturn the cherished beliefs of the Western world? Could it be possible that many traditions beloved of generations will be exposed as the empty window dressing of a bankrupt religion? The point is approaching in this investigation where many doctrines and dogmas beloved of historical Christianity will be exposed as fables.

In the previous chapter it was shown that the number 40 could, quite justifiably, be held up as a candidate for the position of the 'next jigsaw piece.' It is not a prime. But whereas the number 7 for the Moon (Christ) pertains to *identity*, the number 40 is derived from the Moon's *motion*. If the Moon can have a number for motion, what about the Sun? Can the Sun move?

Although it is in the centre of our solar system and by definition stationary relative to Earth, the Sun does have *internal* motion. The Sun has a core, a radiative zone, a convection zone and a photosphere: each of these is in motion. This suggests the number 4.

Four is already associated with motion in many things around us, for example the four seasons and the four forces of the universe: gravity, electromagnetism and the weak and strong nuclear forces. In nature, the four chemical bases of DNA imbue cells with their self-replicating power. In the Bible the number four is also associated with power and motion. The book of Ezekiel records a vision of the transportable throne of a deity featuring *four* Cherubim (winged angels) each with circling eyes forming wheels.

Furthermore, the 400:1 ratio of the Sun to its partner the Moon comprises *four* hundreds. In all of the above examples the number four was equally indicative of *power* as of motion. It is evidently the number for both. The four numbers: 11, 7, 40 and 4 suggest a preliminary 'grid' shape:

The Moon's albedo **7**   **11** The sunspot cycle

Moon: 173 day cycle **40**   **4** Motion of the Sun

Adding headings produces the following:

	(Christ) **MOON**	(The Father) **SUN**
Signs		
Identity number	**7**	**11**
Motion number	**40**	**4**

A further Bible clue for the number 4 is Bullinger's analysis of the number 11 in connection with 'the Father.' The only exception to multiples of 11 for the Father was the multiple of 4 in the book of Luke, a likely indication of a secondary number.

The languages of the Bible offer further confirmation. In any search for a hidden meaning in Greek or Hebrew it is necessary to be aware of the number values assigned to letters and words in those languages, according to their position in the alphabet. This is known as the *gematria value*. One further confirmation for the Father's secondary number 4 is that the *gematria* for the Greek word Father (αβα) is 1 + 2 + 1.

Assessing the result so far, adding up the four numbers of this 'grid' gives the total of 62, or twice 31, and it contains two deities: the Father and Jesus Christ. Of the number 31 Bullinger wrote:

Number in Scripture, page 265
THIRTY-ONE

"The Hebrew expression of this is אֵל, El, the name of God, and its signification as a number or factor would be Deity."

This deity number as 31 is also confirmed by Dr. M. Mahan:

Palmoni, or - the Numerals of Scripture, page 90

"As to his name: Elijah, spelled in full, is 31, the number of Deity…"

Therefore a further confirmation of 11 as the Father's principal number is that the 11th prime number is 31, the number of deity. The result so far could be called a first, or basic, 'grid.'

Multiplying its numbers has an interesting effect. What emerges is an object built by Moses.

8

A PICTURE PATTERN

"All great truths begin as blasphemies." George Bernard Shaw

Just as an accountant will make a first attempt to reconcile a balance sheet, the set of numbers so far obtained as a 'grid' can be scrutinised to see if in some way they 'add up.' Can they be shown to complete a pattern, a message or even a *picture?* So far the following 'grid' has been deduced:

	Son	Father
Signs:	**MOON**	**SUN**
Identity number:	**7**	**11**
Motion/power no.	**40**	**4**

Apart from the obvious theme of 'family' might there be a hidden message or picture in this 'grid'? One could begin by asking the following:

1. Does it have internal patterns?
2. Does it suggest any other numbers that may offer meaning?
3. Can it 'self-validate?'
4. Does it suggest any new idea or insight?
5. Could it conceivably picture anything in the Bible?

To enlarge firstly on the third point of *self-validation*: can this grid of four numbers provide some indicator of its own completeness? Perhaps it is not complete. An internal combustion engine will not start if parts are missing, so is there a way to 'start' this Grid? Perhaps that could be done by a simple calculation based on its digits. Could that produce a self-characteristic result? Firstly, this Grid does have an internal pattern because the product of the vertical multiplication 11×4 produces the same total as the addition of the numbers of the bottom line, 44 in each case. So this gives two self-characteristic results. There are also the numbers 28 and 280 derived in the following manner:

7　　　　　**11**

(11×4 = 44) RESULT

40　　　　**4**

(40+4 = 44) RESULT

280
RESULT

28
RESULT

Furthermore, a self-validating function is also evident, because there is (as noted in the previous chapter) a simple arithmetic function performed on *all* four numbers that produces a result characteristic of the Grid itself. The four numbers of the Grid total 62. This is 2×31 and *31 is the number of deity*. Because the pattern contains 2 deities a total value of 2×31 is characteristic of it! Thus this four-numbered grid can self-validate in several ways.

So regarding the first four questions posed earlier, it is evident that there are patterns and that further numbers are suggested. There is also a sense in which this 'grid' balances, or self-validates. But what of the fifth question concerning a picture: is a picture emerging? Because - all of a sudden - we have 28 and 280 surrounding this 'grid.' Do these numbers occur in the Bible?

Yes, they are found *surrounding* the Tabernacle of Moses in Exodus 26 verses 2-3:

> *"...the Tabernacle with ten curtains...one curtain shall be eight and twenty cubits."*

So that ancient Tabernacle was surrounded by 28 and 280! Note the following points:

a. The Tabernacle had **eleven** covering curtains

b. Teaching was conducted on a **seven** day cycle

c. The Children of Israel were led by Moses through the wilderness for **forty** years

This is the layout of that tabernacle:

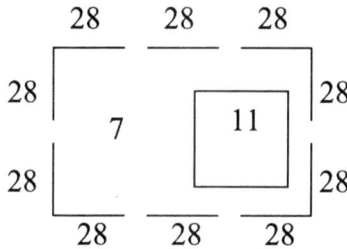

The above perimeter fence contains the numbers 28 and 280, so the tabernacle is surrounded by them - just like the emerging Grid! The four-numbered 'grid' is not yet complete but the next chapter contains the seeds of a solution.

9

A 19 YEAR MIDDLE EAST CYCLE
"Common sense is the collection of prejudices
acquired by age eighteen." Albert Einstein

In recent decades the Arab-Israeli conflict has made the Middle East the world's most volatile region. It has seen major invasions, assassinations, oil embargoes and pre-emptive strikes on nuclear facilities. It has spawned the phenomenon of the suicide bomber. Now a Shiite empire led by Iran is in prospect. This is occurring in the region that has the greatest potential to plunge the world into a species-endangering nuclear war.

These events appear to have been regulated according to a 19 year clock. Strangely enough, the Moon acts as a clock running according to a 19 year cycle[1]. But in terms of events, here is the evidence of a nineteen year cycle:

    1910 The first Jewish Kibbutz was founded.
    1929 The Arab slaughter of Jews in Jerusalem.
    1948 The declaration of a Jewish State
    1967 The Six Day War
    2005 The withdrawal from Gaza after thirty-eight years (2 × 19)
    A full description of these events is given in Appendix III

The Bible is first and foremost a book describing the histories of the Israelites. The ancient Israelites were a family of Middle Eastern tribes. They had a calendar system preserved to the present day by the Orthodox Jews. It works by this same time cycle of 19 years associated with the Moon. This calendar system, known as the *Intercalated Hebrew Calendar*, determines the dates of the seven 'Annual Holy Days' first mentioned in Leviticus chapter twenty-three. These seven days encapsulate a total of nineteen in all. They are determined according to an Earth-Moon cycle of 19 years.

These 19 days feature in both the Old and New Testaments. First mentioned in Leviticus, they were observed by Christ and the early Church as mentioned in several New Testament passages, in particular *Acts*. The system is therefore integral to the entire Bible. These days are as follows:

Passover	1	
Days of unleavened bread	7	(first and last are days of rest)
Day of Pentecost	1	
Day of Trumpets	1	
Day of Atonement	1	
Feast of Tabernacles	7	(first is a day of rest)
Last Great Day	1	(day of rest)
Total number of days	19	

There are 7 annual holy days, or days of rest, incorporated within this cycle of 19 days. The system, as regulated by the Intercalated Hebrew Calendar, is explained further in the chapter *The 1711171 System*. Thus the number 19 is shown to be foundational to the Bible. Can such a fundamental number help in completing the Grid?

This number 19 is used in a system controlling the way the God of the Bible is supposed to be worshipped, so might it be a factor in identifying who or what he really is?

	Christ	Father
Signs:	**MOON**	**SUN**
Identity number:	**7**	**11**
Motion/power number:	**40**	**4**

Only two personalities occupy the top row and their numbers total eighteen.

Historical Christianity would contend that there is a missing third personality. That idea of a Trinity[2] of gods or personages or facets of God is not based on the Bible, but only on manmade traditions.

The Bible does however mention a spirit essence associated with the Father and the Son. Clearly this essence is very important as it is mentioned over three hundred times. Even if one were to assume that it is not a personality, could it still be important enough to warrant its own column, a third column, in an emergent Grid? Supposing that this is the case, the next gambit would be to experiment with a further column for this phenomenon of the spirit:

	Spirit	Christ MOON	Father SUN
Signs:			
Identity number:		7	11
Motion/power no.		40	4

If the 'spirit' really is, in Bible terms, a person then the question arises as to whether its signifying number is larger than 11 or smaller than 7. A larger number would imply a person more important than the Father: a smaller implies a person less important than Christ. Neither of these ideas fit the Trinitarians' notion of God, nor that of the Bible itself. In the Bible 'the spirit' is compared to fire, wind, water and a dove. These symbols suggest something akin to a force or essence from God. Consider this description of the first appearance of the spirit in the Church age:

"...a sound from heaven as of a rushing mighty wind...cloven tongues like as of fire, and it sat upon each of them." Acts 2:2-3

This describes the appearance of a spirit essence, the holy spirit, on the heads of the apostles. This spirit force is generally characterised in the Bible as a means by which deities act at distance, rendering them omnipresent. The flames on the heads of the disciples were the visual manifestation of a force, or essence or power sent from God. Though Biblically not a person, it may still have an identity number because it is so strongly associated with the deities numbered as 7 and 11.

One further lead could be the idea that disciples are 'unified by the spirit.' They could be expected to be of one mind! *Such a unifying factor could be most logically represented by the number 1*. Using that idea in conjunction with the number 19 obtains the following:

	Spirit	Christit MOON	Father SUN	
Signs:		**MOON**	**SUN**	
Identity number:	**1**	**7**	**11**	= 19
Motion/power no.		**40**	**4**	

This gives a line of 1 7 11: an abbreviated version of the 1711171 holy day pattern! That pattern of 19 days is regulated by a 19 year cycle and this top line now also totals 19. Moreover, a further fit is possible when the 11 year cycle of sunspots is considered. This cycle is more accurately measured as 11.1 years and the string 1711171 has 111 at its centre. The main findings can now be summarised:

The Grid top row (of identity numbers) 1 + 7 + 11 = 19

The total of holy days 1 + 7 + 1 + 1 + 1 + 7 + 1 is also = 19

Years of Sun/Earth/Moon motion regulating 1711171 = 19

The number 19 here appears to denote a set, or family, comprising everyone and everything that is, in Bible terms, God. This group 'trademark' attaches to both the Grid top line and the system of holy days, the Bible's own (and only) authentic pattern of worship days.

These days total nineteen in number. Nineteen is the 8th prime number. If 19 be a trademark of sorts then *8 is its shorthand form*. It was a giant figure of 8 that appeared across planet Earth in 2003.

Due to the addition of 'the spirit' in this emerging Grid a third vertical column has been created with a description and number, but not yet any symbol. But the Sun and the Moon relate to one another in a special way, as they produce a 'special effect' when acting in concert. This mystical phenomenon is called a *solar eclipse*. Could the solar eclipse be a clue to further Grid numbers or symbols?

Further information is given concerning the 19 year cycle in Appendix II.

1. The 19 year cycle of the Earth-Moon System is called the Metonic cycle, or Enneadecaeteris, and is a common multiple of the tropical year and the synodic month; 19 tropical years fit 235 synodic months to within two hours accuracy. The 76 year cycle (19 × 4) has a mean year of 365.25 days exactly.

2. Theologians agree that the New Testament does not contain an explicit doctrine of the Trinity. The passage found in Matthew 28:19 *"...baptizing them in the name of the Father, and of the Son, and of the Holy Ghost"* is described as "...strange; it was not the way of Jesus to make such formulas" by the Schaff-Herzog Encyclopedia of Religious Knowledge; it goes on to state that "...the formal authenticity of Matthew 28:19 must be disputed..." and that, "...the New Testament knows only baptism in the name of Jesus." Another passage more commonly known to be of dubious authenticity is that of 1John5:7 *"For there are three that bear record in heaven, **the Father, the Word and the Holy Ghost, and these three are one. And there are three that bear witness in earth**, the spirit, and the water, and the blood..."* The words in bold should be scored through in every King James Version. These were inserted in mediaeval times and are never used to by scholars in supporting the Trinity. Deliberate fraud should serve to alert the diligent seeker of truth. Sacramentum Mundi, an Encyclopedia of Theology, states: "There is no systematic doctrine of the 'immanent' Trinity in the New Testament. The nearest to such a proposition is the baptismal formula of Matthew 28:19, though it must be noted that modern exegesis does not count this saying among the *ipsissima verba* of Jesus." David Kemball-Cook states on page 159 of *Is God a Trinity?* "The Trinity doctrine cannot be found in the Bible."

10

TONGUES OF GOLD

"...a fire infolding itself, and a brightness was about it..." Ezekiel 1:4

The sky is almost empty. Grey clouds hang on the horizon in the gathering gloom. Dark shadows race across the distant hillsides as voles and field mice burrow down in obeisance as the darkness gathers pace. Nature falls silent before a mystical majesty.

It is now more than eighty minutes since the full Moon began its collision with the Sun, slowly swallowing it up like a serpent consuming a prey as large as itself in the most entrancing spectacle in the solar system[1]. The power and glory of a star 109 Earth-diameters in size is about to go fleetingly on display. Diamonds flash on the edge of the Moon's black disc as it slips into exact alignment with the Sun. There is a momentary flicker of regal crimson before numerous yellow tendrils of fire appear amidst a ghostly glow. Time seems to halt for over three minutes during a mesmerising vision of shimmering loops of flame the size of Jupiter, erupting and falling back in the silent sky as they have done for millions of years; flailing tongues of gold one hundred thousand miles in length. Suddenly, a piercing coal of light appears on the shoulder of the Moon's charcoal disc and a wedding ring of blinding brilliance forms as the two orbs part.

At other times the visible Sun's turbulent surface is afflicted with dark roving Earth-sized spots, from which gargantuan flares erupt. Controlled by their attendant magnetic field loops, these bear a striking resemblance to human hair. The blazing flames of the Sun bring to mind the description of a vision of a deity that appears in the Bible:

Ezekiel 1:4 "...a fire infolding itself, and a brightness was about it, and out of the midst thereof as the colour of amber..."

Dramatic things happen when '7 & 11' work in conjunction during an eclipse. The combining of these two numbers within the fabric of the Bible is also highly evocative. Although verse and chapter numbers are not explicitly stated in the original text, those that have been added closely follow the underlying structure and natural breakpoints. Some calamitous events on Earth appear to be coded by chapter and verse[2] using the numbers 7 and 11.

The Sun and the Moon are the two most imposing symbols that nature offers. A solar eclipse reveals to the human eye the vast Corona, a halo of particles streaming out from the Sun. At 2,000,000° Celsius it is far hotter than the Sun's surface which is a mere 8,000°. The Corona seems almost to have a life of its own. Radiating particles out beyond the great gas planets, the Corona forms a protective bubble around the entire solar system, blocking many dangerous emissions from deep space[3]. The particles of the Corona also interact with the magnetic field of the Earth to create a protective bow-wave around it as it traverses the solar system.

These particles streaming out from the Sun constitute a *solar wind*. Thus flames and 'wind' are characteristic of the Sun's Corona. These were the same manifestations that were described in the Bible concerning the spirit:

"...a sound from heaven as of a rushing mighty wind...cloven tongues like as of fire, and it sat upon each of them." Acts 2:2-3

The sound of *wind* and the sight of *flames* were the first visual signs of the spirit at the beginning of the Church age. The proposed third grid column provisionally has a name (the spirit), the number 1 and perhaps now also a symbol - the Corona! The heavenly mating game revealing this Corona is suggestive of a theme that will become supremely important in this discovery – sexuality!

The spirit described in the Bible is a force, or essence, not a person yet it is still linked to God. In nature the Corona is clearly linked to the Sun. The Corona has a 'wind' and flames. *The missing symbol must surely be the Corona.* Gathering all other implied or derived information produces the following:

	(Spirit)	(Christ)	(The Father)
	CORONA	**MOON**	**SUN**

Identity: **1**   **7**   **11**   =   19

Motion: **40**   **4**

                         280   28

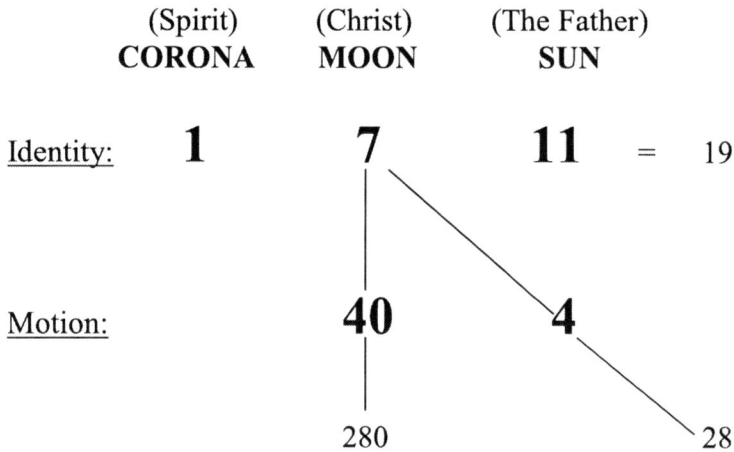

This leaves a sixth Grid number still to be identified. Once found it causes everything to fall neatly into place. An idea then emerges, a theme that is at once so very obvious and logical.

But it is a proposal that establishment religion brands as blasphemy.

1. The greater distances of the larger planets from the Sun mean that any eclipse created by their moons would lack the visual impact of our own.

2. Genesis 7:11 records a Great Flood, and Genesis 11:7 records the scattering at the Tower of Babel, built in response to it. In the book of Exodus 11 judgments fell upon Egypt over 7 chapters.

3. "...the solar wind is linked to the Sun's magnetic field and itself affects the amount of cosmic radiation hitting the Earth's atmosphere..." www.cern courier.com

11

AN ABSTRACT MACHINE
*"Science without religion is lame,
religion without science is blind." Albert Einstein*

Perusing Bullinger's book *Number in Scripture* it took this author less than two minutes to realise that 1, 7 and 11 probably formed a set. It then took two years to complete the Grid and find the *seventh* number.

For many months a sixth number had been lacking, yet there had been a strange number flagged by the events of 2003, the year of the initial discovery. Did that number have anything to do with the Grid? By 2005 an emergent Grid of five numbers was still stuck at the stage illustrated at the end of the previous chapter. It could be likened to a chariot with a missing wheel, or a food grinder with a missing handle. A component part to this 'abstract number machine' was needed and 2003 had been a year in which the number 44 was strangely and repeatedly highlighted. Firstly, consider how that number appeared in connection with the events of the year:

> 44 days - over which 8 world-wide power cuts occurred in 2003
> 44 days - the activities of a magician (This is explained later)
> 44 in total - the place positions of the worldwide figure of 8

For a Bible clue or idea, in the earlier chapter *Prime numbers of the Bible* it was shown how the number 44 appears as the exception to the rule (of prime numbers) in the Torah.

However, the number 44 makes an appearance in *Number in Scripture*, the significance of which Bullinger could have had no inkling. It suggests a meaning that has some fascinating, but some may say blasphemous, implications:

44 occurrences of the Greek word *sperma* (seed) in the Bible
44 as the total of Hebrew gematria of father and mother, 3 & 41

The Hebrew gematria of child is 44! Unless that is a coincidence (a possibility soon to be eliminated here) then 44 has something to do with *reproduction*. It was a father and son pattern involving 11 and 7 that triggered this whole investigation. Some further findings to do with the number 44 reinforce the idea of reproduction:

44 in number – autosomes (not chromosomes) in human DNA
44 as a factor – 1:44 is the relative size of the human ovum & spermatozoon cells

There is no time to ponder here the breathtaking implications of finding a reproductive number tagged, in two clear ways, onto the human reproductive system. But the discoveries concerning this number leave no doubt that it has something to do with reproduction. With a father and son at the core of the Grid does it not make sense to try this astonishing *number of reproduction,* 44, to see if it completes a pattern? With this number as the sixth number of the Grid the jigsaw is almost complete:

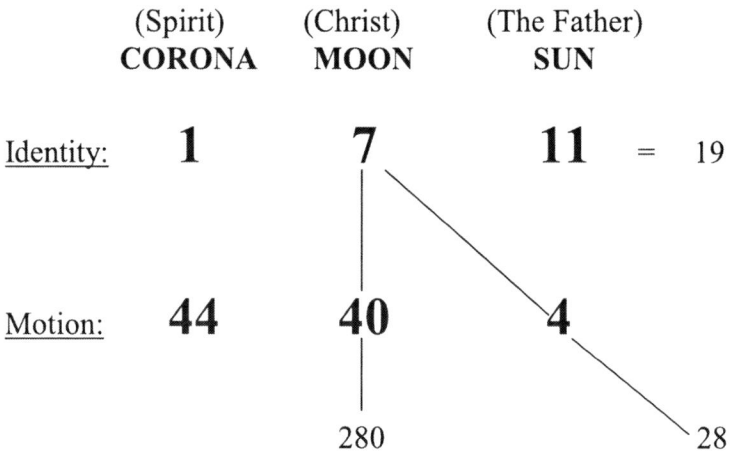

	(Spirit) CORONA	(Christ) MOON	(The Father) SUN	
Identity:	1	7	11	= 19
Motion:	44	40	4	
		280	28	

This solution opens many doors, including a well marked route to a seventh and final number. Before that, there is another indicator concerning 44, because placing it on the second line of this new Grid is consistent with the Bible's structure. The 44th book is *The Acts of the Apostles*, a book of actions and *motion and power*. It is also the book in which flames of fire - corresponding to the Corona - appeared on the heads of the apostles, thus fixing 44 in the left hand column of the Grid!

Of these six proposed Grid numbers two are primes and each of these represents a deity. Clearly primes are of fundamental importance. Their operation was seen in the preliminary 'grid' of four. Those four numbers totaled 62, or 2×31, where 31 is the 11th prime number, the number of deity. That amounted to a simple self-validation: two deities matching two 31's. It was an arithmetic operation in which the emergent 'grid' produced a result that was characteristic of itself.

Might there be a similar process available for the new six-numbered Grid, also involving primes, that has as its product or sum another result *characteristic of itself?* Firstly, check the total of this new six numbered Grid:

$$1 + 7 + 11 + 44 + 40 + 4 = 107$$

It is the prime number 107. Immediately there is a link to a previous observation. Prime numbers emerged in the Torah/Psalms pattern shown in the earlier chapter *Prime Numbers of the Bible*. The reader will recall that these sets produce prime numbers in all but one case, the exception being the last comprised of 44 psalms. Here once again is that pattern:

Genesis	Psalms 1 – 41,	41 chapters; 41 is the 13th prime
Exodus	Psalms 42 – 72,	31 chapters; 31 is the 11th prime
Leviticus	Psalms 73 – 89,	17 chapters; 17 is the 7th prime
Numbers	Psalms 90 – 106,	17 chapters; 17 is the 7th prime
Deuteronomy	Psalms **107** – 150,	44 chapters; **44** is not a prime

The fifth set begins with the 107th Psalm, as highlighted above. This is an extraordinary fact because, as is now evident, a final solution for the Grid may be seen in the number 44 which produces a total of 107. The discovery could not have unfolded in a different order. The number 1 did arise in the author's mind before the number 40 but both had to be confirmed before 44 could be tested. It is inconceivable that anyone solving the six-numbered Grid conundrum could do so, and not confirm 44 last. That is why the interplay of the two numbers 44 and 107 in respect of Psalms/Deuteronomy is so striking, because as the Grid reveals itself the number 44 inevitably leads to 107. Apparently, no other passages of the Bible contain numbers coded in this way.

However, there is a much more staggering observation. The Grid is surrounded by 28's, just like the Tabernacle of Moses. The teaching number of the Bible is proven to be 28. So it is an astonishing fact that the number 107 is the 28th prime number! **In this way the Bible's unique and hidden teaching number 28 functions as the *validation* number of the Grid.**

Also, as previously shown, 28 is *characteristic of the Grid:*

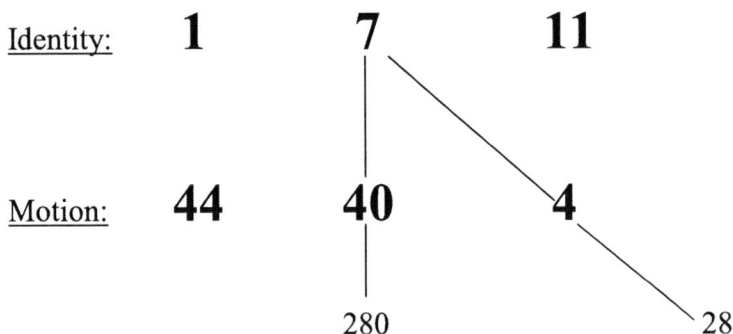

Identity: **1**   **7**   **11**

Motion: **44**   **40**   **4**

                        280   28

The idea of a 'validating' number being used to confirm a larger number (or set of numbers) is not uncommon. Before electronic reading machines were introduced, bank employees would type account numbers manually. In order to accept an account number, a computer would perform a standard calculation on the first seven digits. If this produced the eighth number correctly, according to what had been typed, all eight digits were accepted as a valid account number. But if the *calculated* eighth digit differed from the actual eighth digit typed, the account number was rejected. In that way wrongly typed account numbers could be intercepted. In searching for other factors that may bind the Grid together, an intriguing possibility arises when we ask why this set of six numbers possesses outside columns with the same product:

$$1 \times 44 = 44 \quad \textbf{but so does} \quad 11 \times 4$$

If the 44's of the two outside columns were to be linked in some way then a circular pattern would be the *simplest diagrammatic resolution*. This can be illustrated by 44's in a circle as follows:

The use of 44's can convey the idea of a spirit essence moving in a circle. The phenomenon of a spirit essence moving in a circular manner is described in the first chapter of Ezekiel. The book begins with a vision of the throne of God, supported by four living creatures using four identical wheels:

> *"And the living creatures ran and returned as the appearance of a flash of lightning...they four had one likeness: and their appearance and their work was like as it were a wheel in the middle of a wheel... Withersoever the spirit was to go they went...for the spirit of the living creature was in the wheels."* Ezekiel 1:14-20

In this description there are wheels within wheels (such wheels are mentioned ten times in Ezekiel chapter one) at the four corners supporting a throne. The impression is of a carriage with four wheels located at each corner, each affixed to a winged creature that is powered by a spirit essence spinning in circles. It is thus shown that there is a Bible precedent for a 'spirit essence' of some sort to spin in circles.

This brings a fascinating possibility into view. The sum of the bottom line of the Grid 44 40 4 is 88. But the total of the products of the two outside columns 1×44 and 11×4 is also 88. What could this mean? Little imagination is needed to extrapolate from this a line and a circle, where the diameter of the circle is equal to the length of the line:

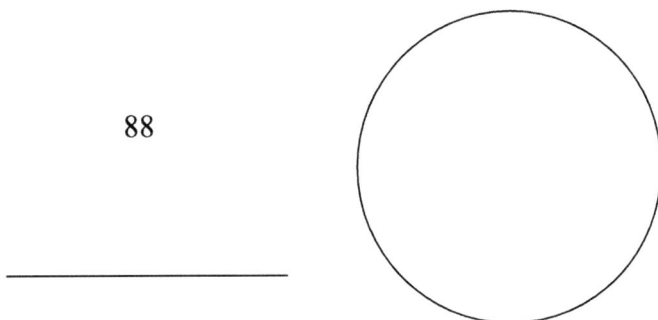

88

If the line is made equal to 1 the circle would be equal to pi (3.14159…). A 'diameter' line and circle suggest the number pi because pi is the ratio between the circumference of a circle and its diameter. Here are the first 31 digits of pi:

3.1415926535897932384626643383279

These numbers begin with a three and a one, suggesting the deity number. The successive digits of pi as they appear are for ever new, never repeating in any discernable pattern.

A further clue for circles and therefore pi is seen in the fact that the Sun and the Moon, both of which appear circular and on which the whole discovery is based, each occupy $1/720^{th}$ of the circle of the sky (half of one degree), or $1/360^{th}$ in all. From ancient times there are 360 degrees to the circle.

There is one more compelling clue for pi. The number 28 multiplied by pi almost produces a Grid derived total:

$$28 \times 3.141592... = 87.96462...$$

Almost[2] 88 is obtained, the value of the line and the circle. There therefore appears to be a relationship between pi, the Grid and the Bible's teaching number 28.

It will be shown that the Grid functions as a machine with circles, an abstract machine for traveling down the digital highway of pi. Unlike any other set of numbers it can stop at strategically and symmetrically positioned metaphorical 'lay byes.' With this ability the Grid bursts onto the pages of history as the first ever pattern in pi.

But before that story is told, there is a deeply disturbing vision within pi's *first* 31 digits. It will be shown in the next chapter how this mysterious self-contained 31 digit 'deity sequence' in pi is seen to match perfectly with the first chapter of the Bible[3]. As already shown, 31 is the number of deity.

One particular fact drew the author's attention to the possibility of a match with the deity number: Genesis chapter one falls into 31 verses.

1. The popular approximation for pi 22/7 is in itself a clue for a link to the Grid as 7 and 11are easily seen within it.

2. The discrepancy of 0.03548… is just short of a 28th.

3. Some will claim that the Bible gives the value of pi as 3.0. If that were true there would be little sense in hoping for any link between the true value of pi and the Bible. That claim is spurious and the point is covered in detail in the later chapter *That They May Be Taken.*

12

THE DEITY STRING

"A man may imagine things that are false,
but he can only understand things that are true." Isaac Newton

Not only the Bible, but the languages in which it is written, would have had to be inspired for the discoveries shown here to be possible. Could another race inhabiting a distance place in our galaxy have accomplished this, or might someone *outside* of the universe have put it all in place?

Some readers may be disappointed by this suggestion. They may feel that ideas of alien life forms have been impressed upon them only for the argument to switch, in effect, to God. Many are repulsed by crucifixes, candles, cathedrals and incense, and perplexed by mosques and Ramadan. Religion for them may have negative associations: wars, inquisitions, dark ages, crusades, witch-hunts and burnings at the stake. Science goes out of the window with the Christian fundamentalist (and non-biblical) assertion that the universe is a mere 6,000 years old; and the Catholic notion of communion bread turning into the literal human flesh of Jesus is outrageous.

Yet this book will incontrovertibly prove that powers in an ancient universe and foreign to this world have implanted, in eleven different passages of the Bible, the Grid system. These powers are *strangers to this world*. Their thinking and their values are *alien* to this world. The Bible predicts that the armies of the world will one day fight against them.

Clearly their intervention on earth will not be welcomed. The major Christian denominations claim to follow Christ but have long departed from the Sermon on the Mount. The prophecies of the

Bible reveal, as is later shown, that Christ at a second coming would be regarded by mankind as an intruder - *as an alien.*

Religions based on the names of God the Father and Jesus Christ for centuries foisted gross deception on the world. They have fomented carnage and genocide throughout the ages, right down into the present day. Most of these groups do not represent the God 'brand' they claim as their own. The real God has been hidden from the world, but now he can be glimpsed in the first few digits of pi:

3.14159265358979323**8**462643383279

The three eights are highlighted. In a book about pi (*The Joy of pi* by David Blatner) the 8's seemed misaligned to the author. This drew them to his attention. The well known 888 gematria of the Greek name Jesus is very simply derived. In the original Greek the name rendered Jesus has six letters: **Ιησους**. These letters have the gematria of 10, 8, 200, 70, 400, 200: total 888 (*Number in Scripture*, E.W. Bullinger, page 203).

A pattern can be seen in the following way: Jesus lived 33 years as a man and his role as the Christ was to teach. It is has been shown that the number 28 is associated with the theme of teaching in the Bible. These numbers will be seen to fit the following idea. Firstly, the 31 digits of pi with the three highlighted 8's enclose the name Jesus Christ:

The man: The teacher from God:
Digits total 33   Digits total 28

<u>3.1</u>4159265358979323**8**462643383279
Deity   six digits=Jesus seven digits=Christ

(Ιησους)   **(χριστος)**
6 Greek letters 7 Greek letters

Totalling the six digits 979323 that correspond to the letters of Jesus in Greek (Ιησους) will yield his human lifespan, 33 years. Totalling the seven digits 4626433 that correspond to his seven lettered title Christ (χριστος) produces 28, the number of the teacher from God. But what of the other great personality, the Father: where might he appear?

From the Grid it is seen that 'the Father' corresponds to the Sun. However, the circular Corona emanates from the Sun. The part of the Grid pertaining to the Father is as follows:

If the Corona emanates from the Sun, then the implication is that the spirit emanates from the Father. The circular motion implied by the Grid suggests that the spirit and the Father are one and the same. This power of the spirit emanates from his person because he is the power source. Four facts are worth summing up here:

1. That the ratio of the diameter of the human ovum to a spermatozoon is 1/44. The detailed evidence for this is given in the chapter *The ovum and the spermatozoon*.
2. That the number 88 is encoded in the human genome where there are found 44 autosomes and 2 sex chromosomes.
3. That the Greek world *sperma* (seed) is mentioned in the New Testament 44 times. (Bullinger, *Number in Scripture, p.29*)
4. That the 44th book of the Bible is *Acts*, the book in which the spirit was given to the New Testament Church.

Plainly the number 44 is the number of reproduction. The Father has encoded his purpose for humans in the Grid: *He is reproducing himself.* That is why humans are made in his image. This great Being denoted by the number 11 is having a family through the process encoded by the number 44. The Father is the senior of the two personalities and therefore comes first in pi. This how he is represented:

$$3.14159265358979323846264338327 9$$

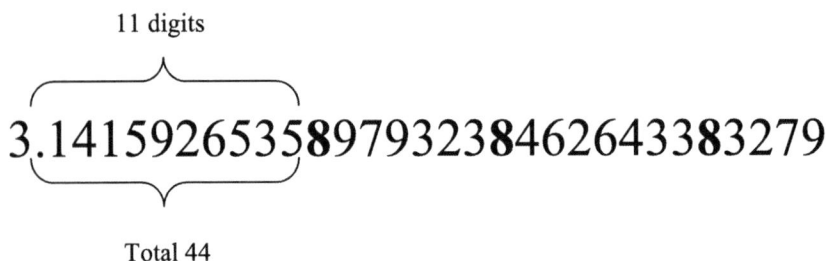

The Father is doing what all parents and grandparents have done: he is reproducing himself, and his reproducing spirit is represented by the number 44.

Having accounted for the first twenty-seven digits of the Deity String, four remain. The clue leading from the Grid to pi was a cryptic suggestion of a circle, so using the same idea:

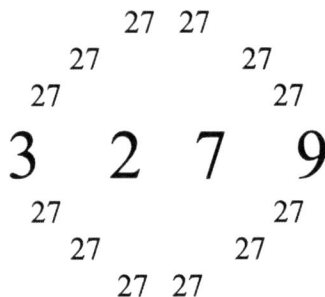

```
          27  27
       27          27
    27                27
    3    2    7    9
      27              27
        27          27
          27  27
```

The number 27 is obtained by multiplication of the outside numbers, suggesting the 27 in the centre. If 27 is the top line in a miniature 'grid' the remaining numbers can form the bottom line:

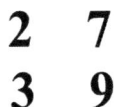

2 7
3 9

These numbers can account for the 27 books of the New Testament & 39 books of the Old Testament. They constitute the 66 spirit-inspired books of the Bible (For the total of 66 a proof is given in the later chapter *A Secret Architecture*).

Thus the first 31 digits of pi give a complete representation of God. There is also a secondary pattern reinforcing the place of the Bible in the overall plan:

Total enclosed 66 (books)

$$3.141592\ \underline{6}53589793238\underline{4}62643383279$$

However, the primary pattern displays four segments:

$$3.141592653\underline{5}8979323\underline{8}462643\underline{3}83279$$

The totals of these segments reflect the two personalities of the Grid in the following way:

	Sum Total	*Greater Factor*	*Smaller Factor*
Segment one:	44	**11**	4
Segment two:	33	**11**	3
Segment three:	28	**7**	4
Segment four:	21	**7**	3

The product (multiplication) of the four numbers 4, 3, 4 and 3 in the *Smaller Factor* column is 144, a key number later in this discovery. The Deities are each represented twice in the *Greater Factor* column by 11 and 7. Furthermore, the *Smaller Factor* column reflects the 'three to the power of four law' discussed in the later chapter *Numbers: Figment or Fact.*

Finally, there are 150 Psalms. These were instrumental in revealing the use of prime numbers (see the earlier chapter *Prime numbers in the Bible*) and confirming the key numbers 44 and 107. The Deity String of pi also totals 150.

In these many ways it is shown that the two creating Deities and their Bible are encoded in the first 31 digits of pi.

13

MESMERIC EYES

"Fie, fie, you counterfeit. You puppet, you!" William Shakespeare

Weeks before the Pope was plunged into darkness by a power cut in 2003, some strange events were unfolding around the world.

In the late summer of 2003 there was the fasting of the magician David Blaine. Caged in a Plexiglas box suspended by crane in front of Tower Bridge the tattooed semi-naked Blaine gazed out at the boats, the crowds and the world's television media. For 44 days in his fourth great public endurance test, Blaine drank water from a tube and urinated into bottles as his body slowly wasted away. Then at the *exact mid-point of his fast* the lights in Rome went out.

In the *Daily Mail* of the September 7th 2003 Olivia Stewart-Liberty recounted her London excursion with David Blaine shortly before the start of his fast. At a restaurant on the Edgware Road it became apparent that despite the best efforts of Blaine's PR to persuade the driver Ken to join the group, he was insisting on staying outside in the car. Sensing a challenge to his authority Blaine went to fetch him in. *"Don't let me sit opposite him. Don't let me sit opposite him"* Ken muttered, shaking his head. He shuddered, sat down and bowed his head. *"It's not what he said, It was his eyes. Horrible. He compelled me with his eyes."* From where does Blaine summon up his strange powers?

It had been a frustrating morning of traffic jams and at best stilted conversation. Olivia Stewart-Liberty concluded that David Blaine could freeze out anyone, as when he reportedly sat through a live three minute GMTV interview with Eamonn Holmes, fixing him with a fish-eyed stare, not uttering a word. Holmes gave his own account of the interview in a Daily Mail piece of May 2nd 2006 in which he recalled the staring eyes, the moody persona, the silence, and the moment Blaine raised the palm of his hand to reveal a drawing of an eye. Holmes said, *"What's that - what's the eye all about?"* and got his first verbal response: *"Protection."* Holmes asked *"Protection from what?"* *"From death"* Blaine had replied.

His first major stunt in 1998 was to be buried for seven days and nights. Two years later he had been encased in ice for sixty-one hours, attempting to remain there, like Christ in his grave, for three days and three nights. Then in 2002, in a stunt reminiscent of the temptation of Christ, Blaine stood on top of a ninety foot pinnacle for thirty-five hours. The top was a mere twenty-two inches wide.

When at last his car reached Oxford Circus the action began. Blaine got out and engaged passers by with some of his most impressive tricks. A playing card was thrown through a plate glass window to become stuck to the inside surface of the glass. A gasp went up from the crowd. A sceptical onlooker muttered *"mumbo-jumbo!"* He was invited to hold the torn shreds of a playing card in his hand. Blaine asked if he could feel the card getting hot. The elderly man nodded vigorously. *"It's hot"* he cried and opened his hand. The card was back in one piece and the man, practically in tears, scurried off.

Little more than six weeks later the thirty year old Blaine had lost fifty-nine pounds in weight. At the end of his 44 day ordeal of fasting he staggered from his Plexiglas box into an icy October Thames breeze. A crowd of 10,000 stood shivering by Tower Bridge ready to greet him under the spotlight of the world's television media. Gone were the egg throwing odd-balls, the sandwich-flying model airplane eccentrics and 'flashers' who had strutted their stuff by the big bridge during the interminable days and nights of the fast. Blinking under the glare of powerful arc-lamps at just before 10.00pm the sobbing, shivering 30-year-old Blaine, master of the black art of publicity, grasped a microphone to whine in his adenoidal drawl:

> *"I've learned more in that little box than I've learned in years. I've learned how important it is to have a sense of humour because nothing makes sense anyway. I've learned how strong we are as human beings. Most importantly I've learned to*

70

appreciate all the simple things in life – a smile from a strange one or a loved one; the sunrise, the sunset; everything that God has given us."

Blaine had succeeded in his 44 day fast where many had predicted failure.

Two and a half years later Blaine entered an eight foot high glass sphere filled with salt water outside the Lincoln centre in New York. Blaine had been coached for months by Navy Seals and had lost fifty pounds in weight for the event. He had been training to hold his breath. In this latest 'drowned alive' stunt Blaine was fed by tube with a hose for air and remained inside the sphere, against all medical advice, for seven days and nights. Experts had said he risked hypothermia, nerve damage and even permanent brain damage.

Despite having blacked out during training only three days before the start of his ordeal, Blaine was to attempt to break the current world record of eight minutes and fifty-eight seconds for holding breath under water. This he would do while freeing himself from chains, Houdini-style. At the end of his seven days in the sphere Blaine duly had chains attached, psyched himself up for thirty minutes and jettisoned his air pipe. He lasted seven minutes and eight seconds before his assistants plucked him from a watery grave.

In a subsequent stunt in November 2006 Blaine was shackled inside a giant gyroscope of two near-identically sized steel circles. For two days he was rotated in two directions eight times a minute, at the end of which he successfully freed himself from the shackles to fall twelve metres through a wooden stage and limp to a taxi.

Could these stunts have any conscious or unconscious meaning? Was the stunt with the glass sphere some type of baptism; a birth from a glassy womb? And then what was the meaning of the two near-identical circles? Perhaps Blaine consciously or unconsciously exhibits a messiah complex. Was the pinnacle stunt in New York really a reference to the Temple pinnacle on which the Devil taunted Christ? One televised trick performed in front of a studio audience saw Blaine cutting off the top of his ear, a likely allusion to the

71

biblical incident where the Peter the disciple cut off the ear of the High Priest's servant.

Another trick allegedly performed by Blaine on U-tube is *levitation*.

Who or what is guiding David Blaine? The mesmeric Blaine, the global Blaine, the powerful Blaine: but is this man powerful enough on his own to coordinate eight giant power cuts around the world, to plunge millions of workers and commuters into darkness and chaos, to bring huge cities to a standstill; to provoke anxious and angry commentary from the Mayor of London, the Prime minister of the UK, the President of the USA and disrupt the 2003 preparations for the succession of the Pope?

14

THE SIGNAL

"I would like to see some evidence of extraterrestrial life. I have always believed we are not alone in this universe." Arthur C. Clarke

Most people around the world in 2003 were as usual preoccupied with work and money, or the lack of it, and the various pastimes and entertainments of which David Blaine and his 44 day fast by Tower Bridge was merely one.

No one seemed to notice that David Blaine's fourth endurance test overlapped symmetrically with a succession of huge power cuts, also occurring over 44 days on four different continents. People had their tax demands and house moves, weddings and funerals and would not have considered it likely that somewhere in the trillions of miles of the cold, empty void beginning 110,000 metres above sea level that a source superior to man had somehow sent us a signal. Especially a signal no one had noticed.

Yet at one level the signal certainly was noticed: it began in Manhattan. The first power cut plunged New York and other cities across America including Detroit - and some right up in Canada - into chaos and darkness. It was ranked as the biggest power cut in history. But the seventh power cut was even larger: it came 44 days later. Proclaimed the greatest ever, the eighth power cut disrupted Pope John Paul's inauguration of 31 new cardinals, blacking out the entire Italian peninsula. The humbling of New York and Rome made front page news around the world although the six intervening power cuts, large and inexplicable as they were, did not (all were reported and are detailed later). The world carried on without recognising any pattern.

Before dismissing the suggestion of a signal from another world, or another dimension, one must consider the evidence that other-world entities have intervened already to deliberately impede the progress of mankind. This they might have done to preserve our species by blocking the development of key technologies that could have led to our destruction. In doing this they may have been following an agenda: a predetermined *program.*

Take the innovation of the steam engine, Hero's turbine, first demonstrated at around 50BC. The principle of steam power was inexplicably shelved only to re-appear millennia later as the key to the Industrial Revolution. No explanation has ever been offered for its disappearance. And why did it take a uniformly brilliant and numerate species many thousands of years to begin taking the first faltering steps towards industrialisation? Once having done so, how did man progress from a universal dependence on the horse and buggy to the atomic bomb in fifty years, to a Concorde aircraft in under seventy years and to a moon landing in under eighty years? This hugely uneven rate of progress requires an explanation.

Soon after the power cuts of 2003 and David Blaine's fast, another highly unusual and perhaps unique event took place between October 19th and November 4th of that year. Eleven giant solar flares[1] erupted from dark earth-sized spots moving across the face of the Sun. Enormous damage to satellites and electrical equipment around the world would have resulted, had these been directed towards Earth. Yet 99% of the inhabitants of our planet had no idea that Earth's closest star had experienced, according to numerous scientific websites, the most violent storm of activity in 144 years. Is it possible that events of the year 2003 were signaling something to us, a message the like of which has never been seen before?

The year 2003 was a significant year in the world of publishing with the appearance of *The Da Vinci Code,* author Dan Brown's fourth book. That bombshell publication was strongly opposed by the Catholic Church. It may have begun a cultural revolution in the way many people regard the Christian religion. It became one of the best-selling books of all time and was published in 44 languages. *The Da Vinci Code* was launched in the UK on the first day of the seventh month of 2003.

Many critics of establishment religion were emboldened by Dan Brown's fourth book. More ammunition was to come their way in

July 2007 with the first ever observation of water on a body outside of the solar system. At a distance of 376 trillion miles (64 light years) the planet known as HD189733b orbits a star in the constellation of *Vulpecula the Fox*. Finding such a body can only serve to encourage the scientists running the S.E.T.I. program whose dream of finding alien life forms seems to draw closer. According to the so called *Drake Equation* there are at least 200 civilisations populating the over 200 billion stars of our galaxy, the Milky Way. Not everyone accepts this claim. In a lecture at Caltech the late author Michael Crichton stated *"The Drake equation cannot be tested and therefore S.E.T.I. is not science. S.E.T.I. is undoubtedly a religion."*

Yet the question looms ever larger, alluded to in one way or another almost daily in the media and in science circles: when if ever will we discover alien life? However, the Signal of 2003 and the Genesis Grid discovery now render the S.E.T.I. program obsolete. A new vista of knowledge has been uncovered since the strange events of 2003. Three numbers unlocked this Aladdin's cave of arithmetic treasure; the numbers provided by the events of the year: 7, 11 and 44. These then led to pi.

A new world record for calculating pi was announced by computer scientist Fabrice Bellard on January 6th 2010 when nearly 2.7 trillion decimal places[2] were obtained. No pattern or system in pi was seen or even expected. In an article *The Mountains of Pi* which first appeared in the March 1992 edition of *The New Yorker* Dr Richard Preston wrote:

> *"If a deep and beautiful design hides in the digits of pi, no one knows what it is, and no-one has ever been able to see it by staring at the digits. Among mathematicians there is a nearly universal feeling that it will never be possible, in principle, for an inhabitant of our finite universe to discover the system in the digits of pi."*

Finally a pattern in pi (perhaps better expressed as a pattern devised according to pi, embedded in both the Bible and the fabric of the universe) has been found. Now it is possible to glimpse at least one part of a system, because pi is now shown to contain the Grid in distant and *self-validating,* or self-characteristic, positions.

The structure of the first chapter of Genesis has been shown to match the first 31 digits of pi. The world has been looking at those 31 digits, unseeing and unknowing, for 400 years. This book offers proof that the 31 verses of Genesis are based on the first 31 digits of pi. Yet the history of the discovery of pi leaves no possibility of any man possessing the knowledge of the decimals of pi circa 1400 B.C., when Moses wrote the book of Genesis.

Here is the history concerning the development of man's knowledge of pi:

Ancient Egypt	$[(8d/9)]^2$ (equivalent to 3.1605)
200 B.C. Archimedes	3 & 10/71, or 3 & 1/7
263 A.D. Liu Hui	3.141014
5th century Aryabhata	3.1416
500 A.D. Zu Changzhi, China	3.1415926
1200 Fibonacci, Italy	3.141818
1424 Ad-din Janshid Kashani	3.1415926535897932
1430 Al-Kashi, Samarkand	14 decimal places
1593 Adriaen van Rooman, Netherlands	15 decimal places
1615 Ludolph van Ceulen, Netherlands	35 decimal places
1949 The ENIAC computer	2037 decimal places
1961 IBM, New York	100,000 decimal places
1994 Chudnovsky Brothers, New York	100 billion decimal places
2002 Professor Yasumasa Kanada, Tokyo	1.24 trillion decimal places
2010 Fabrice Bellard, France	2.7 trillion decimal places

It is evident that the builders of the Great Pyramid at Giza knew pi to the *equivalent* of two decimal places, as did the builders of the Mexican Pyramid of the Sun at Teotihuacan. This is not such a great feat if one can experiment with a draughtsman's compass on a very flat surface. But even 31 decimal places for the ancient world would have been a total impossibility. In the next two chapters, it will be shown that the Grid pattern appears in perfect positions 115 million decimal places along pi's highway of digits.

The world's religions have no knowledge of whoever or whatever buried this Grid pattern in the Bible. These beings are behind *the program* of events set to occur on Earth. Major doctrines and dogmas of historical Christianity are brought down in flames by the discovery of the Grid. The Grid carries a message about an ancient universe, not the imaginary 6,000 year old universe of evangelical 'Christianity.' A message concerning an unknown God that establishment religion has rejected, whose system of worship it vilifies.

In 2003 a starting gun was fired: a gigantic figure of eight across the globe was drawn by eight power cuts in rapid succession; these included the two largest ever recorded. The Northern Rock debacle of 2007, both a symbol of and a catalyst for world economic upheaval, is linked with perfect precision to the power cuts four years earlier. It will be shown that these events are part of a program, a countdown of events.

The late author Arthur C. Clarke, whose book inspired the film classic *2001: A Space Odyssey*, expressed this 90th birthday wish:

> *"I would like to see some evidence of extraterrestrial life. I have always believed we are not alone in this universe, but we are still waiting for E.T. to call us, or give us some sort of sign…"*

A signal from somewhere was received by this planet in 2003, the year when the world turned a corner it never meant to turn.

1. At least eleven giant X-class solar flares erupted from the surface of the Sun over a 387 hour period; some websites such as the NASA site list twelve. As two were of the borderline X1 category, there were differences in measurements. This outburst on the surface of the Sun was unprecedented. The last flare was the largest ever recorded at a magnitude of X45, over twice the previous record. Remarkably, all this occurred when the 11-year sunspot cycle was already in decline. See www.weathermaine.com for the list of eleven incidents and www.spaceref.com. The BBC website www.bbc.co.uk reported: "Solar scientists have confirmed that Tuesday's explosion on the Sun was, by far, the biggest flare ever recorded, capping an energetic solar period." The event 144 years earlier on September 2nd 1859 is discussed on www.science.nasa.gov. And www.firstscience.com relates that "Even 144 years ago, many of Earth's inhabitants realised something momentous had just occurred. Within hours, telegraph wires in both the United States and Europe spontaneously shorted out…"

2. Pi was first calculated beyond a trillion places (to 1.24 trillion) on a Hitachi supercomputer in a 400 hour long session during September 2002, supervised by team leader Professor Yasumasa Kanada of Tokyo University.

15

A SELF-SIMILAR GAP

"A pattern within pi is not something that even God could arrange."
Professor Ian Stewart, Warwick University

Pi has passed into our popular culture. Doctor Spock of Star Trek defeated an out of control computer by ordering it to calculate pi to the last digit. Kate Bush composed a song using the digits of pi.

In the USA on March 14th 2006 World Pi Day was declared in honour of the number that 3.14 and which has *"beguiled and bewildered successive generations of numerate scholars since the days of ancient Babylon"* according to the British newspaper *The Independent*. The number pi is classified as both irrational (not a simple fraction) and transcendental (not arising from ordinary algebraic expressions). Its digits never repeat in such a way as to produce any discernable pattern, yet finding the Grid in pi at positions characteristic of itself would constitute a pattern.

The Grid provides three clues pointing to the number pi, as explained in the earlier chapter *An Abstract Machine*; therefore searching within pi for a Grid pattern is an obvious next step. But any string of numbers can be found amongst the digits of pi (up to an apparent maximum of 11 digits, and if this was ever proved then pi has a law involving the number 11).

Logically, because any small string – like 12345 for example – can be found repeatedly in pi, it is only in its positions, and the gap between those positions, that a pattern can be seen. For example the string 12345 appears in pi as follows:

1st appearance, beginning in position number 49,702

2nd appearance, beginning in position number 181,676

The gap between these positions is 131,974

This result shows no pattern at all. One can conduct endless searches like this without finding any pattern. But suddenly, when the Grid is searched, there is a pattern! Here are the first occurrences in pi of the top line (1-7-11) and the bottom line (44-40-4) of the Grid:

Position number in pi of the top Grid line 1-7-11 is   4,802

Pos. number in pi of the bottom Grid line 44-40-4 is 218,802

In this result, the last three digits are the same in each case. Imagine trying to guess the last three digits of your car's milometer! Furthermore, the gap between the two pi positions contains a factor that is characteristic of the Grid:

The gap comprises **214,000**. This equals **2000 × 107**.

Of this result one might say, 'now try and see if your (old fashioned) three digit number plate matches your milometer!' The number 107 has been seen before: it is the total of the Grid. The Grid was *validated* when it was found that its total, 107, appeared 28th in the series of prime numbers:

2, 3, 5, 7, 11, 13, 17, 19, 23, 29, 31, 37, 41, 43, 47, 53, 59, 61, 67, 71, 73, 79, 83, 89, 97, 101, 103, **107**

The number 28 was already shown to have singular importance vis-à-vis the Bible due to its repeated association with the theme of teaching and teachers. It is also strongly implied by the Grid:

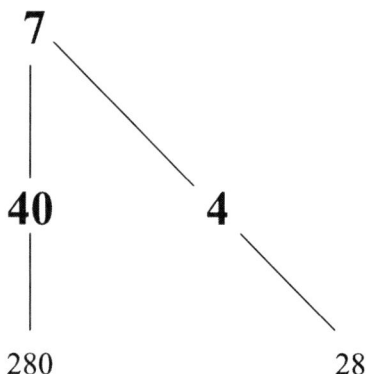

The 107-28 link is, at first, an internal validation. The most complex part of this validation system, the number 107, appears in the gap between the Grid positions in pi. The number 107 is ten times more difficult to find than 28 and is therefore more impressive as a confirmation. Also, in terms of numerical simplicity, it might be said that the outcome of 2000 × 107 is the second best result; the 'best' being 1000 × 107 one might suppose.

Yet there is a case to be made that the result of 2000, containing the factor 2, is the most *fitting* outcome of all obvious permutations. This duality has been a constant theme throughout:

1. Two deities (Father/Son) and their symbols
2. Two grid rows (Identity and power/motion)
3. Two columns of 44 (1 × 44 and 11 × 4)
4. Two lines of 88 as clues to pi (circle and bottom line)
5. Two forms of the pi string 171144404 show a result
6. Two positions in pi show a result, 1st and 2nd, but apparently none after this
7. Two segments encode a name (see the chapter *The Deity String*)
8. Two segments of pi match two parts of Genesis one (see *The Great Breach*)
9. Two numbers 11/44 in Genesis 1:28 (see *Genesis: Game, Set & Match*)
10. Two commands in Genesis 1:28 of the same pattern (*Genesis: Game, Set & Match*)
11. Two sub-segments encode a name (see *The Deity String*)
12. Two solid symbols (Sun/Moon) matched by two ephemeral (see *The Rainbow*)
13. Two mounts (Sinai/Olives) contain Grid numbers (see *The Two Mounts*)
14. Two occurrences of 11 and 7 in factorizing the four segments of the Deity String (see chapter of that name).

The Grid is derived from the Bible in which many things are also dual in nature, or are seen to appear in pairs:

1. The first Adam and the second Adam (I Corinthians 15:45)
2. Cain and Abel (Genesis 4:8)

3. The unrighteous line & the righteous line (Genesis ch. 4 & 5)
4. The 'Church in the wilderness' & the N.T. (Acts 7:38)
5. The Old and New Testaments
6. The Old and New Covenants
7. The Promised Land, physical/spiritual (Hebrews 3 & 11)
8. The first Elijah and the Elijah to come (Matthew 17:11)
9. The physical Jerusalem and the New Jerusalem (Rev 21:2)
10. The two deities, Jesus Christ and the Father (John1:1-14)
11. Two genealogies of Christ (Matt 1, Luke 3)

Of pairs in scripture Bullinger says the following:

"It is impossible to even to name the vast number of things which are introduced to us in pairs, so that the one may teach concerning the other by way of contrast or difference." page 96, Number in Scripture.

The finding 2000×107 performs the function of a fitting, elegant and economical Grid confirmation. In the next chapter, the full Grid-string 171144404 also appears in pi at positions characteristic of the Grid.

16

THREE DEFINING NUMBERS OF THE GRID IN Pi

"It's as if pi has been waiting for billions of years for ten-fingered mathematicians with fast computers to come along." Carl Sagan, Contact

It has been demonstrated that the manner in which the Grid appears in pi constitutes a pattern. So far only the positions of the Grid lines 1711 and 44404 have been shown. A further pattern is seen when searching with a continuous string of all Grid numbers. Here are the search results for the full string of the Grid, 171144404. The first two occurrences within pi are:

First appearance of 171144404: position 44,196,605   *[begins with grid number 44]*

This is equal to the product of four primes: $643 \times 233 \times 59 \times 5$

In the series of primes these are: 117th 51st 17th 3rd

[The largest prime positional 117 begins with the Grid no. 11]

(Prime positional here means the position number that a prime occupies in the series of prime numbers, e.g. 7 would be 4th)

Second appearance of 171144404: position 115,130,585 *[begins with grid number 11]*

This is composed of two primes: $23,026,117 \times 5$

In the series of primes these are: $449,817^{th}$ and 3rd

[The largest prime positional 449817 begins with the Grid no. 44]

Note here the interchange between 11 and 44 highlighted in bold. Then note this discovery in respect of the gap:

The gap between these positions is 70,933,980 *[begins with 7]* This factors into **seven** primes: 431, 211, 13, 5, 3, 2 and 2.

These findings can be summarised as follows:

a. The number for position one starts with 44 and its largest prime positional starts with the number 11. For position two it is the opposite way round: it starts with 11 and its largest prime positional starts with 44.

b. The gap number begins with 7 and factorises into 7 prime numbers.

c. The position numbers and their gap each begin with a different Grid number. These are the three *defining numbers* of the Grid **7, 11** and **44** as shown below:

44196605   position 1
115130585   position 2
70933980   the gap

Here, for the second time, the manner in which the Grid string appears in pi is undeniably *characteristic of itself.* The two *rather different ways* in which the Grid appears in pi, firstly by way of the two lines of the Grid and secondly in respect of the full string, appear to be independent of one another. A definition of independent events is as follows:

> *"If either of the events A and B can occur without being affected by the other, then the two events are independent."* Advanced Level Statistics, fourth edition, J. Crawshaw & J. Chambers.

It is hard to see how one appearance of the Grid in pi could be affected by, or be in anyway causing, the other. Yet both are characteristic of the Grid: one invokes the number 107 while the other displays the three defining numbers of the Grid, as well as the 11:44 interplay. Each discovery uses the principle of 'position one, position two, gap.'

The next pi discovery, the *Deity String* of 31 digits, is far removed in character from the above discoveries yet contains the 11:44 pattern. It highlights the teaching number 28 in an utterly astounding way.

17

SEGMENTS IN GENESIS

"Belief in the truth commences with the doubting of all those' truths' we once believed." Friedrich Nietzsche

Could Moses, the author of the first book of the Bible named Genesis, have known the first 31 digits of pi? Here is proof that either he did or that he was guided by some higher power. Of the four segments with their partitioning 8's in the 31 digit *Deity String* of pi (shown in the earlier chapter *The Deity String*), an exact match between these segments and the 31 verses of Genesis chapter one can be demonstrated:

verses
1 2 3 4 5 6 7 8 9 10 11 12 13 14 15 16 17 18 19 20 21 22 23 24 25 26 27 28 29 30 31

3.1415926535897932384626433832 79

Segments: one   two   three   four

The digits of segment four, 3279, represent the 39 books of the Old Testament and the 27 books of the New Testament. The three partitioning 8's can be distributed as follows:

314159265535**8** 979323**8** 4626433**8** 3279

The corresponding verse numbers of Genesis chapter one would then fit in this way:

verses 1-12   verses 13-19   verses 20-27   vs. 28-31
314159265358 9793238 46264338 3279

That distribution follows precisely the actual structure of Genesis chapter one, as follows:

Segment one: twelve verses describing the initial perfect creative act and its undoing (this destructive phase is explained in the chapter *The Great Breach*). That is followed by three days of restorative activity, culminating in the creation of living matter containing seeds. Verse 11 makes the first mention of seeds; this corresponds to the 11 of the 'reproducing Father.'

Segment two: seven verses describe the establishing of times and seasons via the Sun and Moon. The 19 year cycle of the Sun-Moon-Earth system corresponds to verse 19, completing the segment.

Segment three: eight verses comprising the creation of fish, birds, land animals and mankind made in the image of God. All living things are described, with man in dominion over them.

Segment four: four verses beginning with the very first instruction from God to his human protégés; in verse 28, the teaching number. The first command to mankind is in two parts: *"Be fruitful, and multiply…and have dominion…"* The dual nature of these instructions is analysed in the chapter *Genesis: Game, Set & Match.*

These four segments correspond perfectly with the segments of pi previously shown. The discovery can be summarised as follows:

3.14159265358 9793238 46264338 3279

Segments:	one	two	three	four

CREATION & THE FIRST SEEDS	THE TWO GREAT LIGHTS	CREATURES & MANKIND	THE FIRST TEACHING

The chance of obtaining by luck this matching permutation of four segments of the correct length and in the correct order within a string of 31 digits is vanishingly small. Inspiration can be the only reasonable explanation for the manner in which these thirty-one foundational verses of the Bible match against digits of pi. A comparison between the verses of Genesis chapter one and the numbers of the Grid provides further confirmation of the relationship between the Grid and the Bible:

Verse 1:   *In the beginning God created the heaven and the earth.*
This corresponds to the Grid number one representing the 'spirit of the Father.' This spirit had been the agency of creation in Genesis 1:1.

Verse 7:   *And God made the firmament, and divided the waters which were under the firmament from the waters which were above the firmament: and it was so.*
This corresponds to the Grid number 7 which is Christ's identity number. This 7th verse describes a most mysterious matter that today is rarely discussed: the dividing of the 'ocean above' and 'the ocean' below. The Hebrew word in both cases is MAYIM and means 'waters', not vapors or clouds, thus suggestive of a giant bubble or canopy of water in the sky. This 'canopy' of water was apparently a giant bubble of suspended liquid, braced against a huge atmospheric pressure. This greater pressure could have extended the lifespan of mammals[1] while the water barrier shielded life from cosmic rays (distorting today's radio-dating results accordingly). Rainbows would have been impossible with such an 'ocean above' prior to the flood. The *subsequent* appearance of rainbows, following that Great Deluge partly facilitated by a collapse of the water canopy, was an automatic result. The link between this mysterious canopy and Christ's identity number 7 is seen in the fact that the canopy concealed, and later revealed, the rainbow of 7 colours.

Verse 11:   *And God said, let the earth bring forth grass, the herb yielding seed, and the fruit tree yielding seed, and the fruit tree yielding after his kind, whose seed is in itself, upon the earth: and it was so.*

89

The Grid number of the Father's identity is signalled by this verse 11 that makes the very first mention of seeds, a theme of reproduction.

Verse 16: *And God made two great lights; the greater light to rule the day, and the lesser light to rule the night: he made the stars also.*
Sixteen is the number of power (4) squared, concentrated, or accentuated and corresponds to 4 in the Grid. The power theme is dominant in this verse as it describes the creation of the Two Great Lights.

Verse 19: *The evening and the morning were the fourth day.*
The 19 year time cycle previously discussed presents an excellent match with the contents of this section.

Verse 28: *And God blessed them, and God said unto them, Be fruitful, and multiply, and replenish the earth, and subdue it: and have dominion over the fish of the sea, and over the fowl of the air, and over every living thing that moveth upon the earth.*
The teaching number 28 contains the *first ever recorded instruction from God to man*. The dual command seen here is to do firstly with reproduction, the principal theme of the Grid.

On six counts the first chapter of Genesis confirms what is known about Grid numbers 1, 4, 7 and 11 and the Grid's associated numbers 19 and 28. In matching pi to Genesis chapter one there is another spine tingling realisation: the decimal point in pi matches the greatest division or 'split' in the entirety of scripture.

---

1. Certain surgical procedures are more successful under increased atmospheric pressure. Oxygen uptake is far more rapid, see www.cja-jca.org

18

THE GREAT BREACH

"Here the first [verse] speaks of perfection and order. The second of ruin and desolation..." E.W. Bullinger, Number in Scripture, page 93

It has been shown that there is an astounding match between the first chapter of Genesis and pi. The clue for this was that Genesis chapter one falls into 31 verses, the number of deity.

If a match with the *digits* of pi is demonstrated, what could the significance of its decimal point be? The decimal point is a prominent feature of pi. It separates the whole number 3 on one side and an endless stream of 'irrational' digits on the other (1415926535...). Matching the decimal point to Genesis reveals one of the most vital truths encoded in the Bible, because it transpires that there is a little known but highly illuminating sub-text at the beginning of Genesis. A dramatic breach is contained within the first two verses of the Bible, a cataclysmic divide within the scriptures that is strongly signalled by the Hebrew meanings. There is *no other comparable case* of such an enormous and abrupt switch from one state to another to be found in all of scripture:

"In the beginning God created the heavens and the earth."
Genesis 1:1

Yet this immediately switches to:

"And the earth was without form and void; and the darkness was upon the face of the deep. And the spirit of God moved upon the face of the waters." Genesis 1:2

In these first two verses of the Bible, the perfect creation implied by the character of the Hebrew words used in verse one is completely usurped by the chaos and 'sin' reflected in the words of verse two. Bullinger highlighted the phenomenon in *Number in Scripture*:

ORDER: *Of verse one*: *"Here the first [verse] speaks of perfection and order..."*

CHAOS: *Of verse two*: *"...The second [verse] of ruin and desolation..."*

(Page 93 of *Number in Scripture*)

The Hebrew *bara* (created) of verse one implies a perfect work, but the Hebrew *tohu* (without form) and *bohu* (void) in verse two strongly suggest *sin*[1] as the explanation for their appearance. There is compelling evidence of a change of state. A comparison of the number of Hebrew letters used is revealing:

Genesis 1:1 *"In the beginning God created the heavens and the earth."*

28 Hebrew letters (The teaching number, with the factor 7)

Genesis 1:2 *"And the earth was without form and void; and the darkness was upon the face of the deep. And the spirit of God moved upon the face of the waters."*

52 Hebrew letters (With the factor 13, the number of rebellion)

In verse one the factor 7 denotes perfection and completeness[2]. Abruptly cutting across that theme, the factor 13 in verse two suggests apostasy and rebellion. All English translations have been entirely inadequate in failing to reflect the magnitude of the change of state described in the Hebrew. It is as if a forward moving car was suddenly slammed into reverse gear.

The magnitude of the switch is increasingly obvious when one considers the definitions of the key words used in each verse:

'created'

bara – "A prim. root; (absol.) to *create*; (qualifies) to *cut* down (a wood), *select, feed* (as formative processes):- choose, create (creator), cut down, *dispatch, do, make (fat)." Strong's Exhaustive Concordance*

'without form'

tohu – "from an unused root meaning to lie waste; a desolation (of surface), i.e. desert; figuratively a worthless thing; adverb in vain:- confusion, empty place, without form, nothing, (thing of) nought, vain, vanity, waste, wilderness." *Strong's*

'void'

bohu – "from an unused root (meaning to *be empty*); a vacuity, i.e. (superficially) an undistinguishable *ruin:*- emptiness, void." *Strong's*

Could there have been chaos at the original creation? Consider this verse from Isaiah:

> *"For thus said the Lord that created the heavens; God himself that formed the earth and made it; he hath established it, **he created it not in vain (tohu), he formed it to be inhabited..."** Isaiah 45:18*

Clearly God did not create the conditions described in verse two of the Bible. The idea that God made a chaotic universe in a state of ruin and decay is illogical and unbiblical. How then did it become so quickly ruined? The subsequent use of the words *tohu* and *bohu* provides a further layer of knowledge. Consider their use in Jeremiah:

> *"I beheld the earth and lo it was without form (tohu) and void (bohu); and the heavens, and they had no light." Jeremiah 4:23*

This verse gives a description of an 'evil from the north': destruction, cities laid waste; great deception, wickedness and God's anger on a people who refuse his correction and who have *"faces harder than rock."* The words *tohu* and *bohu* are associated here with sin. Consider these further examples of sin associated with *tohu*:

> *"...city of confusion (tohu) is broken down." Isaiah 24:10*
(The context: haughty people, the earth is defiled, spiritual laws are broken; treachery.)

> *"...he shall stretch out upon it the line of confusion (tohu)." Isaiah 34:11*

(The context: indignation of God on all nations, armies are slaughtered; stinking corpses.)

"...their molten images are wind and confusion (tohu)."
Isaiah 41:29

(The context: punishment from the north, no wise men; the people are abominable.)

In every case *tohu* is associated with sin. The cataclysmic event in the second verse of Genesis is therefore linked to that theme. A terrible ruction occurred that made necessary the *renewal* of an existing planet. This is alluded to later on in the Bible:

"Thy sendeth forth thy spirit, they are created: and thou **renewest** *the face of the earth." Psalm 104:30*

The reference is to the *face* of the earth, not its foundations. The foundations of the Earth were once laid by God. Nobody had rebelled at that time:

"Where were you when I laid the foundations of the earth...7 When the morning stars sang together, and **all** *the sons of God shouted for joy." Job 38:4*

This is a reference to humanoid figures who inhabit the same plane of existence as God. They are called angels, the created sons of God and pre-existed our present world. Consider again the meanings of the Hebrew *tohu* and *bohu*: "Waste, desolation, desert, worthless thing, vain, confusion, empty, undistinguishable, ruin, emptiness, void." These two words *tohu* and *bohu* always occur as a result of sin. They could never describe what God directly creates by his own spirit. So who sinned?

It is true that God created conditions for others to commit evil:

"I form the light and create darkness: I make peace and create evil [Strongs - bad, evil, adversity, affliction]: I the Lord do all these things." Isaiah 45:7

This passage is used to try and explain the chaos of the Bible's second verse, but it is quite insufficient as it *cannot explain the sin factor.* Neither does it explain the switch from one state to another different state, or the highly meaningful numerical contrast (28 letters/52 letters) of verses 1 and 2. A far more comprehensive explanation is required. The *Gap Theory*[3], as it has become known, fits perfectly. No other viable theory is available.

In the second verse of Genesis chapter one, destruction had been brought about by the sin of the fallen angels. This is the only biblical explanation. That event was referred to by Christ:

"I beheld Satan as lightning fall from heaven" Luke 10:18

One can compare the clear and obvious breach between the Bible's first two verses to the number pi. Is there a dichotomy in pi, some great split or division, in its first 31 digits? It should now be obvious that **the great breach of Genesis chapter one corresponds with the decimal point in pi.**

Therefore, not only are Grid and associated numbers encoded and segments of the *Deity String* matched to events, but even the decimal point - the great breach of pi - has a concrete meaning. It represents the most cataclysmic rupture in the history of the universe: Satan's fall to Earth. Occult forces appeared from that time.

The explanation as to why our world is so weighed down with grief and insecurity is encoded in the first two digits of pi. A true understanding of scripture will cause us to see that occult forces rule the world of today. Paul attributed this to:

"...the god of this world..." (1Corinthians 4:4)

...before whom our leaders are little more than pawns, pushed around a giant chessboard.

In a belated attempt to avert humankind's greatest calamity, Neville Chamberlain told Hitler in 1938 that he had come a long way to see him. Hitler is said to have retorted *"I fell from Heaven."*

1. *"Sin is the transgression of the law."* I John 3:4, or more literally 'the transgression of law' as the word <u>the</u> is added by the King James translators.

2. Bullinger says of the number seven: "In the Hebrew seven is…from the root **savah**, to be full or satisfied, have enough of. Hence the meaning of this word 'seven' is dominated by this root, for on the seventh day God rested from the work of Creation. It was full and complete, and good and perfect…It is seven therefore that stamps with perfection and completeness."

3. The Gap Theory was popularized in the Schofield Reference Bible of 1909, although it had first been proposed in modern times by William Buckland and Thomas Chalmers in the early 19th century. Critics of the theory have argued that as all was pronounced by God as *very good* at the end of six days of creation, an evil Satan could not have been in existence until his appearance as a serpent in Genesis chapter three. According to this reasoning, an evil Satan must therefore have been made by God at about that point. The truth that Satan *became* evil through his own choice is evident from the combined accounts of Isaiah 14 and Ezekiel 28, in particular Ez.28:15 *"Thou (Lucifer) was perfect in thy ways from the day thou want created until iniquity (lawlessness) was found in you…17…I will cast thee to the ground…"* Satan is composed of spirit. What was pronounced very good in Eden was only the new physical creation.

19

GENESIS: GAME, SET & MATCH

"If such numerals were to elude casual observation ... [it] would be decided evidence of a supernatural design..." M. Mahan, Palmoni

Much has already been shown here about the structure of the Bible that is new knowledge. The next insight must surely move the most ardent skeptic.

Books have been written before about patterns in the Bible, some fair and some fraudulent. None ever offered claims as staggering as those now offered up in this chapter. This next discovery in Genesis chapter one hugely reinforces the finding that:

1. The Father's secret identity number is 11 and

2. His number of reproduction 44.

It is worth first of all reviewing the manner in which the six numbers of the Grid suggest pi as the seventh. A pattern of numbers would be of little use to anyone if it has no meaning. A bar code can be read by machine and encodes a price, a useful meaning. The symbols of the Grid, the Sun and the Moon, provide vital insights concerning the relative power and position of the two Deities. This they do while ruling out any personhood for the 'holy spirit.'

The six numbers of the Grid convey an amazing amount of information, yet the number six is strongly linked to man (created on the sixth day) and his incompleteness. Therefore it would seem strange that a Grid purporting to encode the identity and nature of God would be comprised of only six numbers. A seventh number for the Grid would seem imperative.

Clues for pi are in abundance. For example, an obvious thing about the six Grid numbers is the outside columns. They both possess the same product, 44. Drawing a circle around the Grid was the simplest way to link these columns. The sum of the 44's of the circle is the same as the sum of the bottom line of the Grid, 88. This hints at a circle as it is (almost) the product of pi and the teaching number 28, the number that had validated the Grid by its total 107 (the 28th prime number). The truth of the idea is confirmed when pi is searched and Grid lines found in self-validating positions.

The circle solution also has one fascinating spin-off. It appears to shed further light on the question of who, or what, is the 'holy spirit.' Pictorially, the Grid suggests that the spirit emanates from the Father, a perfectly scriptural idea. Then there is the sensational realisation that the number 44 is a reproduction number (see the chapter *The ovum and the Spermatozoon*). The Father's column is unified with the reproduction/spirit column via the circle:

$$
\begin{array}{ccccc}
 & & 44 & 44 & \\
 & 44 & & & 44 \\
44 & & & & 44 \\
\mathbf{1} & & \mathbf{7} & & \mathbf{11} \\
\times & & & & \times \\
\mathbf{44} & & \mathbf{40} & & \mathbf{4} \\
 & 44 & & & 44 \\
 & & 44 & & 44 \\
 & & 44 & 44 &
\end{array}
$$

The theme is obvious: there is a reproducing Father. A shorthand expression for that idea would be 11/44. Then it is seen that the first chapter of Genesis comprises 31 verses, the deity number. The structure of that passage is based on four segments. Those segments correspond to four segments in pi. The first 11 digits of pi 3.1415926535 total 44. This represents the Father, 11, and his reproducing spirit, 44. Then the 11/44 pair was seen to occur as an alternating pattern in the full Grid-string 171144404 in pi. (That 11/44 pattern is also prominent in the frequencies of related musical notes, described in the later chapter *Grid Vibrations*.) Evidently,

Genesis chapter one is built on pi. But the 11/44 pattern makes a further dramatic appearance in its Hebrew letters:

Genesis chapter one:

Verse	(Hebrew letters)
1	**28**
2	52
3	23
4	45
5	49
6	44
7	65
8	39
9	52
10	49
11	69
12	67
13	22
14	76
15	37
16	80
17	33
18	51
19	22
20	57
21	89
22	52
23	22
24	56
25	69
26	84
27	50
28	**88**
29	83
30	69
31	50

The teaching number is 28. It is in the 28th verse that the *first ever recorded command from God to man* appears. At this point it is worth recapping that the number 44 encodes the theme of reproduction in the following ways:

1. New Testament occurrences of the Greek word for seed, *sperma:* 44 times
2. The Hebrew gematria for father(3), mother(41) and child(44) suggests 3 + 41 = 44
3. The human spermatozoon/ovum diameter ratio of 1:44
4. The human genetic package: 44 autosomes and 2 sex chromosomes[1]

Now consider, in this next discovery, the strength of the manner in which the key number 44 appears within the all important first chapter of Genesis. The first instruction to man was given in verse 28 *but this instruction was dual.* If we ignore the English punctuation of the numerous translations in use (there is no punctuation in the Hebrew) we can see *two* distinct instructions, the first of which is:

Multiply and replenish the earth

And the second:

Subdue and have dominion

Here are these two commands in the original Hebrew.

Set 1, command one:

1 2 3 4 5	6 7 8	9 10 11 12 13	14 15 16 17 18	19 20 21
ויברך	אתם	ויאמר	אלהים	להם
word *one*	*two*	*three*	*four*	*five*
and-he-is-blessing	them	and-he-is-saying	Elohim[God]	to-them

22 23 24 25 26	27 28 29	30 31 32 33	34 35 36 37 38 39 40	41 42 43 **44**
אלהים	פרו	ורבו	ומלאו	את–הארץ
six	*seven*	*eight*	*nine*	*ten* *eleven*
Elohim[God]	be-fruitful	and-increase	and-fulfill	» the-earth[2]

Without punctuation the King James Version of this section reads:

> *"And God blessed them and God said unto them be fruitful and multiply and replenish the earth."*

The passage contains **11 Hebrew words with 44 characters**. The second half of verse below 28 follows precisely the same 11/44 pattern:

Set 2, command two:

1 2 3 4 5	6 7 8 9	10 11 12 13	14 15 16	17 8 19 20 21
וכבשה	ורדו	בדגת	הים	ובעוף
word *one*	*two*	*three*	*four*	*five*
and-subdue-her!	And-sway!	in-fish-of	the-sea	and-in-flyer-of

22 23 24 25 26	27 28 29 30	31 32 33	34 35 36 37 38 39 40	41 42 43 **44**
השמים	ובכל	חיה –	הרמשת	על – הארץ
six	*seven*	*eight*	*nine*	*ten* *eleven*
the-heavens	and-in-all	living-one	the-one-moving	on the-land

Without punctuation the King James Version of this section reads:

> *"...and subdue it and have dominion over the fish of the sea and over the fowl of the air and over every living thing that moveth upon the earth."*

So the 11/44 pattern appears twice in verse 28 of the Bible. This is the first recorded instruction from God to man. The pattern 11/44 represents the first 11 digits of pi that total 44, identifying the Father and his reproducing spirit: 31415926535 where 11 digits total 44. The Bible confirms that the Grid and the two Deities and their word are woven into the first 31 digits of pi. Those 31 digits also confirm the Grid. The Grid's remote appearance in pi confirms itself and the 11:44 pattern, again found here in the 28th verse of Genesis (see the earlier chapter *Three defining numbers of the Grid in pi*). In the light of all this, who could be surprised that the first conversions (spiritual *begettal*, not birth) in the Church occurred in the 44th book of the Bible, *Acts of the Apostles*. This was the book in which flames of fire (the Corona) appeared on the heads of the Apostles, at the giving of the holy spirit.

Thus a multifaceted and interlocking proof is offered for the Grid and its defining numbers 7, 11 and 44. Who could fabricate such a watertight case? Two seemingly unrelated sets of data, one from ancient manuscripts and the other derived from modern computer calculations of the digits of pi, have been found to correspond precisely in multiple ways.

Pi is defined by the simplest shape in the universe. It is part of the fabric of our universe and of basic truth itself. It could no more be altered than 2 + 2 could be made to equal 5. Therefore the only way for such a precise fit between the two sets of data would be for a Higher Power to inspire the Bible and its languages (with their gematria values) in such a way as to ensure a fit.

Several more ways in which the Grid is encoded in the Bible are yet to be presented, but even at this stage it is obvious that this discovery amounts to proof of the inspiration of the Bible.

We need not doubt the statement made in chapter one concerning those who have intervened on this planet:

> *"their finger print is 44 and their calling card is pi."*

1. Where the 2 sex chromosomes are a key to the individual's *identity*, but the human frame is built in both cases on the 44 autosomes.

2. This is the layout used in *The Interlinear Bible*, where the symbol >> is inserted instead of a meaning in the translation space beneath a two lettered word. The words 'the-earth' appear directly under the corresponding Hebrew four lettered word. The two words joined by a hyphen at the end of set 1 follow the same pattern of words ten and eleven in set 2 (clearly two words) where a translation for word ten is shown: the meaning "on."

20

THE TWO MOUNTS

"No man ever believes that the Bible means what it says:
He is always convinced that it says what he means." George Bernard Shaw

The Grid and its associated number patterns are spread across the entirety of the Bible from Genesis to Revelation, in eleven passages. It can be seen in the two main passages of scripture that dispense the moral law.

The 'Sermon on the Mount' comprising Matthew chapters five, six and seven is considered by some to be the most important passage in Bible. It is Jesus Christ's detailed exposition of moral principles and practices based on, and derived from, the Ten Commandments. This was not the first Sermon on the Mount, nor was its message entirely new. It merely expounded the original law given on the other mount: Mount Sinai. For example, in Matthew chapter five:

> *"...it was said unto them of old time, thou shalt not commit adultery: 28 But I say unto you, that whosoever looketh on a woman to lust after her commits adultery with her already in his heart."*

Here Christ refers back to the sixth Commandment and then magnifies its intent, intensifying the degree of obedience required. Outward conformity is no longer sufficient: it must now be of the heart. So the law has its basic form, but also a higher expression – its fullest *intent*. There is a similar principle operating in respect of the Bible's hidden numerical structure. God is an architect applying the same principles across different layers of his great construction, the Bible. It should not be surprising that the development of matters ordinal (law) should proceed in a manner rather similar to those that are numerical (the Grid and its associated patterns).

To illustrate this principle further, it can be shown that a number can appear in a basic form, such as 44, but can also be related to a higher number; another 'version' of that number. This higher version expresses a greater depth or range of meaning than the lower version. The greater expression of number 44 is evidently 88. For the teaching number 28 it is obviously 280. So in its basic form, 44 denotes reproduction, but in its greater expression 88 it points to the *end result* of that: eternal life as pictured by the digits of pi to which it points. The following numbers and their meanings can be deduced as follows:

BASIC FORM	GREATER EXPRESSION
11 The Father	111 The Father's plan
28 The teaching function	280 The teaching program
44 Reproduction	88 Eternal life

With regard to the 111 pattern, this is seen in the centre of the 1711171 holy day cycle. The holy days denoted by the three 1's, Pentecost, Trumpets and Atonement, are *shadows* of yet future events.

If the Grid is integral to the Bible, one might reasonably hope for evidence to that effect within its most fundamental passages. This could include the giving of the Ten Commandments from Mount Sinai and its New Testament equivalent, the Sermon on the Mount. Upon investigation this is found to be the case. Firstly, the Sermon on the Mount comprises 111 verses. The Grid number of reproduction, 44, is also flagged by verse 44 of Matthew chapter five (the 44th verse of the entire discourse), which contains the following action step:

> *"But I say unto you, love your enemies, bless them that curse you, do good to them that hate you, and pray for them that despitefully use you, and persecute you."*

Christ gives the reason for this in verse 45: *"That you may be the children of your Father which is in heaven..."* Nowhere else in this Sermon on the Mount is the parent/child connection so directly stated. No other part of the discourse links sonship to a mode of thought and behaviour. *The required action step is in the 44th verse.*

The reader will by now be aware that the Grid suggests a far more literal concept of sonship (as do the scriptures) than the *adoption* idea put out by mainstream religion. The main pattern of the Grid is the circle of 44's that leads to pi. This number 44 is the biological reproduction number, as previously shown. The Grid contains a Father and a Son and its main theme is reproduction. As before in this investigation, one could ask if God would really be incapable of something humans take for granted: *the ability to beget, through impregnation of a life-giving seed, one's own literal children.*

The structure of the three chapters comprising the Sermon on the Mount is astonishing in one more respect. There are 7 threats of eternal death. These are found in verses positions 13, 20, 22, 29, 102, 105, and 109. But there are 7 offers of eternal life and these are in verse positions 3, 5, 8, 9, 10, 12 and 19. The position numbers for the 7 verses pertaining to eternal life total 66, matching the total number of books of the Bible. The 7 verses pertaining to eternal death total 400, the identity tag of the 'Two Great Lights' who are to judge on matters of eternal life, and eternal death (Revelation chapter twenty).

So the defining numbers of the Grid, 7, 11 and 44, are unmistakably stamped on Christ's first, longest and most important dissertation, the Sermon on the Mount. But what of the Ten Commandments that appear for the first time in Exodus chapter twenty?

The Ten Commandments can fall into two sets. The first four pertain to the relationship between God and man, while the remaining six regulate relationships between man and fellowman. But it can be argued that the fifth commandment (honour your father and mother) also pertains to, and can affect, one's relationship with God and should appear in the first set to make a set of five. The two sets in short form are then as follows:

SET ONE: 1. Thou shalt have no other Gods before me
2. Thou shalt not make unto thee any graven image
3. Thou shalt not take the name of …God in vain
4. Remember the Sabbath day to keep it holy
5. Honour thy father and mother

SET TWO: 6. Thou shalt not kill
7. Thou shalt not commit adultery
8. Thou shalt not steal
9. Thou shalt not bear false witness
10. Thou shalt not covet

From these two sets of five a further proof of the Grid emerges. The first five commandments are comprised of 111 words in the *original Hebrew text*. These five commandments are also made up of 535 Hebrew letters, or characters, where 535 is 5 × 107, the total of the Grid! Thus the average length of these commandments is 107. But the number 535 has appeared before. It was a distinct sub-segment within the 11 digits of the Father's pi string 3.1415926**535** where the sum of 5+3+5 is 13, the reversal of 31.

The above occurrence of 13 within the Father's string is discussed in the chapter *And then there were Two*. But here is seen an average of 107 Hebrew characters for each of the five commandments. The Grid total is 107, the 28th prime number.

But what of the second set of five commandments? In the Sermon on the Mount the command to love neighbours (including enemies) is the *action requirement* of verse 44 linked to sonship (*"…that you may be the sons of your Father…"* verse 45). That same link between the number 44 and the 'loving neighbours' concept is clearly seen in this second set of five Commandments. Concern for neighbours is the preserve of this set, also marked with 44: they are comprised of 88 Hebrew characters.

Thus the Grid is stamped upon the words spoken on the two Mounts. There is now a super-abundance of proof that the Grid is real.

21

JOHN CONFIRMS THE GRID
"...that disciple whom Jesus loved..." John 21:7

The book of John is the odd one out of its set, the four Gospels. It lacks genealogies and has fewer historical data than the other books, but *it encodes the Grid in its entirety.*

John does not feature a lengthy denunciation of the Pharisees, as do the other three gospel writers. Also, the threats of eternal damnation (destruction) present in the other three books are largely absent in this the fourth gospel. These odd and distinct qualities of *John* vis-à-vis the other three gospels support a 3 + 1 pattern, or a 3^4 law, as discussed in the later chapter *Numbers: Figment or Fact.*

John has a style distinct from the other three gospel writers. His book possesses a profundity that sets it apart. It also goes much further than the other three books in one key area: Christ stated (in Matthew 11:27 & Luke 10:22) that he had come to reveal the Father, the most important personality of all. If men had known of this great personality it would not have been necessary to reveal Him. Most of this revealing of the Father occurs in *John.* One aspect of this revealing is described in Luke 10:21 – 22:

> *"In that hour Jesus rejoiced in spirit, and said, I thank thee O Father, Lord of heaven and earth, that thou hast hid these things from the wise and prudent, and has revealed them unto babes: even so, Father, for so it seemed good in thy sight. v22 All things are delivered to me of my Father: and no man knoweth who the son is, but the Father; and who the Father is, but the son, and he to whom the son will **reveal** him."*

The process of revealing the Father is also described in Matthew chapter eleven in very similar terms: twice in all. But it is in John that the Grid is encoded in its entirety by way of the revealing of the mysterious Father. This revealing has not been to the *"wise and prudent"* but to *"babes."* The most educated classes of society were and are, by and large, excluded. From such, indeed from almost everybody, the true nature of God remains hidden in this present age.

There is a sense in which the *revealing* of the Father is linked to the number 107. It was shown earlier that Deuteronomy, the *odd one out* of its group of five books, yielded a finding concerning the number 107, the Grid total. John is also *odd* within its group (the four gospels) and it too contains a hidden match with the Grid. In fact there is a 'duality' operating, because the same pattern of Grid confirmation is seen in both the Old Testament (Deuteronomy) and the New Testament (John). In each case they were the *favourites of Jesus Christ*: the book of Deuteronomy which Christ quoted from exclusively during the Temptation (Matthew chapter four); and the man John, the disciple Christ especially loved (John 19:26).

Why was Deuteronomy Jesus Christ's favourite book? Could it be that it was more reflective of the Father's nature than any other within the Old Testament (the only Bible Jesus had)?

It is worth recalling that in the case of *Deuteronomy* the last (sixth) Grid number to be deduced, 44, was represented by a corresponding set of 44 Psalms. That set began with Psalm 107. And the Grid was finally confirmed by finding 44, resulting in a total of 107. So this 44/107 sequence is seen in Deuteronomy, but also in John. As with Deuteronomy, the book of John differs sharply in character from the others in its set. It too *strongly features the number 107* as will now be shown.

The realisation that there exist two God-beings is a fact fundamental to both the Grid and the truth concerning Jesus' revealing of the Father in John. The whole point of that revelation concerning the Father was this: *He is another person that nobody had any inkling of before!* Old Testament Israel had no idea about any 'Father' superior in authority to the God leading them out of Egypt. The being Moses interacted with was the 'Word' (Gk. *Logos*) of John chapter one, the pre-existent Christ. This was the same being that had followed the Israelites in the wilderness, signifying himself as a pillar of fire by night and of cloud in the day. He was not the Father.

The pre-existence of Christ as a God-being is a fact of great importance attested to by several Old Testament passages. The account of Melchizedek in Genesis chapter fourteen concerns an eternal being, the *"priest of the most high God"* where the Father is referred to as God. Of this priest the Apostle Paul wrote:

> *"Without father, without mother, without descent, having neither beginning of days nor end of life; but made like unto the Son of God; abideth a priest continually." Hebrews 7:3*

To accept that Christ and Melchizedek are one and the same is to accept that Christ acts as a priest in relation to the Father – that he worships the Father! Hence Christ's statement in John: *"My Father is greater than I."* This does not lie comfortably with false trinitarian reasoning. But just as John reveals the Father and contains the Grid of 107 in its entirety, so does also the Grid reveal the Father. It shows his distinctness, as a personality, from Christ. The Grid shows that the Father has determined reproduction as his theme or purpose.

Melchizedek was a human manifestation of 'the Word', the 'Logos' of John chapter one. In human form he interacted with Abraham. Paul understood that the *"Rock that followed them"* in the wilderness was Christ (1Cor10:4). This was the same being that had addressed Moses from the burning bush (Exodus 3:2), a *messenger* (Greek *aggelos*) from the Father. This was not an angelic messenger with wings as is sometimes supposed, but that greater Messenger.

But why does it matter what kind of messenger spoke in the burning bush? It matters because the whole question of the nature and purpose of God hinges upon it! Jesus Christ, 'the Word' of John chapter one, eventually became a human messenger: the *messenger of the New Covenant*. The 'Word' had always been a messenger! (John 1:1 *"In the beginning was the Word..."* the 'Logos' – spokesman, or 'messenger.') During his pre-existence he had dealt directly with Moses and the Israelites. Here is the proof: Moses had been dealing with a God-being, which is why at the burning bush he was *required to remove his shoes. No one ever did that for a winged angel*. Compare this incident with what actually happens when humans encounter an angelic being:

> *"...when I [John] had heard and seen, I fell down to worship before the feet of the angel which shewed me these things. v9 Then said he unto me, See thou do it not: for I am thy fellow servant..." Revelation 22:9*

John was rebuked for trying to bow down to an angel, thought of commonly as a winged messenger. He wasn't asked to take his shoes off as Moses had been. Neither did this angel (messenger) say 'this is holy ground.' Contrast this with the passage in Exodus where Moses confronted another great being:

> *"...God called to him out of the midst of the bush, and said, Moses, Moses. And he said here am I. And he said, draw not nigh hither: put off thy shoes from off thy feet, for the place whereon thy standest is holy ground. Moreover he said, I am the God of thy father, the God of Abraham...Moses hid his face; for he was afraid to look upon God."*

This account shows the stark difference between Moses' experience and that of John in Revelation. John did not hide his face; neither was he told to put off his shoes. The Angel did not say *"I am the God of Abraham"* or anyone else, neither was he told that he was on holy ground. Also, he was not afraid to look but he was chided for bowing down to 'a messenger.' In *Acts* 7:31 Stephen recounted that Moses had heard *"the voice of the Lord"* not merely the voice of an angel (messenger). Deuteronomy 5:4 reiterates that it had been a God-being speaking with Moses:

> *"The Lord talked with you [Moses] face to face in the mount..."*

In the later chapter *And then there were Two* the issue of the 'twoness' of God is further discussed. This truth fits the symbols of the Grid, the Sun and the Moon, and their status as monuments at creation. It fits the correct understanding of the nature of God as a family, not the Trinity heresy; this is why it's important. In support of this, an analysis of the key passage in Exodus chapter thirty-three pertaining to Christ's 'pre-existence' is contained in Appendix X.

Christ came to Earth as a human to reveal the Father. Below, again, is the breakdown of the 280 mentions of "the Father" listed by Bullinger:

Number in Scripture
Page 28: words in the New Testament

"The Father" occurs in Matthew	44 times	*(4 * 11)*
Mark	22	*(2 * 11)*
Luke	16	*(4 * 4)*
John	121	*(11 *11)*
The rest of the N.T.	77	*(7 * 11)*

Bullinger correctly tallies occurrences of the phrase 'the Father' at 121 (11²) times in John's gospel. This is the most <u>accentuated appearance</u> of the number 11. *Thus the Father's Grid number 11 is most strongly affirmed in the book that makes greatest mention of him,* the book of John. The book in which he is revealed far more than any other. But the tally shown above uses 'The Father' singular. This is a different total to the occurrences of merely 'Father' or 'my Father's' which in the four gospels are as follows:

Matthew	42
Mark	5
Luke	18
John	107

The total times that the Father occurs in conversation (107) differs from the count for the use of his name (121). Using the King James Version[1], it is found that counting occurrences of the actual *mention of the person* of the Father is 107. To do this one must include the expression "my Father's" in the count (as in John 14:1 *"In my Father's house are many mansions"*). Then should be taken into the count statements in which the name (heavenly) 'Father' occurs *twice in one mention of him,* for example:

John 10:15 *"As the Father knoweth me, even so I know the Father."*

111

In the above statement Christ *broaches the subject* of his Father once (in one breath as it were) but uses the name twice. Thus one sees that the Father is <u>broached in conversation</u> 107 times in the gospel of John. A list of all 107 instances is given in Appendix IV.

This Grid verification in John is reinforced by the manner in which the 107 verses are distributed. They form successive *sets* according to the numbers of the Grid. Each set reflects the meaning of that part of the Grid. The match is seen when the 107 verses are grouped so as to fit Grid numbers ordered by the column of the Grid in the following manner:

Column	Column	Column
1	**7**	**11**
44	**40**	**4**

These three Grid columns can be collapsed into one line in the following manner:

Column	Column	Column

1   44   7   40   11   4

This arrangement[2] of the Grid facilitates a match with John. The following correspondence between the 107 mentions of the Father in John, taken in the order given, and the numbers of the Grid then emerges:

Grid number 1: begettal.
Corresponds to Father-statement number 1: chapter 1:14 "...the only begotten of the Father."

Grid number 44: predominantly begettal. *Corresponds to Father-statements 2-45 inclusive.* The section begins on the origin of Christ from the intimate bosom of the Father, and ends on what can be understood as the complete security of the 'womb of the Church' for the true follower.

Grid number 7: identity of Christ. *Corresponds to Father statements 46 – 52 inclusive.* Christ begins by identifying himself in the Father and then twice in this section by his works.

Grid number 40: trials and testing of Christ. *Corresponds to Father statements 53 – 92 inclusive.* This begins with a reference to Christ's biggest trial, "this hour"; then further mention that "his hour had come" i.e. the greatest trial; mention of the "comforter" by which trials may be endured; Christ and his followers to be hated; the section ends on the trial of no longer seeing Christ and the burden of being spoken to in parables, soon to be ended.

Grid number 11: the Father's identity. *Corresponds to Father, statements 93 - 103 inclusive.* This section contains themes showing from whence Christ came, i.e. the Father, and identifying the Father's true worshippers; concluding on the fact that the world cannot identify the Father.

Grid number 4: the Father's power. *Corresponds to Father, statements 104 - 107 inclusive.* This shows the power to resurrect and the power to have actual sons (not adopted) to do his work. Indeed, an equivalence between Christ's sonship with the Father and (potentially) our own is inferred (item 106) here.

This close correlation between successive sections of the 107 statements and the Grid can be summarised in the following way:

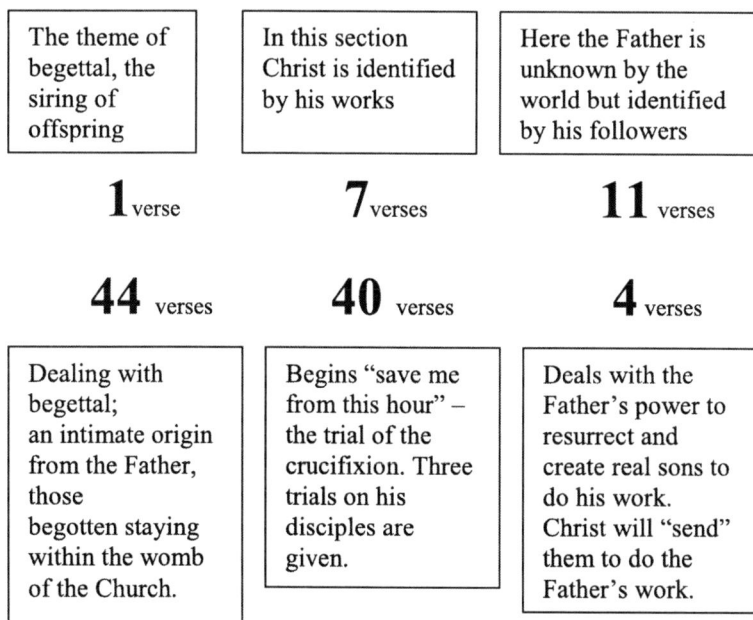

The theme of begettal, the siring of offspring	In this section Christ is identified by his works	Here the Father is unknown by the world but identified by his followers
1 verse	**7** verses	**11** verses
44 verses	**40** verses	**4** verses
Dealing with begettal; an intimate origin from the Father, those begotten staying within the womb of the Church.	Begins "save me from this hour" – the trial of the crucifixion. Three trials on his disciples are given.	Deals with the Father's power to resurrect and create real sons to do his work. Christ will "send" them to do the Father's work.

In *John,* the statements concerning the Father confirm the Grid *in its entirety.* How fitting that within the gospel written by the *disciple Christ loved* - and in which he most fully revealed the Father - is seen the complete pattern of the Grid.

The 107 occurrences are listed in full in Appendix III.

1.The King James Version is almost a 'word for word' translation where an additional English word is added for clarity are in *italics.* This means that the number of times words such as 'Father' are mentioned will tally with the original Greek.

2. The string 144740114 does not occur within the first 200 million digits of pi, the maximum searchable on the Internet at the time of writing.

22

COMMONPLACE PATTERNS

*"Nothing occurs at random, but everything for
a reason and by necessity." Leucippus (5th century B.C.)*

The Grid, the inspired Bible and the imprinting of the number 44 on the human reproductive system are now facts that rather cruelly undermine cherished beliefs held by both scientists and the churches.

Questions such as the origin of life and the evolution of man require re-examination in the light of evidence here presented. As for the dogmas of historical Christianity, these could always be challenged with a galaxy of evidence: geological, astronomical and biblical. Traditions of pagan origin adopted by Christendom do not withstand scrutiny and neither does a 6,000 year old Earth.

With regard to the question of origins, if God is all powerful he could have created humans by any method or system he determined, marking their reproductive system with the number 44. He could have accomplished this through Darwinian evolution, but did he? That is a matter requiring a further book. The transition from belief in a flat Earth to a globe contradicted many assumptions in the middle ages. Sailors were no longer in fear of falling off the edge of the world once the new truth had been assimilated.

As for the revelations of the present book, patterns in pi aside, they are not new truth, but a *reversion to old truth*. The Bible is now proved to be inspired.

It may be claimed by some that the findings of a Bible pattern in pi are coincidental and could be easily replicated in other ways. For example, finding the 7, 11 and 44 set of numbers in the first and second positions of 171144404 in pi is not special and presumably could be repeated in other positions in pi. Yet if these kinds of self-characteristic patterns are easy to find then subsequent pi positions should yield similar results, but they don't. There appears to be no Grid patterns beyond position one and two in pi. Yet along the 1.24 trillion known places of pi there may be about 5,000,000 examples of 1711 and 44404 and 600 occurrences of the full string 171144404. Here is a challenge: can a meaningful comparable pattern be found in any of the Grid positions in pi, *other than those already found in positions one and two in each case?*

If God is supreme in power, he will have either foreseen the eventuality of the discovery of the Grid, or he will have actually engineered its discovery. Has he arranged matters in such a way that (following a principle familiar in this investigation) it is the *first occurrences* of the Grid in pi that are important, and no other? The author lacks the wherewithal to demonstrate this conclusively for the full string, as only the first 200 million digits can be searched using the Internet at the time of writing. Those with the necessary resources are invited to find subsequent Grid patterns in pi of comparable improbability to those now discovered. Certainly the first fifty appearances (shown below) of the strings 1711 and 44404 offer little encouragement.

Apparently there exists no Grid pattern in pi comparable to those already shown for 1711 and 44404. Seemingly there is nothing beyond Grid positions one and two. The low probability of finding anything like the 2000 × 107 gap (214,000 shown below) is illustrated by the following where, for example, the second appearance of 1711 in pi is at position number 6,323 and so on:

Pi position no.	1711	44404	the gap
1 (the only pattern)	4,802	218,802	214,000
2	6,323	279,765	273,442
3	8,140	411,202	403,062
4	23,588	447,654	424,066
5	29,847	557,385	527,538
6	35,632	562,054	526,422
7	50,444	787,731	737,287

8	61,461	826,262	764,801
9	62,028	934,974	872,946
10	62,141	1,210,736	1,148,595
11	89,027	1,264,273	1,175,246
12	99,869	1,374,026	1,274,157
13	117,421	1,458,740	1,341,319
14	124,503	1,507,037	1,382,534
15	133,894	1,705,383	1,571,489
16	134,694	1,737,713	1,603,019
17	155,832	1,976,293	1,820,461
18	159,662	2,161,766	2,002,104
19	163,699	2,270,876	2,107,177
20	168,997	2,290,879	2,121,882
21	175,507	2,317,122	2,141,615
22	181,721	2,321,705	2,139,984
23	190,126	2,368,949	2,178,823
24	201,433	2,380,019	2,178,586
25	203,391	2,534,962	2,331,571
26	208,995	2,843,157	2,634,162
27	210,556	2,913,577	2,703,021
28	219,454	2,939,073	2,719,619
29	244,892	3,334,910	3,090,018
30	251,156	3,434,419	3,183,263
31	255,943	3,442,191	3,186,248
32	293,363	3,471,717	3,178,354
33	300,825	3,783,288	3,482,463
34	305,252	3,787,399	3,482,147
35	306,466	3,979,400	3,672,934
36	309,280	4,124,599	3,815,319
37	314,210	4,224,218	3,910,008
38	331,675	4,429,976	4,098,301
39	356,054	4,493,528	4,137,474
40	375,124	4,542,286	4,167,162
41	377,663	4,556,105	4,178,442
42	386,724	4,994,711	4,607,987
43	421,539	5,108,652	4,687,113
44	441,890	5,251,576	4,809,686
45	457,700	5,286,739	4,829,039

46	458,921	5,293,937	4,835,016
47	458,956	5,320,353	4,861,397
48	465,120	5,372,866	4,907,746
49	474,012	5,527,991	5,053,979
50	490,208	5,694,865	5,204,657

No patterns can be seen here, except in the first line.

Of the full string 171144404, this occurs only twice in the first 200,000,000 positions of pi, so the compilation of a similar table for that pattern is not yet possible. But this too shows remarkable Grid similarities in the positions[1] one and two.

As already shown, the full Grid pattern 171144404 appears at positions one and two in a manner that is self-characteristic due to the 'appearance' of 7, 11 and 44. In reality, because the digits of pi are never ending every possibility could eventually play itself out. So eventually logic dictates that you would find the pattern you want if you go far enough along pi. But the impossible thing to explain is: *why do the astonishing gap-patterns occur at the very first coupling of 171144404 and the very first appearance of 1711 and 44404?*

That thought begs the question now addressed in section two: could all this have been brought about merely by another intelligent life form from another world? Or must it have been inspired by a universe-creating power? Can this question be answered conclusively?

Further Grid verifications are given in Appendix V.

1. There is only a 9.5% chance of any 9 digit string such as 171144404 appearing just once within the first 200 million places of pi. In the 2.7 trillion places of pi so far calculated the string could therefore be expected about 1300 times.

PART TWO: WHO AND WHY?

23

TIDES THAT ENCODE THE GRID
*"What manner of man is this, that even the wind
and the sea obey him?" Mark 4:41*

If the schemes of mere mortals cannot account for the astonishing structures in the Bible then one must look elsewhere for an explanation. Here, in part two, further evidence is gathered to cast light on the following question: have other physical life forms been responsible for the observed Grid phenomena, or is the Grid so fundamental to everything that it demands another explanation? In successive chapters this will be conclusively answered.

There is a further point of proof concerning the Grid discovery, relating to the Moon and its tides. Probably many secrets are entombed in the rocks of Moon. It is far lighter than its volume would suggest yet its movement controls the tides of our vast oceans. Its mass is a mere 1/81 of planet Earth (to the nearest whole number). This number 81 also happens to be the total for the number of stable elements in the periodic table, the basis of all science. In the later chapter *Numbers: Figment or Fact* the number 81 is investigated in some depth.

With respect to the Moon's motion it has already been shown that the lunar eclipse cycle of 173 days confirms its secondary Grid number 40 (as 173 is the 40th prime number). But there is a further Grid number encoded by the Moon, because in the procession of the ocean tides the number 88 can be clearly seen. High and low tides occur at 12.44 hour intervals and so high tides are therefore

119

advanced by a period of 0.88 hours in each 24 hour period, giving an actual 24.88 hour double cycle (Thus a 9 a.m. tide returns at 9.53 a.m. the following day).

This discrepancy of 88 in a double cycle matches the double appearance of 44 in the line and circle pattern. The symbolism of the Grid, where the Moon pictures Jesus Christ, is also apt. The cleansing action of the Moon's tides, removing sewage and refreshing coastal areas, offers a fitting metaphor for Christ's preaching: *"Now you are clean through the word which I have spoken unto you."* (John 15:3).

Christ walked on the Sea of Galilee (Matthew 14:25) and subdued it, illustrating his power to control the demonic forces abroad in the world. In walking on the water Christ also enacted the role of the Moon - his own Grid symbol - in controlling these turbulent waters, themselves a metaphor for the dark demonic powers[1] of the Earth. In Bible prophecy it is the fearsome Beast Power of Revelation chapter thirteen that emerges, in symbol, from the sea.

So it is shown that the Grid is encoded by the tides caused by the Moon, as well as its luminosity and the 173 day cycle. The number corresponding to each twelve hour tide is 44, the sixth Grid number. But there is another important match between the Moon's motion and the Grid. The top line of the Grid 1 7 11 is the abbreviated version of the 1711171 pattern of 19 annual holy days. These days are regulated by the 19 year cycle of the Moon's motion! This gives four points of Grid confirmation by the Moon.

Thus the entire Grid is indirectly encoded by the Moon itself, as the Christ/Moon identity number is shown to be 7, and the corresponding power and motion number 40. The Moon's motion regulates and subdues the wild turbulent oceans, a metaphor for the demonic powers abroad in the Earth. In so doing earth's satellite encodes, through the timing of the tides, the number 44.

It was the number 44 that completed the Grid of six numbers. It has been prominent throughout this narrative in the following ways:

120

The period of the worldwide power cuts, in days	44
The duration of a magician's fast, in days	44
The position totals of the eight power cuts	44
The occurrences of the Greek Sperma, seed	44
The gematria in Hebrew, father/mother/child	44
The autosomes responsible for the human frame	44
The relative size of the human ovum vs. sperm	44
Last set of psalms corresponding to Deuteronomy	44
Acts the 44th book; flames representing the spirit	44

It was stated in chapter one of those who sent a signal in the year 2003:

"Their finger print is 44…"

---

1. In biblical symbolism the power of the sea is essentially evil. *Jude* warns of the fate of the incorrigibly wicked, likening them to *"…waves of the sea foaming out their own shame…for whom the blackness of darkness is reserved forever."*

24

GRID VIBRATIONS

"... the spirit of God vibrating upon the face of the waters..."
Genesis 1:2, Interlinear Bible

Can it be shown that the powers behind the universe have cunningly encoded their signature within its laws as a Monet or Constable might on a painting? If their finger print is 44, as stated in the introduction, and their calling card is pi then what is their signature?

An answer is found through a detailed investigation into the vibrations that create musical notes. Ethelbert Bullinger found patterns involving 7 and 11 in musical scales and structures:

> *"On examining the above [chart of frequencies] it will be at once seen that the number eleven is stamped upon music; and we may say seven also, for there are seven notes of the scale..."*
> *Page 16, Number in Scripture.*

The major and minor scales are built on 7 notes before they repeat. What is not generally realised is that the notes of most instruments are tuned to frequencies that relate to one another in ratios of 11. All musical instruments are constructed with the human ear in mind, so a dog whistle would not qualify as such. The frequency of notes in the major scales was determined in 1839 in Stuttgart[1]. There have been some subtle refinements since then; however, the fundamental factor of 440 vibrations per second for A4 (above middle C) is preserved to this day, as are also its octave relatives 880 (A5), 220(A3), 110(A2), 55(A1) and 27.5(A0).

The most expressive instrument devised, also possessed of the greatest range of frequencies, is the grand piano. This is evident from the enormity of the repertoire for this instrument. From the most humble models to the most expensive Steinway model D, or Yamaha equivalent, this instrument possesses 88 notes. Although dogs hear higher frequencies than man, and whales much lower, it can be seen that the range of useful notes for musicians is 88 in number. These find their embodiment in both the finest and least expensive pianos made. George Bernard Shaw wrote:

> *"The pianoforte is the most important of all musical instruments; its invention was to music what the invention of the printing press was to poetry."*

Certain refinements in the tuning of the piano have been introduced since 1839. The refinements were developed to compensate for the "stretch factor" (inharmonicity due to stiffness of the string) and to permit the transposition of music from one key to another. Twelve semitone intervals are contained within each octave cycle on the piano. Experiments on various stringed instruments have been made to try and vary this number but with quite unsatisfactory results, so it would appear that the numbers 7 and 11 are fundamental to the production of music that has meaning to the human brain: a brain that seems wired in that particular way and able to differentiate, through the ear, notes in a range of 88.

The Grid and its associated numbers appear in the 88 notes of the piano. The three musical Grid numbers appear prominently and it is particularly striking that the 11th note from the bass-end of the piano has 7 × 7 vibrations per second, or at least the modern value 48.999 which is remarkably close. In an interesting reversal, the 49th note from the base (7 × 7) turns out to be the pivotally important A4 'Concert Pitch' of 440c.p.s. (11× 40) of the piano tuning fork to which orchestras tune up. The factor 11 is therefore represented, as is the Grid number 40.

In these several ways 7 and 11 are strongly linked to one another in music and to the Grid number 44 as shown below:

Position from bottom A:		Note name	Cycles per second	Approximates to
1	(bottom A) A0		27.5	exact (half 5 ×11)
4		C1	32.703	33 (3× 11)
7		D1sharp	38.891	39 (3 ×13)
11		G1	48.999	49 (7 × 7)
12		G1sharp	51.913	52 (4 × 13)
19		D2sharp	77.782	78 (6 × 13)
40	(middle C) C4		261.63	264 under old scheme
44		E4	329.63	330 under old scheme
49 (7×7)		A4	440.00	exact (44 × 11)
88		C8	4186.00	exact (2 × 7 ×13 × 23)

values from www.vibrationdata.com

It is a sobering fact that the three defining numbers of the Grid, 7, 11 and 44, are inexorably linked together within the greatest and most expressive musical instrument man has devised. The piano possesses 88 keys: the number of the line and the circle encoded in the Grid, where both the totals of bottom line numbers and the products of the outside columns equals 88.

In the scheme of everything, the laws of music as they relate to the human ear appear to harmonize with all other laws including the spiritual. The six Grid numbers – the seventh unseen – are woven into a creation brought into being for man. This is the *Strong Anthropic Principle* in action: the idea that the universe and its laws were fashioned with human life in mind; that mankind was made last but thought of first. The number 6 is characteristic of man and his works which, without the spirit (represented by 1), cannot be complete. Bullinger said of the number 6:

Page 150, Number in Scripture:

"...man was created on the sixth day...six days were appointed unto him to labour...the sixth commandment relates to the worst sin, - murder...Cain's descendents are given only as far as the sixth generation..."

With regard to the workings of the piano, it can be seen that its operating principle hints at a quality peculiar to God, that of infinite power. This is because the force of a hammer blow at the point of contact with the string is theoretically infinite[2]. But would God go to so much trouble with the laws of nature, with the way copper wires vibrate and the functioning of our inner ears, all for the sake of a pattern in the piano? There is overwhelming evidence that he did.

Musical intervals also harmonise with the Grid. It can be shown that the characteristics of certain musical intervals correspond to Grid numbers. For example, an interval of seven semitones (i.e. C to G), form a perfect fifth. This interval produces a most harmonious effect, as the name suggests, and seven is the identity number of Christ. Frequently, the six semi-tones of the augmented fourth, i.e. C to F sharp are used in music to evoke sinister or demonic overtones[3] (e.g. Danse Macabre by Saint-Saens, Liszt's Dante Sonata). This six-factor corresponds to the incompleteness of the six numbered Grid (if the seventh number pi is not perceived) just as it does to man, created on the sixth day. In both cases a seventh number (or day) is required for their completion.

A question was posed earlier: does God have a signature? It should be obvious now that there is a numerical signature craftily woven into the fabric of God's handiwork. The seven numbers of the Genesis Grid must constitute that signature.

Further Grid patterns in music are shown in Appendix VI.

1. On page 15 of *Number in Scripture* Bullinger shows the frequencies, or cycles per second, that were adopted in Stuttgart in 1834, later by the Paris Conservatoire in 1859 and by the Society of Arts in 1869. The seven note scale Do Re Mi Fa Sol La Si and back to Do was comprised of the following cycles per second: 264 (which is 24 × 11), 297 (27 × 11), 330 (30 × 11), etc. all with 11 as a factor. The small modifications introduced to certain notes in the 20th century are discussed later. The 440 for A is unaltered as are all other A-notes.

2. This nearly infinite force enables the light-weight hammer to overcome almost any achievable tension in the string and cause it to vibrate. The mathematical formula demonstrating this "infinite" force is simple enough. The horizontal string is stretched between the bridge at the keyboard end and the agraffe at the far end of a grand piano. The rising hammer strikes the string and in so doing creates a microscopic kink in the string. Momentarily two right-angled triangles are thus created back to back, with their points at the bridge and the agraffe. The angles at the bridge and the agraffe are infinitesimally small, both of which can be designated as "theta." Extra tension in the string is produced when the hammer strikes; the more force in the hammer the greater the extra tension. The formula, where "f" is the hammer force and "F" is the extra string tension is: $f = F. \sin (theta)$ and therefore $F = f/\sin (theta)$. At the moment of the strike of the hammer the angle theta is zero and so sin (Theta) will also be zero making "f" infinite. In reality, as "f" tends toward infinity some deformation of the string occurs thus bringing "f" back into the realms of the finite; see the website www.member.aol.com/chang8828/tuning.htm

3. This interval is known as Diabolus in Musica. The Oxford Companion to Music states under the subject of TRITONE: "The interval of the augmented fourth....is relatively difficult to frame in singing....in early ecclesiastical music (fourth to sixteenth century) its melodic use was forbidden....Early theorists called it the Devil in music (*Diabolus in Musica*)....modernist composers have made it a cornerstone of their atonal and polytonal practices."

127

25

NUMBERS: FIGMENT OR FACT

"Look at the cut, look at the stitch, look at the fit..."
Danny Kaye, The Emperor's Clothes.

If the universe is built on numbers, experience would suggest that an 'architect' designed it. But in the face of this argument some will deny that numbers exist at all. According to their reasoning numbers are merely a product of the human imagination, but can this be true?

The idea that numbers have no real existence is a proposition many people will have never considered. How can one set about determining whether or not they do? Before answering that, it is necessary to know what a number is in terms of its everyday usage. In the simplest terms it is a quantity: it could be three fingers or four loaves of bread, with familiar symbols used to communicate the information as 3 or 4.

Number symbols can express a quantity, but they can also express a relationship – or a ratio – between two things. If one tenth of a population is left handed then 1 in 10 or $1/10^{th}$ have that characteristic. While that ratio in living things may fluctuate, the basic elements in nature such as Oxygen, Iron or Sodium combine in fixed amounts. Every laboratory experiment will show an identical result because these ratios were (for any given isotope) determined at 'creation.' Such ratios do not need an observer to confirm them. If all observers died the elements would still obey the same rules in combining one with another. In such a case it is self evident that, for example, the simple ratio of 2:1 in H_2O dictates the existence of numbers. It matters not which symbols are used; in Roman numerals one would say II combines with I.

If one were to say that the universe dictates *ratios* then the matter is beyond dispute. Moreover, those who deny that numbers exist outside of the human mind come up against another fundamental concept in mathematics: the idea of *sets*. Bullinger had noticed groups of *seven* successive elements (by atomic weight) that form groups according to their ability to combine with other elements. These groups, or sets, were illustrated in the following way on page 12 of *Number in Scripture*:

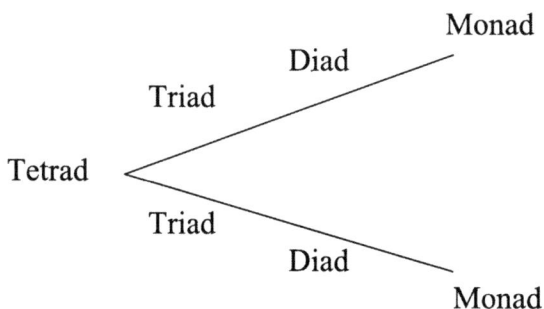

Bullinger noted that monads combine only with one atom of another element, diads with two and so on to create the above pattern of 7. The number 11 was also noticed by Bullinger:

> *"There is an element whose atomic weight is a **multiple of 7** (or very nearly so) for every multiple up to 147, while the majority of the other are either square numbers (or multiples of a square number), **multiples of 11**, or cube numbers." Number in Scripture, page 14*

All this shows that numbers are seen to exist in the physical universe in ratios and as sets. Bullinger was astonished to find sets of 7 and multiples of 11 in the Periodic Table of Elements. But now, due to the Grid discovery, the top line of the Grid can be seen within it in two different ways.

There is an obvious similarity in colour between the elements Gold and Silver to the Sun and the Moon. The jet-black colour of the element Boron and the yellow lustre of Gold correspond to the apparent colours of the Sun's spots and fiery face, as seen from Earth. The table below shows a match for 1 7 11 with two sets of elements who atomic weights are also found to be proportionate to 1, 7 and 11:

Corona *(Fiery)*	Moon *(Silvery)*	Sun *(Golden)*
1	7	11

Oxygen		Silver		Gold	
Rocket Fuel	1	Mirrors	6.74	jewelry/electronics	12.31
r.a.w*.					

Hydrogen		Lithium		Boron	
Rocket Fuel	1	Batteries	6.88	glass making	10.72

The *relative atomic weight** where the weights of the three elements Oxygen, Silver and Gold are divided by sixteen - rebasing Oxygen as 1 - produces numbers very similar to the Hydrogen, Lithium and Boron line and the Grid. Seen that way, the two sets of three have very similar weights. The total of the averages is within 2% of the actual Grid top line total 19:

All six elements:	1	6.81	11.51 (total 19.32)
Grid numbers:	1	7	11 (total 19.00)

The invisible 'spirit' of the Grid left hand column is amply represented by gaseous rocket fuel (Hydrogen and Oxygen, both invisible at room temperature) producing one of the most spectacular displays of power. The fierce behaviour of Lithium in water is in character with the personality of Christ, just as glass-makers' Boron suggests the glassy sea before the throne of God (Revelation 4:6). The extreme electrical conductivity of Gold also suggests a theme of power, and an eternal aspect is hinted at by this, the most beautiful of metals that does not tarnish.

The notion of numbers having no existence outside of the human imagination is a foolish one. Observers are not required in order to make reality real. Take for example the surface area of a globe. The area of a circle with radius of 1 is π. The equivalent sized globe has a surface area of 4π. This astonishing ratio exists without the aid of observers. If all scientists and their imaginations were to perish, with no conscious person left alive in the universe the number 4, or IV (regardless of how it is written) would still exist. This seems blindingly obvious, but the 'naked emperors' have been influential and the sad truth is that most people prefer others to do their thinking for them!

The *Alice in Wonderland* nonsense that numbers have no existence outside of the human mind is nothing more than 'science' propaganda. Some in the science community display the impudence of Goebbels in the face of facts repeatedly encountered in their investigations of the universe: evidence of order, logic and design. Peering through telescopes and microscopes, such scientists could be likened to ants crawling across a skyscraper, denying as they go that there could be any intended symmetry, beauty, or elegance to the object; decrying the idea that there could be any intelligence responsible for the refined surfaces beneath their feet. Every parallel line must be a coincidence; every right-angle merely the product of blind chance.

This approach of denying the obvious, and censoring anyone who ventures to discuss it, has Orwellian overtones. It is as if scientists are trapped in their own Ministry of Truth. But do all scientists really believe in their heart of hearts that all is coincidence, and that the materialistic view must be the end of the investigative process? Many are afraid to speak out, though not all. One such scientist is Professor Brian Josephson, a Nobel Prize winner, who investigates the 'quantum interconnectedness of entangled particles' and telepathy. He leads the *Mind-Matter Unification Project* at Cambridge University. In an interview in the December 2006 *New Scientist* magazine he expresses the view that many fellow scientists suffer from bias:

> *"... [Many scientists have] an irrational bias against unconventional ideas. I've had funding problems due my choice of areas to work in. Lots of scientists believe in telepathy but won't admit it. If you say you accept the paranormal this automatically affects your reputation. I don't have the kind of support network that researchers usually have. It's also hard to*

change how people think... People have vested interests, and their projects and reputations would be threatened if certain things were shown to be true."

If scientists are supposed to be impartial investigators of *the truth* then the above statement, if correct, highlights a scandal to make the Enron and Madoff debacles utterly trivial by comparison. It appears that most scientists are working in emotional, financial and intellectual straight-jackets. Today's science community cannot admit to any possibility of the supernatural. The following comment from Professor Richard Lewontin highlights the existence of a metaphorical 'science dam' holding back evidence from a credulous public:

"...we have a prior commitment, a commitment to materialism...we are forced by our a priori adherence to material causes to...produce material explanations...we cannot allow a Divine foot in the door." From an article Billion and Billions of Demons, The New York Review, January 9th 1997.

What if the universe came into being by a non-material cause? What if it is sustained by an undetectable non-physical substratum, or support mechanism, just as a motion film is sustained by a powered projector or DVD player? Many observations suggest that something like this is likely, but pursuing such ideas in science could be a career-wrecking option. There is an unspoken McCarthyism in science. A glass ceiling exists beyond which the 'scientific mind' *dare not* aspire. Many scientists could be likened to characters in a film who, when noticing that their world flickers at a rate of thirty times a second, are told 'don't ask why, or else.'

Everything that we can see, our guides assure us, came from an impersonal 'big bang' which, by definition was *unintelligent*. Such a universe could be expected to fit the description given by the biologist Richard Dawkins:

"The universe has no design, no purpose, no evil and no good, nothing but pitiless indifference." R. Dawkins, The God Delusion.

Dawkins himself is anything but indifferent towards the God of the Bible whom he considers a *"petty, unjust, unforgiving control freak"* and *"a misogynistic, homophobic, racist, infanticidal megalomaniac and capriciously malevolent bully."* Perhaps there is evil in the universe after all! The logic that one requires goodness of one's own to pass moral judgments on others seems to elude this critic; besides which, God should be commended for his bullying in a Darwinian universe. However, the 'no design' scenario makes little sense. Intricate laws are seen to operate in both organic and inorganic matter and laws require a lawgiver. Such laws often find expression in ratios - in effect, numbers.

Laws are seen in the way numbers (more to the point, *quantities*) behave one to another. This is true at the very simplest level, and the astonishing thing is that the number 11 is preeminent. This is seen when a simple, in fact the simplest possible, number pyramid is assembled. This pattern is known as Pascal's triangle:

$$
\begin{array}{ccccccccc}
 & & & & 1 & & & & \\
 & & & 1 & & 1 & & & \\
 & & 1 & & 2 & & 1 & & \\
 & 1 & & 3 & & 3 & & 1 & \\
1 & & 4 & & 6 & & 4 & & 1
\end{array}
\qquad
\begin{array}{l}
1 = 11^{\circ} \\
1 = 11^{1} \\
1 = 11^{2} \\
1 = 11^{3} \\
1 = 11^{4}
\end{array}
$$

Each line of this most basic of all number constructs totals successive powers of the number 11, such as 11^2 or 11^3. After the fourth power the pattern appears to fail but on closer examination it actually does not. The pattern of increasing powers of 11 continues uninterrupted if we convert the rows of Pascal's Triangle to what one could term their *true quantitative values*. This is awkward to understand at first, but when each number in a row is assigned to successive decimal columns – thousands, hundreds, tens and units – then the true quantitative value can be ascertained. The fifth line of Pascal's Triangle normally looks like this:

$$1 \quad 5 \quad 10 \quad 10 \quad 5 \quad 1$$

The two 10's are the problem, but it is resolved by revaluing them as follows:

1 ×	100,000	=		100,000
5 ×	10,000	=		50,000
10 ×	1,000	=		10,000
10 ×	100	=		1,000
5 ×	10	=		50
1	unit	=		1

Total that, and one can see that the true quantitative value of what first appeared as 15,101,051 (the row 1 5 10 10 5 1) is actually 161,051 or 11×11×11×11×11 or 11^5. So it is seen that the pattern of successive powers of 11 continues without interruption. It is apparent from this that *a law of 11* is in operation. This is the natural outgrowth of the addition of quantities – purely bean counting - in Pascal's Triangle. Therefore the *law of 11* would still be in operation if one used beans on a desk top instead of symbols like 1, 2, and 3.

The Father's number 11 is embedded in 'nature' in one further way. This can be seen in the number 81. The German chemist Peter Plichta PhD noticed that the periodic table contained only 81 stable elements, not the 83 commonly assumed:

> *"...it is clear that the two unstable elements should be omitted from this group of 83. That would then give us a final number of 81 stable elements. In 1981 the paperback German edition of Isaac Asimov's Book of Facts appeared. The author, the world famous writer and chemist, reports on page 93: 'There are only 81 stable elements'..." Page 34 God's Secret Formula*

Plichta then investigated the number 81:

> *"...the reciprocal value of 81 had particularly aroused my interest:*
> *1/81 = 0.01234567901234567901...with the number 8 always missing. The fraction 1/81 can also be represented in an even more unusual way:*
>
> *1/81 = .0123456789 (10) (11) (12) (13)......This idea appeared wonderful to my eyes because now the number 81 appears as a*

reciprocal value for the atomic numbers of the elements. The missing 8 is an illusion..." Page 123, ibid.

Plichta discovered that when the true quantitative values are substituted (the (10) (11) (12) (13)...above) then all numbers from 1 to infinity are seen in the reciprocal value of 81, or 1/81. In this respect the number 81 is uniquely the number of infinity: it is also the correct total number of stable elements[2]. As mentioned in the chapter *Tides that encode the Grid*, the ratio 1/81 relates the mass of the Moon to the Earth.

Unbeknown to anyone, the number of the Father has been 'hidden' in the number 81. His abbreviated number within the Grid is 11. It appeared that its fullest expression was the number 111. But a new aspect to the expansion of 11 appears within the square root of the reciprocal of 81:

$$\sqrt{1/81} = 0.111111111...$$

In this way the Father's number is, uniquely, encoded within the number of stable elements, 81. A further vital law operates within the family of 81 stable elements. Plichta had observed that things of a fundamental nature occurred in *sets of three* and cites the following examples:

All elements fall into three types:	*stable, radioactive or artificial.*
All elements fall into three groups:	*non-metallic, metalloid and metallic.*
Elements have three types of bond:	*simple bonds, double bonds or triple bonds.*
Only three elements form all bonds:	*Carbon, Nitrogen and Oxygen.*
All elements are basically made of:	*protons, neutrons and electrons.*
All matter at room temperature is:	*solid, liquid or gaseous.*
All radiation is of three types:	*alpha, beta or gamma rays*
Water has three components:	*H_3O+, H_2O and H_1O-*

But the threefold law Plichta saw was complemented by a further fourfold law:

"I had previously assumed that everything had a threefold nature, just as all atoms in the universe are composed of

protons, neutrons and electrons. If this law holds good, how can it be that the orbiting electrons possess four quantum numbers?" page 68.

Plichta began to notice numerous examples of fourfold phenomena. There are four chemical bases in DNA, and life is sustained by heat from the Sun generated by the transformation of four protons into the helium isotope of mass four. The Earth has four seasons and the Moon four phases. The fourfold principle also operates at the fundamental level as electrons in their shells are regulated by it:

"They [electrons] are found like sparrows sitting on a wire. Some secret law tells them that in the innermost orbit one pair may sit, the next orbit can take a total of four pairs, the third orbit has nine, and there are [a maximum of] sixteen in the fourth orbit. The mathematical law for this had a quadratic nature since these numbers - 1, 4, 9, 16 – can be described as the squares of 1, 2, 3 and 4." Page 68, ibid.

Plichta synthesized these two laws into a *'three to the power of four law.'* This vital insight offers a remarkable match with his earlier finding of 81 stable elements, because 81 is itself $3 \times 3 \times 3 \times 3$ (*three to the power of four, or 3^4*). This law is certainly seen to operate in nature. DNA, although exhibiting the *four* bases mentioned earlier, is composed of *three* fundamental constituents as Plichta noticed:

"DNA consists entirely of only three chemical components: phosphoric acid (P), sugar (S), and a base (B). This mysterious substance consists of chains of phosphoric acid and sugar molecules in the following arrangement:

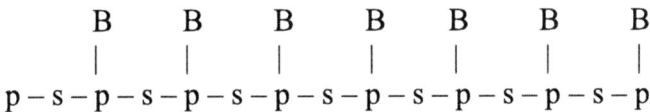

$$
\begin{array}{ccccccc}
B & B & B & B & B & B & B \\
| & | & | & | & | & | & | \\
\end{array}
$$
$$p - s - p - s - p - s - p - s - p - s - p - s - p - s - p$$

A base is attached to each sugar molecule. This comprises any one of four different bases – and nobody knows why there should have to be precisely four – thiamine, adenine, cytosine, guanine." page 110, ibid.

137

This law not only controls living things but inert matter as well. While the fundamental constituents of both the atom (protons, neutrons and electrons) and of living things (phosphoric acid, sugar and base) are *three*fold, the factors that *determine the characteristics* of the atom or molecule are *four*fold. It is the fourfold law controlling the distribution of electrons in their shells that determines the characteristics of an element, as well as its ability to combine with other elements; and it is, again, the fourfold nature of the chemical bases of DNA that determines the characteristics of a living creature.

Plichta's insight concerning the 81 stable elements and the 'three to the power of four law' supports the Grid discovery. This is because while the Grid's 3 aspects of God are fundamental, the bottom line 4's determine their *function*. The *determinant* for reproduction is the number 44 (and for work 40 and for creating 4: all fours). The Deity String in pi also follows a '3^4 law.' This is because the number 3 is fundamental because pi is 3.14159…but there are 4 segments. The first 3 segments relate to God, but the 4th relates to the Word of God. So for human beings - who cannot see God - the fourth (the Bible) is the determinant, telling us what to do. All of these examples follow a $3+1$ or 3^4 pattern.

Not only do the building blocks of the universe and of life itself follow a 3^4 law but so also does the Grid and its attendant patterns. It is not surprising that an investigative chemist such as Plichta, on a search for ultimate truth, found laws that integrate perfectly with the Grid discovery.

See Appendices VII/VIII for further information on the 3^4 law, and mathematical findings regarding the Golden Ratio and Fractals.

1. The observation of a 'law of eleven' as it relates to Pascal's Triangle is more fully explained in the book *God's Secret Formula* by Dr Peter Plichta.

2. *"The stable [non-radioactive] elements in the periodic system are therefore made up of the 1, 2, 3 … 83. There is, however, a problem here of which chemists and physicists are generally not aware: for some strange reason two elements do not exist [in nature, but must be synthesized] – the elements with the atomic numbers 43 [technetium] and 61 [promethium]. They also do not exist outside this planet, at least not anywhere in the solar system…" page 30, ibid.*

26

THE OVUM AND SPERMATOZOON

"...thou hast covered me in my mother's womb...
when I was made in secret..." Psalm 139:13-15

The focus in this second section of *The Genesis Grid* is the question of *"Who and why?"* This necessitates a closer look at the Grid's main and most controversial theme, reproduction. Who is doing the reproducing? Is there a reproductive plan underway involving supernatural beings and if so what is the evidence?

The Grid contains a Father and a Son. Reproduction is evidenced by the number 44 twice in biology, in the gematria values of Hebrew, in the appearances of the Greek word *sperma* and in the scheme of the Grid. It contains a Father and a Son. Linking these ideas invokes many New Testament passages that speak of sonship, implying the existence a family relationship with God. The theme of Christ as an elder brother is found in the New Testament.

Mainstream religion has spread confusion over this simple principle of parenthood. It has claimed that man's potential for a relationship with his creator amounts to mere *adoption*. Describing adoption as once practiced in the Roman Empire, an act as permanent and all encompassing as an adoption could conceivably be, theologians have craftily supplanted the sonship of the scriptures. God is reduced to being an adoption agency; impotent and bereft of the reproductive power displayed by living things on earth. But the Grid affirms what should be obvious from scripture: some kind of *genuine* reproduction on God's part is occurring. That is what it means to have sons and daughters.

There is a close similarity between human reproduction and the spiritual process of begettal and birth alluded to in the Bible.

Theologians seem to overlook the fact that God somehow impregnated Mary to produce Jesus the Christ. In that way, the pre-existent Logos *became* the Son of God in human form, but later restored by the Father after 33 years to *"...the glory which I had with you before the world was"* (John 17:5). His original spirit-composed body was restored, setting a pattern and a precedent for human beings.

Does God lack the resources to beget (or spiritually impregnate) those very few he calls in this darkened age[1] by his spirit? Mainstream religion denies any possibility of true sonship. But if the God of the Bible is an authentic parent, is he not able to exercise the same prerogative as mere humans? Is he unable to take the same course that one's parents and grandparents made when they decided to marry?

The theme of reproduction is encoded in the Grid, the Bible and its languages. The Greek word sperma (seed) occurs 44 times in the New Testament. Bullinger records the fact:

> *Number in Scripture: "(sperma), seed 44 (4 × 11)" Page 39.*

The significance of that number entirely passed Bullinger by. Perhaps he was unaware that in Hebrew (the language of the Old Testament) the *gematria* for child is 44 and that the values for mother and father are 41 and 3 respectively, again coding 44 as a number that pertains to reproduction. The number 44 is encoded twice in the human reproductive system. Precise measurements of the cells involved are not possible and their size can only be given as an approximation. An internet article entitled *What is the size comparison between a human sperm and a human ovum?* by Dr. R. James Swanson of the Biological Sciences faculty, Old Dominion University, illustrates the problem of measuring living cells. The following guidelines were quoted by him:

> *"The normal human sperm has a width range of 2.5 – 3.5 microns (micrometers)...The normal sperm sample will only have between 14 to 20% normal sperm...The immature egg...is about 60 microns diameter...the egg will reach a final diameter of about 120 to 150 microns."* www.madsci.org

These obviously rounded off ranges suggest mid-values for a sperm of 3 microns and for an ovum of 135 microns. Such numbers yield a ratio of 1:45 in width which is, when one considers the imprecise nature of the figures, remarkably close to the postulated ratio of 1:44. An actual measurement of a pair of cells of *3 microns and 132 microns* would yield a ratio of 1:44 precisely. Both the World Health Organisation (1992) and 'Kruger's & Tygerberg's Strict Criteria' agree on a range of 2.5 – 3.5 micrometres, whilst Menkveld *et al* (1990) gave a value of *"approximately 3 microns"* for a normal spermatozoon head (jabsom.hawaii.edu).

In March 2004 a large colour photograph of a human ovum surrounded by spermatozoa appeared in *The Times* of London. Careful measurements of these produced the following figures in millimeters:

Spermatozoa diameters: 2.3, 2.5, 2.0, 2.1, 2.3, 1.9, 2.6, 2.5, 2.5, 2.0, 1.9, 1.9

This yields an average diameter at the widest point of 2.208mm for sperm in the foreground of that photograph. The ovum was then measured at four different angles and found to have an average diameter in this photograph of 97.125mm. This gives a ratio between the diameters of an ovum against a sperm cell of 43.98:1.

The strands of the DNA molecule woven into the structure of our chromosomes contain some precise numbers: prominent amongst them is the number 44. There are 23 pairs of chromosomes in humans of which 22 contain *2 autosomes* each (total 44) and a 23rd pair, comprised of *2 sex chromosomes*. The factors 2 and 44 (implying 88) thus represent the building blocks of the human body. In the Grid, the spinning 44's derived from the two outside columns imply the number 88. The human reproductive system provides a further verification of the number 44. Investigations in the USA in the 1950's discovered that marital relationships are consummated in an average of 88 pelvic thrusts[2].

Here is a powerful challenge to some readers' cherished ideas concerning human origins: either these numerically marked phenomena appeared in the human body randomly due to impersonal resident forces and blind chance, or they were put there.

141

As for the Bible, the 44 occurrences of the Greek word *sperma*, and the appearance of the spirit as Corona-like flames in Acts (the 44th book of the Bible) confirms that God's own idea of sonship is literal. It is now plain that the number for reproduction is 44: *it is the reproduction number of God.* This fact has already been noticed by others in a roundabout way.

In April 2006 the author viewed for the first time the 1997 Darren Aronofsky film *Pi*. It was billed on British T.V. as an intriguing psychological sci-fi thriller. The protagonist Maximillian Cohen played by Sean Gullette was a reclusive mathematics genius trying to find a pattern in pi. He fell into conversation in a cafeteria with a member of an Orthodox Jewish sect, who told him:

> *"You know, Hebrew is all maths. It's all numbers - you know that? The ancient Jews used Hebrew in a numerical system. Each letter's a number. The Hebrew "a" Aleph is 1, "b" Bet is 2, you understand? Look at this: the numbers are interrelated. Take the Hebrew word for father, Av. The Hebrew "a" [he starts writing] Aleph is 1, Bet is 2, that totals 3. Take the Hebrew word for mother, imah - Mem is 40, Aleph is 1; equals 41. Sum of 3 and 41 is 44. Right, now Hebrew word for child [he writes it out] 4, 30 and 10, total 44. The Torah [first five books of the Bible] is just a long string of numbers." Pi, 1997, Darren Aronofsky*

The film did not pursue the reproduction theme and needless to say it did not find a pattern in pi. Both old and new Testaments, Hebrew and Greek, are in agreement about the number 44, a number that is integral to Genesis chapter one (as previously explained in the chapter *Genesis: Game, Set & Match*). The number 44 is contained within the first teaching statement given by God to man, in a command to *reproduce*. Appearing in the 28th verse of the Bible the command encodes the number 44 twice.

In the 44th book of the Bible, *Acts*, the 'tongues of fire' of the spirit, the *seed* of God, first appear on the heads of his disciples. When we add to that the evidence of the 1:44 ratio of the two human reproductive cells - the largest and smallest of human cells - and the 44 autosomes of the human genetic package, it is clear that the number 44 is the primary number of reproduction, both physical and spiritual. But one critical question must be asked: if 'the spirit' as represented by tongues of fire on the Sun and on the heads of the Apostles is the agency of begettal, what is its Bible count for this spirit? That is to say, how often does it appear in the Bible? Surely this could not be a random number if God is the architect of the Bible? The result is quite astounding.

The count for 'holy spirit', 'spirit' and 'spirits' (as the word pertains to God's spirit) provides yet one more confirmation of the number 44, as well as a very powerful confirmation of the Grid and its total 107, the 28th prime number. The complete Bible count for this spirit factor is as follows:

Old Testament 'spirit' pertaining to God	102
New Testament 'spirit' pertaining to God	102
Old Testament 'holy spirit'	3
New Testament 'spirits'[3]	3
New Testament 'holy spirit' ('Holy Ghost', KJV)	94
New Testament 'seven spirits of God'	4
Total	**308**

If the personalities of the Father and Jesus Christ are marked with the numbers 11 and 7, would not the holy spirit - the third factor in the Grid (though not a person) - be marked in some significant way? This is certainly the case because $308 = 7 \times 44$. There is a further stunning observation regarding the count of 308 for the spirit. Because 44 contains the factors 4 and 11, the number 308 inevitably contains 11 and 28 (7×4). Therefore the following is observed:

$$308 = 7 \times 44$$
but also
$$308 = 11 \times 28$$

Thus the three defining numbers of the Grid, 7, 11 and 44 are again verified within scripture. But a crucial byproduct of this discovery is additional proof that the spirit is nothing other than an essence. If it were a person it would presumably be represented by a new *prime number*, not the number 308 which is merely the *product* of the activities of '7 and 11' in their teaching (28) and reproduction (44). Plainly, the spirit is their agency.

The true nature of the spirit is implied by both the circular 44's of the Grid and in one further way: there is a Grid pattern not yet

143

discussed which forms the centre of the 1711171 pattern of 19 days. This can be deduced as follows:

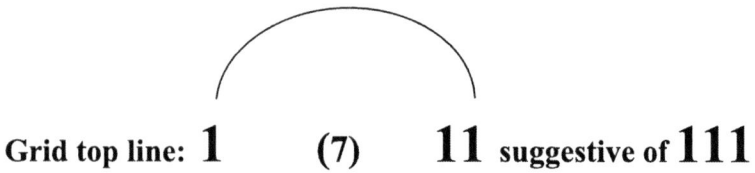

Grid top line: **1**   (7)   **11** suggestive of **111**

The holy day pattern 1711171 contains the number pattern 111 at its centre. Like the number 11 this appears to represent the Father; it is similar to the 11.1 years of the sunspot cycle. The Grid symbolism in this case is 'Sun + Corona' and materially the two are one and the same thing. So, in reality, are the Father and the spirit also one and the same thing? That is the inference drawn from the circling 44's of the outside columns of the Grid. If the Father and his spirit amount to the same thing, if 'particles of God' are streaming out from his person - as the solar wind streams out from the Sun - then utterances of the spirit amount to utterances of the Father. This clarifies certain scriptures that might otherwise be used to promote ideas of the spirit as a person, rendering the Bible self-contradictory. Such scriptures can be understood as prophetic utterances of the Father, for example:

> *"Now the Spirit speaketh expressly, that in the latter times some shall depart from the faith, giving heed to seducing spirits, and doctrines of devils..." I Timothy 4:1*

The use of a capital in 'spirit' indicates the mainstream belief of the translators that it is a person. In fact the only person here is the Father.

Amalgamating the 1 of the spirit with the 11 of the Father, as in the figure above, produces the pattern 111. Regarding this pattern 111, it is explained in the chapter *Subterranean Numbers,* by way of comparison to other pairs (such as 44 and 88), as the 'fullest expression' of the number 11. So by the same token the number 11 is an abbreviated version of 111. The number 111 is therefore a more complete numerical representation of The Father. Therefore it is hardly surprising that the 1711171 pattern, a nineteen day system of worship instituted by God, should have the number 111 at its centre. Scientists will from time to time invoke 'the mind of God.' They may be interested to ponder the evidence here that the pattern 1711171 is straight from His mind.

The number 308 - the total mentions of the spirit in the entire Bible - confirms that the holy spirit is an essence. It confirms the identity numbers 7 and 11 of the two Deities and, by multiplication, their program of teaching (28) and self reproduction (44). The chief Deity is shown to be the ultimate teacher, as in 11 × 28. This is confirmed in John 15:49 where Christ said:

> "...the Father which sent me, he gave me a commandment, what I should say, and what I should speak."

To summarise, this chapter has provided evidence of design in the human reproductive system; it is now marked *three times* with the number 44. It has also been demonstrated that the number of the Father's reproducing spirit is 44, in the Bible and in the Grid.

Both scientific observation and the Grid pattern support the greatest Bible prophecy of all: God is to propagate his *own kind* through the human race.

1. A darkened world in which, long after the resurrection, the great adversary Satan remained *"the god of this world"* 1Cor 4:4).

2. Masters & Johnson.

3.These three scriptures are: I Corinthians 12:10 "...(gifts) to another discerning of spirits..." ; Hebrews 12:9 "...in subjection unto the Father of spirits..." I John 4:1 "Try the spirits whether they be of God..." where in each case the writer could be speaking of the spirit of God, or of the "seven spirits of God."

27

AND THEN THERE WERE TWO

"The testimony of two men is true." John 8:17

A popular assumption from ancient times has been that the Supreme Being is some sort of 'trinity.' Trinities abound in paganism. But it is now shown that the most mysterious being in the universe, or perhaps one should say outside it, is represented by the pi digits 31415926535. This string of 11 digits totals 44 and represents 'the Father'; the one who would at least be 'first among equals' depending on which version of the Trinity is under discussion. Is he part of a Trinity? What can be deduced about him?

It was first shown in the chapter *An Abstract Machine* that the spirit of the Father is represented by the number 44. The Grid possesses outside columns of the same product, 44, but why?

Spirit	*Father*
(Ovum/Spermatozoon)	(Identity/power)
1	**11**
×	×
44	**4**
= 44	**= 44**
(left hand column)	(right hand column)

Are these columns linked and if so what would be the simplest way? If they are linked a circle would the simplest diagrammatic resolution. This idea was illustrated earlier by 44's describing a circle as follows:

<pre>
        44   44
     44   44
  44   44
  1   7   11

    ×   ×

   44   40   4
    44   44
      44   44
        44   44
</pre>

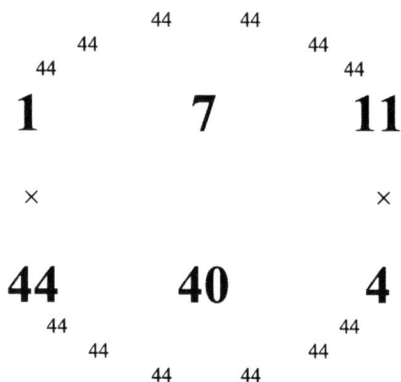

A circle suggests motion, namely the spirit (44) circling. Such a circle possesses a maximum value of 88. The finding that 28 × pi = 88 (almost), combined with the observation of a circle of 88 and line of 88 in the Grid, provided a strong clue for circles. Another fact supporting the circle as a solution is that of flames of fire on the heads of the disciples. These are described in the book of *Acts* (the 44th book) and offer an obvious match with the flames of the circular Corona, itself bearing a fascinating resemblance to the host of spermatozoa surrounding an ovum.

Bullinger asserted that: *"If there be design, there must be significance."* What could be the significance of a circle? One insight offered by the circle of 44's is that the Father and the spirit consist of one and the same thing. He is composed of it and it emanates from him. The circle confirms what common sense would suggest: that the Father is the *power source*.

Because the Father is the chief Deity it is fitting that the deity number 31 is found as a whole number total. It is encoded in the first 8 digits of pi:

The Father's string, 11 digits **3 1 4 1 5 9 2 6   5 3 5**

total 31   total 13

148

Here there is symmetry because totalling the remaining 535 we get 13, which is the *reflection* of 31, the number of deity. The number 13 has appeared before: it is the number of apostasy and rebellion but it is also, as will now be shown, a number associated with Jesus Christ who is the antidote.

Both of the Deities are linked to the number 13. They stated their intention to create mankind in the 26th (2×13) verse of all scripture: *"Let us make man in our image…"* Of the number thirteen Bullinger writes:

> *"The popular explanations do not, so far as we are aware, go further back than the Apostles. But we must go further back to the first occurrence of the number thirteen in order to discover the key to its significance. It occurs first in Genesis 14:4 where we read, 'twelve years they served Chedorlaomer, and the thirteenth year they REBELLED.' Hence every occurrence of the number thirteen, and likewise every multiple of it, stamps that with which it stands in connection with rebellion, apostasy, defection, corruption, disintegration, revolution, or some kindred idea…We see it stamped upon the very forefront of Revelation. For while the opening statement of Genesis 1:1 is composed of seven words and twenty-eight letters (4×7), the second verse consists of fourteen words, but fifty two letters; fifty two being 4×13 tells of some apostasy or rebellion which caused the ruin of which that verse speaks. But it is when we come to gematria that the most wonderful results are seen. These results may be stated thus, briefly: That the names of the Lord's people are multiples of eight, while the names of those that apostatised, or rebelled, or were in any sense his enemies, are multiples of thirteen…No human foresight or arrangement could have secured such a result beforehand; no human powers could have carried it out in such perfection. No matter where we look [in scripture] we find the working of the law without cessation, without a break, without a flaw from beginning to end. Only one conclusion is possible, and that is that the Bible has but one author…" Number in Scripture, pages 205/6*

Here Bullinger contrasts the numbers of the righteous and the unrighteous, these being 8 and 13 respectively, both of which are seen in the Father's string. This is another reason why the discovery of 31 as the total of the first 8 digits of pi is important. Again, this string of 11 digits is seen at the beginning of the first 31 digits of pi as follows:

3.1415926535897932384462643383279

The Father is identified in the following 11 digit segment:

3.1415926535 (totals 44)

with sub-segments totalling:

$$31 \qquad\qquad 13$$

Where $3+1 = 4$ and $1+3 = 4$

Through the two sub-segments of 31 and 13 the number 44 is once more confirmed. Thus two separate confirmations of 44 exist within the first eleven digits of pi, a notable Grid confirmation; but what of the segment 535?

This too matches the Grid in its own way because $535 = 5 \times 107$ where 107 is the Grid total and the 28th prime number. Thus *both the Father and his self-reproducing plan,* so succinctly represented by the Grid, are also both deftly encoded in 31415926535.

The scheme within the 31-digit Deity String is hierarchical: the Father first, the Son second and the written word last. The Father is shown to be the power source discharging the reproductive power of which he himself is composed. That notion is mirrored in the physical realm where every cell of the body contains the genetic signature of the original ovum/spermatozoon package.

This spirit power is shared with Jesus Christ the second Deity, as many scriptures indicate[1]. Thus a Grid containing Father and Son has as its primary theme reproduction, a begettal through this shared spirit. When considering the idea of a hierarchy the opening verses of John's gospel, considered by many to be the most profound prose ever written, are indicative:

"In the beginning was the Word and the Word was with God and the Word was God. The same was in the beginning with God. All things were made by him [the Word]...and the Word was made flesh and dwelt amongst us..." John 1:1-14

From this we may deduce that the God-being who was incarnated as the man Jesus had originally created the universe. Christ referred to that pre-existence in *John*:

"And now O Father glorify me with thine own self with the glory which I had with thee before the world was." John 17:5

But Jesus also said *"My Father is greater than I."* This difference in greatness affirms an obvious hierarchical order. It is vividly illustrated by the symbols given in the 16th verse of the Bible. The pale, scar-faced Moon would seem nothing more than a small dead rock placed next to the massive and resplendent Sun. Yet they appear two of a kind by virtue of their identical width as viewed from the Earth.

The presence of a Grid pattern in pi cannot be due to a choice in presentational method. Admittedly the pattern only appears in the base ten - hundreds, tens and units - but this is the system of the Bible, so there is no choice in that regard. The same can be said of the gematria values as they are built into the languages of the Bible, also according to the base ten using hundreds, tens and units. The discoveries do not work in the base 9. For example, in the base 10 the gematria for 'child' is 44 and mother and father are 41 and 3 respectively, confirming 44. It works in hundreds, tens and units but in base 9 the number 41 becomes 45 (4 nines and 5 units); then added to 3 the total would be expressed as 48, not 44.

When working with the information handed down (the Bible) the patterns are unmistakable. This pattern in pi, and its harmonization via the numbers 7 and 11 with the laws of music, the periodic table, a new scheme in the prime numbers (see the later chapter *The Deity Prime Sequence*) and the scriptures, proves the inspiration of the Bible beyond all doubt or dissimulation. It also proves that we occupy a 'base ten universe.'

As the Grid discovery fits scripture it naturally clashes with humanly devised ideas about God. The trinitarian notion of the Godhead with three co-equal entities is not compatible with the scriptures. An alternative to trinitarianism is monotheism. This asserts that God is one person and that any sense of plurality should be understood as facets, or aspects, of his personality rather than separate entities. The Grid discovery affirms the Bible truth that neither idea is true.

The Grid confirms the biblical truth about this question: that the Father and Christ are separate personalities and the spirit is a shared power that emanates from the Father. The Father is the original power source. Christ used great plainness of language regarding the existence of two separate wills:

> *"...I am not alone, but I and the Father that sent me. 17...the testimony of two men is true. 18 I am one that bears witness of myself, and the Father that sent me beareth witness of me." John 8:16-18*

Christ likens himself and the Father to "two men" who testify, or bear witness, of each other. Is one to believe that Christ was talking to thin air when he prayed to the Father, or that Jesus was the Father masquerading as a human, leaving heaven unattended? Also relevant is the statement by Christ in the garden of Gethsemane:

> *"O my Father, if it be possible, let this cup (crucifixion) pass from me: nevertheless not as I will, but as thou wilt."*

Was this statement nothing more than a charade? Space does not allow for an exhaustive list of all those scriptures offering evidence that Jesus Christ and the Father enjoy an individual existence apart from one another, though the chapter *John confirms the Grid* and Appendix III lists 107 of them. The idea that there could be two personalities in the Godhead, a number that intuitively feels 'unfinished' and not rounded, may seem illogical until one pauses to consider the implications mentioned, in passing, by the nineteenth century theologian M. Mahan:

> *"The number two is the symbol of a transitional or intermediate stage. It is a number of insufficiency or expectancy..." M. Mahan, D.D., page 115, Palmoni, or the Numerals of Scripture.*

According to this, the number two when applied to God seems

revolutionary. Is God in a transitional stage? If so, what is he expecting? This question has already been addressed to some extent and will be developed in later chapters, but it should be obvious that if the two deities are planning an expansion of their group, or family, then they are presently in transition.

The Father's string in pi has two segments. This is not surprising when so many other dualities have emerged:

1. Two Deities (Father/Son)
2. Two Grid rows (identity and power/motion)
3. Two columns of 44 (1 × 44 and 11 × 4)
4. Two lines of 88 as clues to pi (circle and bottom line)
5. Two positions in pi show a result, 1st and 2nd
6. Two numbers 11/44 seen in the prime positional result
7. Two forms of the pi string 171144404 show a result
8. Two halves of pi, Genesis 1:1 & 1:2 (The Breach)
9. Two segments in pi representing Jesus Christ
10. Two segments in the Father's eleven digit string
11. Two Mounts are based on the Grid (as shown later)
12. Two commands, Genesis1:28, of the same pattern (as shown later)
13. Two occurrences of 44 in pi: the total of the first 11 digits, and in the 31:13 reflection of the segments within those 11 digits
14. Two occurrences of 11 and 7 as factors in the four segments of the Deity String.

Is any support for trinitarian ideas seen in the Father's string? Supposing the Father was to be solely represented by the 8 digits 31415926, total 31, could someone or something (else) be represented by the 535? These 3 digits total 13 as do the letters of the name Jesus Christ in the original Greek. The number 13 when associated with Jesus Christ denotes that he is the antidote for sin and apostasy, yet even in that form he is *"in the Father"* (John 10:38). As Christ is *"the firstborn of many brethren"* (Romans 8:29) who is *"bringing many sons to glory"* (Hebrews 2:10) the 535 pattern (totalling13) could be a reflection of the Father's parenthood role for the Word (Christ). But whatever the reason, a third personality as portrayed by the Trinity doctrine is nowhere implied. The conviction of the Trinity tradition would have been a giant barrier in Bullinger's mind to the discovery of anything like the Genesis Grid. It is deeply ingrained in the minds of theologians. This giant obstacle, an implication of *two-ness* in the Godhead, may have deterred him from mounting further investigations into the numbers 7 and 11. Again recall what Bullinger wrote on page 25 of *Number in Scripture*:

> *"Why should it be these two numbers seven and eleven? Why not any other two numbers? Or why two at all? Why not three? We may or may not be able to explain why, but we cannot close our eyes to the fact."*

A further New Testament insight concerning the interchangeability between Christ and 'the spirit' helps answer Bullinger's question *"Why not three?"* What the scriptures say of Christ is noteworthy:

> *"Now the Lord is that spirit."* 2 Corinthians 3:17

In the gospel of John we see the same suggestion of equivalence between the spirit and Christ:

> *"...another comforter...the spirit of truth...v18 I (Christ) will not leave you comfortless: I will come to you."* John 14.16

What seems implied here is that Christ is himself the spirit. Elsewhere it is without doubt the spirit of the Father. This is certainly mysterious, but still it is evident that *the spirit is a shared power* and not a person. As Christ stated:

> *"The testimony of two men is true."* John 8.17

This is Christ's allusion to himself and the Father.

Further information is given concerning God in Appendix X.

---

1. The spirit of the Father is shared with Christ: Galatians 4:6 "...sent forth the spirit of his Son." And 1 Peter 1:11 "...the spirit of Christ that was in them." Evidently the spirit is an essence that is shared by the two Deities.

28

NEVER BEFORE SEEN, THE NUMBER 28

"...the Tabernacle with ten curtains...
one curtain shall be eight and twenty cubits." Exodus 26:2-3

The Bible was once considered to be so important in English society that the version commissioned by King James described it as *"...that inestimable treasure, which excelleth all the riches of the earth."* It remains to this day the world's preeminent religious and historical document. The discovery of number 28 as the secret teaching number in the Bible would certainly be worth a book in its own right, if it were not for the superabundance of discoveries here vying with it.

There are no chapters or verse numbers in the original Bible, yet most of its material falls neatly into clearly discernable sections. For example, the chapter divisions within the four gospels are as plain and obvious as the 31 verses of Genesis chapter one. The evidence that chapter divisions identified by scholars are, for the most part, correct can be seen in the fact that all of the twenty-eighth chapters of the Bible's books closely fit the *teaching theme.*

Excluding the Psalms and Proverbs, where material is not structured in the usual way, eleven books of the Old Testament and two of the New Testament are composed of 28 or more chapters. All of these books have a twenty-eighth chapter either expounding a theme of teaching or containing a doctrine of great importance. The slight exception to this is the book of *Numbers*, where the material is so lengthy it is divided into chapters 28 and 29. The thirteen books following this pattern of 28 are:

Genesis	1Samuel	Isaiah	Matthew
Exodus	Chronicles	Jeremiah	Acts
Numbers	2Chronicles	Ezekiel	
Deuteronomy	Job		

The teaching themes contained in these 28th chapters contrast with material in the preceding and following chapters, as follows:

Genesis 28 The doctrine of tithing was established through Jacob, who receives for his progeny the first promise of national greatness and prosperity; a prophecy foundational to understanding most other prophecies. The adjacent chapters are only narrative.

Exodus 28 Aaron and his sons (the teachers) are set apart. This is preceded by measurements and specification for the tabernacle (a place of teaching) in chapter 27, and followed by offerings and rituals in chapter 29.

Numbers 28 The cycle of holy days is set out; this is the premier teaching system in all of religion. It represents the true religion. The preceding chapter deals with inheritance laws and the appointment of Joshua; the following chapter with offerings and the Feast of Tabernacles.

Deuteronomy 28 The terms of the Old Covenant: its blessings & cursings are expounded here. This chapter is preceded by altar building and followed by comments on the covenant and exhortations to follow it.

1Samuel 28 The deceased prophet Samuel (a teacher) is 'seen' by King Saul in a false vision, due to demonism practiced by the Witch of Endor. This is preceded by the activities of David and Saul and followed by military matters.

1Chronicles 28 Solomon was to build the first Temple, a place of teaching. This is preceded by captains, princes and officers and followed by offerings, thanksgiving and David's death.

2Chronicles 28 False teaching reaches its zenith as Ahaz becomes the only Israelite king to burn his children in sacrifice to a false god. It is preceded by military matters and followed by a cleansing of the house of God.

Job 28 This is Job's exposition on wisdom and understanding. It is preceded by his condemnation of the wicked and the rich, followed by a nostalgic narrative about the way things once were.

Isaiah 28 A vital teaching principle is given in this chapter. The truth on any subject is shown to be spread throughout the scriptures: v13 *"...here a little and there a little, that they might go and fall backward, and be broken, and snared, and taken."* This is preceded by a description of vineyards and followed by judgments on Jerusalem.

Jeremiah 28 This is concerning false prophets and false teachers. It is preceded by the yoke of Nebuchadnezzar and followed by Jeremiah's letter to the captives.

Ezekiel 28 This is all about the source of false teachers, namely Satan, and a description of his downfall. This chapter is preceded by the riches and the fall of Tyrus, and followed by a judgment on Egypt.

Matthew 28 Its last chapter contains an exhortation to teach all nations, the previous chapter describing the crucifixion.

Acts 28 This describes the culmination of Paul's teaching career where people flocked to his rented house to hear him preach on *"...the kingdom of God...no man forbidding him."* This is the last chapter in *Acts*, preceded by an account of Paul's travels.

So it can be seen in the Old Testament that the 28th chapters of books follow the teaching theme consistently. The book of Acts is the thirteenth and final instance of a 28th chapter. Grid numbers are encoded within it. For example, Acts has 28 chapters that fall into 4 sections of 7. They end as follows:

Chapters 1 – 7: The stoning of Stephen, approved by Saul (Paul)

Chapters 8 – 14: The stoning of Paul, though he survives

Chapters 15 – 21: The beating/chaining of Paul before he speaks

Chapters 22 – 28: Paul preaching, no man forbidding him

Acts is the 44th book of the Bible. The spirit, which is marked by the numbers 1 and 44, was first given to the Church in Acts. By virtue of its structure and position Acts encodes the numbers 4, 7, 44 and 28. In this way Grid numbers are encoded in Acts, the book itself containing 28 chapters.

The number 28 is also of importance mathematically. It is classified as a very rare type of number called a *perfect number*. Such numbers are the sum of its several factors, for example $1 \times 2 \times 3 = 6$ but also $1 + 2 + 3 = 6$. The first few perfect numbers are:

6	$(1 + 2 + 3 = 6)$
28	$(1 + 2 + 4 + 7 + 14 = 28)$
496	
8128	
33550336	
8589869056	
137438691328	

More often than not perfect numbers end with 28. As mentioned in the earlier chapter *A Picture Pattern* the number 28 was prominent in the construction of the Tabernacle (a temporary dwelling, or tent) in Exodus. Two sets of curtains were to be constructed: one set to used as a perimeter fence, the other as a covering for the Tabernacle itself. This ancient design is in complete accord with the Grid:

The perimeter
Exodus 26:2 "...ten curtains of fine twined linen...v3 The length of one curtain shall be eight and twenty cubits, and the breadth of one curtain four cubits..."

The Tabernacle within
Exodus 26:7 "And thou shalt make curtains of goats' hair to be a covering upon the tabernacle: eleven curtains shall thou make. v8 The length of one curtain shall be thirty cubits, and the breadth of one curtain four cubits: and the eleven curtains shall be all one measure."

These measurements can be simply illustrated:

10 curtains 28×4 cubits, as fencing

11 curtains 30×4 cubits as a covering

This 'spiritual teaching centre' was to be surrounded by curtains of 28 cubits long. These are 10 in number giving a total end to end of 280 cubits. The 'two rows of three' of the ten curtains forming the perimeter fence of necessity resemble *the layout of the Grid* (three by two), as any other arrangement would create a corridor of 4 by 1:

The above perimeter fence contains the numbers 28 and 280 which that appeared earlier, 'going around' the Grid:

The Old Testament Tabernacle is therefore based on the Grid: both are 'surrounded' by the number 28. Its *holy of holies* is covered by 11 curtains, the Father's secret identity number in the Genesis Grid.

The 10 curtains of 28 cubits length surround this teaching centre, from which the number 280 derives. The teaching program

159

conducted regularly in and around the Tabernacle was based on the Sabbath, the seventh day, coding the number 7. Even the number 44 is present in a cryptic way, as the *two categories* of curtain are *each* 4 cubits in diameter. Furthermore, the two rectangles each offer four sides, suggesting 4 and 4. Thus the three defining numbers of the Grid can be seen in the Tabernacle of God, surrounded by the number 28.

The number 28 is encoded in one more way in the Tabernacle. It could be almost a point of complaint that the whole edifice is made needlessly complicated by the use of curtains of odd lengths. These lengths were 30 cubits for the 11 curtains covering the holy of holies and 28 cubits for the perimeter curtains. But there is a reason for a length of 30, because when the number of square cubits (as in the sense of 'square feet') for the two different types of curtain is calculated, a familiar pattern emerges:

Holy of Holy's curtains:
$$11 \times 30 \times 4 = 1320 = \mathbf{40} \times 33$$

Perimeter :

$10 \times 28 \times 4$

$= 1120$

Which equals

$\mathbf{40} \times 28$

Digits total 33   *Digits total 28*

3.14159265358979323846264338

Jesus 28   Christ 33

No book or website in the world, at the time of writing, discusses the number 28 in the Bible. Here *the power of 28* has been expounded for the very first time. Is it any wonder that the Bible has been called 'the book nobody knows'? The Bible is a coded book, but the time has now come for more of its mysteries to be uncovered. As the prophet Daniel wrote:

"At the time of the end men shall run to and fro, and knowledge shall increase." Dan 12:4

Is knowledge of the Bible increasing today? Here is a summary of key observations (incorporating Bullinger's unwitting testimony shown earlier) of the number 28:

The Father: mentioned in the N.T. (Bullinger)	280
Total teaching statements, four gospels combined	280
Warnings of false teachers, four gospels combined	28
The wave-sheaf symbol (Christ 'offered)	28
The heave offering (prophetic of Christ)	28
"A son" in respect of Jesus Christ; New Testament	28
The Lamb, Jesus Christ; New Testament	28
Beings in Revelation Ch. 4, surrounding the throne	28
"As it is written" in the New Testament	28
Genesis 1:1, comprised of 28 Heb. letters	28
Genesis 1:28 man's first instruction from God	28
The ass (only animal likened by birth to man)	28

The teaching number of God, 28, is embedded in scripture in multiple ways. It is, in and of itself, a world-first discovery. It has been unearthed at a pivotal time in history when knowledge of every kind is rapidly increasing.

REVELATION ENCODES THE GRID

"The most beautiful emotion we can experience is the mysterious." Albert Einstein

The book of Revelation is the 66th book of the Bible. Its 22 chapters provide further confirmations of the Grid. In this second section of *The Genesis Grid* the question *who and why* can be answered convincingly by a further revealing of the Grid pattern in scripture, as well as the manner in which it is encoded. The true nature of the 'spirit' identified by the Grid number 1 is here brought into even sharper focus in this the last book of the Bible.

In Revelation the principle is again seen that an important number can appear as a factor within a larger number that is also part of the same scheme. For example, the number 28 and its scaled-up version, 280, are both clearly represented in the four gospels. Similarly, the meaning of the number 44 is expressed more forcefully by 88.

The 'spirit' is a subject of particular interest in Revelation. Uniquely in that book, it features four times as the *"seven spirits of God."* The first such mention is made in the 4th verse of Revelation and the fourth and final mention in its 88th verse. The four occurrences of the 'seven spirits' in Revelation are as follows:

Fourth verse Rev 1:4 "John, to the seven churches....from him that is, and which was, and which is to come, and from the seven spirits....before his throne."

Fiftieth verse Rev 3:1 "And unto the angel of the church in Sardis write; these things saith he that hath the seven spirits of God and the seven stars; I know thy works..."

Seventy-sixth verse Rev 4:5 "And out of the throne proceeded lightnings, and thunderings and voices: and there were seven lamps...the seven spirits of God."

Eighty-eighth verse of the book *Rev 5:6 "...in the midst of the throne and of the four beasts, and in the midst of the elders, stood a Lamb as it had been slain, having seven horns and seven eyes, which are the seven spirits of God sent forth into all the earth."*

In this there is the series of numbers 4, 50, 76 and 88. Could this relate in some way to the Grid? At first glance verse numbers 50 and 76 do not seem to, but a fit becomes apparent when we set these positions against the natural musical frequencies discussed in the chapter *Grid vibrations*. The appearance of 88 notes of the piano was a clue to the presence of Grid numbers in music, now amply demonstrated, and is a further clue here.

Using the 'verse position pattern' of 4-50-76-88 of the seven spirits of God we find that, by comparing their pattern to keyboard positions, the top line of the Grid is represented in the following way:

4th note - C   1   (bottom C, the fourth note up)

50th note - B flat   7   (the seventh note in a descending melodic minor scale, from C)

76th note – C   1   (C is the 1st note in this scale)

88th note – C   1   (C is the 1st note in this scale)

Thus the top line of the Grid, the identity line 1-7-11, is encoded via the seven spirits of God. But the pattern ends on the 88th verse, coding the number 44 of which it is the fullest expression. Thus the defining numbers of the Grid, 7, 11 and 44 are encoded via the Seven Spirits of God.

There is a further way in which the book of Revelation encodes the numbers 7 and 11 by virtue of its structure. Revelation has 22 chapters falling into two halves of 11 chapters each. This can easily be shown to be the case, because a clear and uninterrupted narrative thread runs through the first 11 chapters; it is then broken at the beginning of chapter twelve.

Evidently, the second half of Revelation has an entirely different structure to the first. Unlike the first set of 11 chapters, there is little in the way of any chronological ordering, or chapter linking, within the second set of 11. The manner in which information is organized is quite different. Moreover, the *seventh chapter of Revelation is an inset* (a Sabbath rest from punishments!) and the book of Revelation as a whole is comprised of *two sections of eleven chapters* each. Thus the numbers 7 and 11 are unmistakably marked on the book of Revelation by virtue of its structure.

One further number is encoded in Revelation, which is the number 400. The book of Revelation as in its original inspired text would appear to fall into 400 'verses.' The allocation of material in the King James Version produces 404 verses. Some few of these are unnaturally short and appear to cut a single statement in half. The following pairs of verses (with English punctuation removed) strongly indicate that 404 may be too many divisions:

Chapter 1 verses 10 to 11: *"I was in the spirit on the Lord's day and heard behind me a great voice as of a trumpet v11 saying I am Alpha and Omega the first and the Last..."*

Chapter 2: verse 2 to 3: *"I know thy works and thy labour and thy patience and how thy canst not bear them that are evil and though hast tried them that that say they are apostles and are not and hast found them liars v3 and has born and hast patience and for my name's sake has laboured and has not fainted. v4 Nevertheless I have somewhat against thee..."*

Chapter 13 verses 9 and 10: *"And all that dwell upon the earth shall worship him whose names are not written in the book of life of the Lamb slain from the foundation of the earth v9 If any man hath an ear let him hear v10 he that leadeth into captivity shall go into captivity he that killeth with the sword must be killed by the sword here is the patience and faith of the saints."*

<u>Chapter 22:3 and 4</u>: *"And there shall be no more curse but the throne of God and of the Lamb shall be in it and his servants shall serve him v4 and they shall see his face and his name shall be on their foreheads v5 And there shall be no night there…"*

In each of these four cases, two verses could more logically be condensed into one. The number 400 appeared first in Genesis as the Israelites' slavery in Egypt. The Sun and the Moon are introduced in the 16th verse of Genesis. These two bodies have a width/distance ratio of 1:400 as first explained in the chapter *Heavenly Bodies*. Apparently, the end of the Bible is marked in this way with the number 400.

With its last book built on the number 400 the Bible would be brought back full circle to the Two Great Lights of Genesis. Thus the Bible ends as it begins, with *The Lamb* and his God: 7 and 11.

30

THE PATTERN OF A THRONE

"...and there was a rainbow round about the throne..." Revelation 4:2

It has been shown that the Grid is encoded in such diverse places as the tides of the oceans, the notes of the musical scale, the reproductive system of the human body, the elements of which the universe consists, the Sun and the Moon and in eleven Bible passages.

In the last biblical occurrence of the Grid pattern, within the book of Revelation, there is an odd development. It is a departure from all other scripture. All of a sudden, the spirit is referred to as 'the seven spirits of God.' This is a new concept and it is quite a mystery: how is it that something represented by the number one is possessed of seven facets?

The seven colours of the Rainbow (red, orange, yellow, green, blue, indigo and violet) may provide an answer to this question. This is an obvious line of enquiry, because a *rainbow is described as being around the throne of God!* (Revelation 4:2). Is it possible that the top line of the Grid, **1 7 11**, shows the pattern of that throne in the following way?

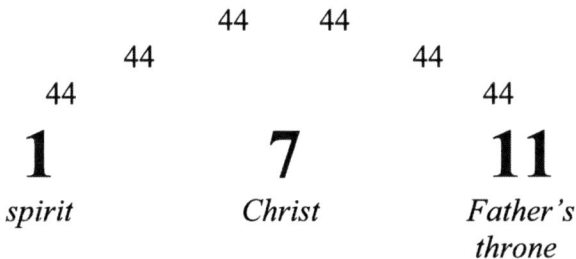

44   44

44   44

44   44

1   **7**   **11**

spirit   *Christ*   *Father's throne*

Christ is sitting on the Father's right hand. If the symbol of the seven spirits of God is the rainbow, which is merely guesswork at this point, are these deities surrounded by one? If so this might conflict with what has been established already about the Father's spirit. It was always a circle of 44's, not a circle of 7's.

However, a resolution to the Rainbow question is possible. A simple formula is found through the product (multiplication) of 44 and 7. This gives 308, the total count for the holy spirit across the entire Bible! As shown in the earlier chapter *The Ovum and the Spermatozoon,* the number of times the spirit is mentioned in the Bible is 308:

Old Testament 'spirit' pertaining to God	102
New Testament 'spirit' pertaining to God	102
Old Testament 'holy spirit'	3
New Testament 'spirits'[3]	3
New Testament 'holy spirit' ('Holy Ghost', KJV)	94
New Testament 'seven spirits of God'	4
Total	**308**

The Rainbow around the throne is a pictorial representation of the spirit. Thus it is seen that Christ and the Father are seated together and encircled by the spirit (44 and the 7 facets of the Rainbow) that emanates from the Father.

The rainbow hypothesis fits another key fact, because the Grid is 'surrounded' by the number 28. In the description of the throne of God in Revelation chapter four there are 24 elders and 4 living creatures, a total of 28 beings. With this fact in mind a further insight is possible: it is the entire Grid, and not the top line only, that pictures the throne of God.

Within the Grid the two deities 7 and 11 are surrounded by circling 44's containing the six Grid numbers. But yet again they are all, by implication, surrounded by the derivative 28's. This is reminiscent of the Tabernacle of Moses with the holy of holies covered by 11 curtains, surrounded by 10 screens of 28 cubits with teaching on the 7th day. The layout is as follows:

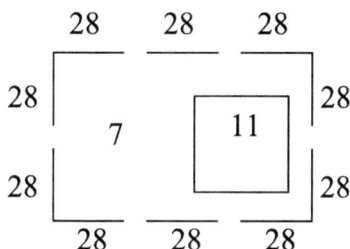

The discovery of the 308 occurrences of the spirit throughout the Bible is a weighty and dramatic addition to the many proofs of the Grid. The number 308 can now take its rightful place:

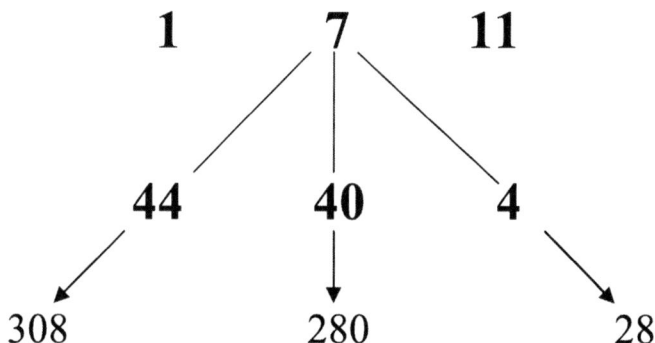

This is proof par excellence. As with the bottom line of the Grid, this result enjoys addition from right to left. Was it ever likely that 308 occurrences of the spirit would randomly emerge to complete the derived 28's in this way?

There is another sobering fact that relates to the pattern, because in 2005 the oldest ever version of the New Testament in its original language was found in Egypt. This 1700 year old document gave the 'number of the beast' not as 666, but 616; the three derivative Grid numbers 308, 280 and 28 total 616. That human leader, marked with a number and typed in prophecy as a beast, will be a Satanic counterfeit.

Because the circle of 44's within the Grid is surrounded by 28's, another way in which this discovery can be illustrated is by a circle surrounding a smaller a circle:

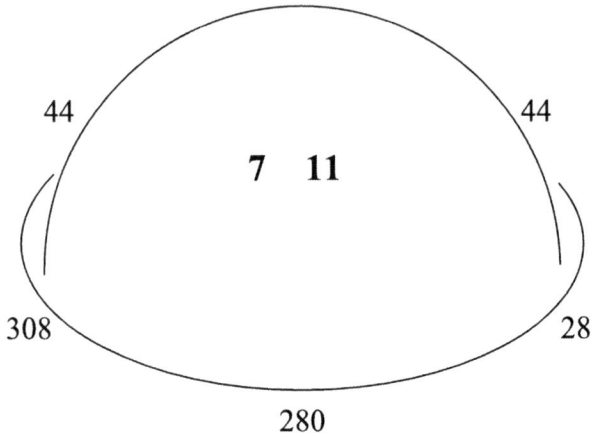

The Grid with its completed pattern of associated numbers holds together in more than one way, because 1 × 308 equals 308, but also 11 × 28 also equals 308:

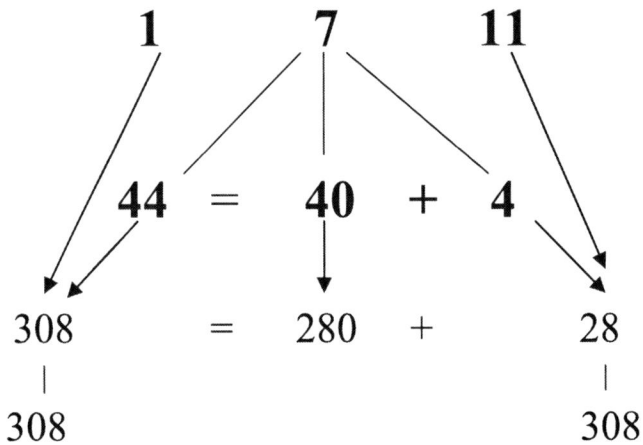

This symmetry is again suggestive of a circle within a circle, as the original inner circle of 44 is now surrounded by a greater circle of 308. Yet the two are linked and would coincide, sharing as they do the common factor 44. But is it a noteworthy fact that the total of 308 is garnered from across the *entire Bible*, not just one section. It unifies all those other totals derived merely from relevant sets of books or testaments.

So what are we to make of these numbers involving the spirit? Initially this essence was represented by 1: the number of unity. As the reproductive power of the Father it is 44 because that is the action, as in the 44th book, Acts. The total number of appearances of God's spirit in the entire Bible, 308, has the factors 44, 28, 11 and 7. Just what is this spirit? Could a seven coloured Rainbow fit the scheme laid out in this book? It transpires that it does.

But realise this: the Rainbow is circular like *the Sun, Moon and Corona!* Are there four symbols all of a sudden? If there are now to be *four* circular symbols, not merely the original three, could there then be a *fourth column* to the Grid? If so, this would surely upset the validation so neatly proffered by the Grid total of 107, would it not? Regardless of that problem, for which a solution does exist, the appearance of a fourth circular symbol virtually demands the existence of a fourth column to the Grid as below:

CORONA	MOON	SUN	RAINBOW
◯	◯	◯	◯
1	7	11	?
44	40	4	?

These symbols follow the 3+1 pattern because there are three phenomena composed of atoms, but a fourth consists of refracted photons that produce seven colours of the rainbow. Could it be possible that a fourth pair of simple numbers, again following a 3+1 pattern, is available and that they could meaningfully represent the Rainbow? Such a pair of numbers could quickly come to mind: the numbers -1 and -7, or minus one and minus seven. The minus signs alter completely the character of the numbers. Their negativity can be said to correspond to the ephemeral photons of the Rainbow, whereas as the atoms of the first three symbols are equivalent to the positive integers of the original Grid.

The result of this arrangement is quite astounding, because it suggests an image of the throne of God. But firstly, a new symmetry emerges because the top and bottom lines of the expanded Grid have totals that are reversible: 18 and 81:

1	7	11	-1	=	18
44	40	4	-7	=	81

This idea appears to offer not only symmetry but an additional, alternative, *validation* to that of the original Grid total 107. Furthermore, in the expanded bottom line, which had been denoted as 'power and motion' we see that the previous total of 88 drops to 81, the number of *creation* – the 81 stable elements! But there is another insight involving the top line: the number 18. This is the identity line of the Grid.

Man is made in the image of God. The evidence for the existence of nine fundamental human personality types is compelling. Furthermore, research has found that each person will usually be found to possess just one other sub-dominant aspect of personality derived from one main type. This gives 18 basic categories, according to the Enneagram system[1] of personality analysis. This system is probably the most thorough attempt to categorize and explain human types. God may have envisaged eighteen basic 'versions' or models of himself. Eighteen is also the sum of 7 and 11. The reader will recall that the multiplication of the two outside columns of the original Grid produced 44 twice. In this expanded Grid below the equivalent total is 308, because the two outside columns now have products of + 44 and +7 = +308 as follows:

1	7	11	-1
×			×
44	40	4	-7

(Note that the two minus signs in the right hand column cancel one another out to produce an ordinary 7, or +7)

As the total Bible count for 'the spirit' is 308 (where 7 × 44 = 308) what now presents itself is a pattern of the two Deities of heaven, the Father with Christ at his right hand and a rainbow above and behind them, as follows:

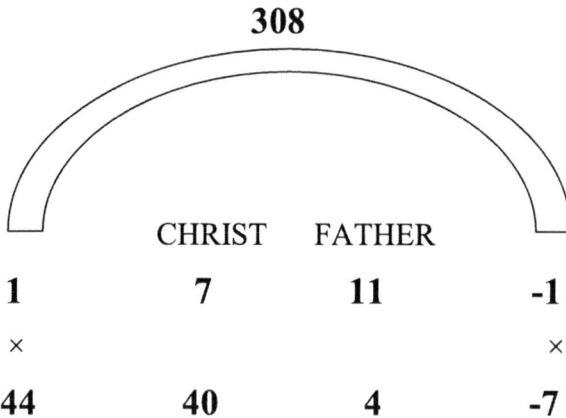

308

CHRIST   FATHER

1	7	11	-1
×			×
44	40	4	-7

The bottom line now totals 81, representing the 81 stable elements of the universe. The two Creators here preside over their creation. (The 'three to the power of four law' was always evident in the throne and its surroundings. Revelation describes its four levels: the Father, the Son, the 4 beasts and 24 elders - 28 in all - and the angelic realm.)

The other discovery concerning the top line total of 18 (7 +11) suggests God as the Creator of the 18 human personality types, casting new light on the statement in Genesis 1 verse 26:

"Let us make man in our image, after our likeness…"

The Grid is not only a representation of God's throne but of his master plan, as represented by the 1711171 holy day system of religious observance. This is the only authentic system of Bible worship; there is no other. The top line of the Grid is an abbreviation of this pattern. At the end of the fulfillment of that plan for mankind pictured by the 1711171 pattern of holy days, following the final destruction of the incorrigibly wicked (Malachi 4:1-3 & Revelation 21:8), the throne of God is to come down literally to Earth as described in Revelation:

> *"And I saw a new heaven and a new earth: for the first heaven [sky] and first earth were passed away; and there was no more sea. v2 And I John saw the holy city, new Jerusalem, coming down from God out of heaven...v23 And the city had no need of the sun, neither of the moon, to shine in it: for the glory of God did lighten it, and the Lamb [Christ] is the light thereof...the throne of God and of the Lamb shall be in it..." Revelation 21*

In this vision, the sea, symbol of the satanic forces of the Earth, is done away. The two deities render their secret symbols, the Sun and the Moon, obsolete. Their light is not needed any more. They are to be supplanted by the Two Great Luminaries that they have silently represented down through the millennia, to a blinded, unseeing world. Trinitarians should note that there is no mention of any holy spirit 'person' or any third entity casting light in this New Jerusalem.

Despite the earth-shattering discovery of the Genesis Grid, pointless billions of dollars will still be poured into the futile search for life on other planets. Men will still crave evidence that life sprang spontaneously from dead matter. They have never found evidence for that idea on Earth. Their defense against truth is as always to ignore it, if they can; to deny it any currency - though that may no longer be possible.

The only alien is God himself. There are no aliens, and the search for them on Mars and other places should be pronounced over and finished with. In the light of these present discoveries S.E.T.I., the Search for Extra-Terrestrial Life, is no more - it is a dead end.

---

1. See the Riso-Hudson Enneagram Type Indicator on enneagraminstitute.com

PART THREE: OUTCOMES

31

A BRIEF HISTORY OF HISTORY

"The longer you look back, the further
you can look forward." Winston S. Churchill

Most people in a school, an office or a gymnasium would utter a knowing chuckle - or throw a worried glance - if someone spoke the number '666.' Nearly everyone in the western world has heard of Revelation's beast and what is generally believed to be the number of his name; but what of the number 444? Consider the following pattern from the Grid:

1   **7**   **11**

44   **40**   **4** is suggestive of **444**

The Grid bottom line relates to power and motion. It suggests the pattern 444. Strangely enough, this number 444 is seen in events of enormous import both ancient and modern. For example the website www.freewebs.com refers to:

> *"...the exactly 444-week-long duration of the Rome-Berlin Axis [led by Mussolini and Hitler] from 26th October 1936 to 28th April 1945 when Mussolini died."*

The same number 444 was linked to the Tehran US embassy siege during the presidency of Jimmy Carter. That siege lasted 444 days, not weeks. These are well known and widely quoted figures. The *Daily Mail* of March 27th 2007 also quoted 444 days as the length of the November 1979 Tehran siege. Whether weeks or days, it is a remarkable pattern, but what could it mean? It does have a meaning going back to ancient times.

World changing events signalled by 444 involve changes in *the power of nations*. Developments influenced by what is sometimes called the occult can mimic or counterfeit[1] themes found in the Bible. Hitler's interest in the occult is fairly well known. It should be no surprise that the Axis powers were led to enact a Bible related number.

A further example of the counterfeiting occurrence of Bible related numbers would be the ordination ceremony held by John Paul II in 2003. The ceremony had been blacked out by the eighth and last of the worldwide power cuts of 2003, when ordaining 31 new cardinals. That pope went on to live exactly 31,000 days[2]. Rome's pontiffs have headed an organization that promotes Christ in name, though not in deed.

Sinister forces are once more moving the world towards a crisis. The decline in US power and prestige are its precursor, though there have been fluctuations in America's fortunes before. They rebounded following the Tehran siege of 1979, after which President Reagan decided to ally with Saddam Hussein. Reagan backed Saddam in his decade-long war against Iran, selling arms to both sides. If that was a bid to hold back Islamic fundamentalism then it has manifestly failed.

There was an ugly rumour that American diplomats gave tacit approval for Saddam's invasion of Kuwait, purely to engineer the outcome of a permanent American military presence in Saudi Arabia. The Gulf War would then have been a mere side show. This cannot be proved, but what is known is that the Saudi ruling family have been feeling threatened by their subjects for many years and have exceptionally close ties to the Bush family, whose fortune was made in oil. Bin Laden, a Saudi national, began campaigning for the removal of American troops from what he regards as his holy land. Alliances with and a dependence upon nations with no natural affinity with the US have proved to be a snare. Globalisation has led to vast financial imbalances and now China's financial reserves hold huge sway over America's fortunes.

The malevolent forces behind the wartime Axis powers have not disappeared. Ancient Bible prophecies lay out a pattern of seven revivals in the geographical area of Europe, as will be explained in this chapter. The 'Holy Roman Empire' was to undergo seven resurrections. The Axis powers fulfilled the sixth, but the Tehran siege set in motion a course of events leading up to the emergence of the seventh and final revival. This is the link between 444 weeks and 444 days: the Axis powers, lasting 444 weeks, were a *fulfillment* of the sixth resurrection of the Holy Roman Empire in Europe whereas Tehran, with 444 days and its fundamentalist aftermath, is to be the *catalyst for the seventh.*

Against the backdrop of prophecy, the link between Tehran and the EU is evidently this: it is looking increasingly likely that an Iranian controlled union of nations will provoke Europe to invade the Middle East. Iran has repeatedly threatened to close the Straits of Hormuz through which 40% of world oil supplies flow. It is racing to produce materiel that would make an atomic bomb possible by 2011. The 444 day Tehran siege was the portal from which Iran has emerged as a world power, having broken the American dominance and control over their country through the Shah.

Daniel chapter eleven contains the longest prophecy in the Bible. It describes a 'King of the South' who will 'push' at a European power:

> *"And the King of the South pushed against the King of the North and the King of the North came against him like a whirlwind...."*

The precursor of this soon coming mega-crisis was history's greatest event, World War Two. One can determine the identity of the two kings mentioned above by combining related passages of the Bible. These describe successive world kingdoms in the area of Europe and the Middle East. The 'holy' Roman Empire has seen many revivals and these are portrayed in the prophecies as 'horns.' A horn is a Bible symbol for power. One other major factor is a religious force *"speaking great things."* This is pictured in *Daniel* as an additional horn *"stouter than its fellows."* Which institution in the area of the

Holy Roman Empire has been the stoutest, has enjoyed the greatest permanence and continuity, while overseeing and controlling successive political revivals? Only the Catholic Church could fulfill that prophecy.

Here is a summary of actual events matched against the combined prophecies of Daniel chapter 7 and Revelation chapters 13 and 17. These passages share a scheme built around eleven prophetic horns:

FIRST FOUR HORNS

THE ROMAN EMPIRE, started 31B.C. Rev 13: 1-2 A beast with 7 heads and 10 horns (also Daniel 7:7)

THE ROMAN EMPIRE, fell 476 A.D. Dan 7:3 The beast suffers a *"deadly wound"*

Horns 1, 2 & 3

VANDALS, HERULI, OSTROGOTHS The first three of the Roman beast's horns Dan 7:8

Horn 4 (stouter)

THE PAPACY (the image of the 'holy' Roman Empire that *"causes many as would not worship the beast to be put to death"* Rev13.

An important key is to realise that the 'little horn, stouter than its fellows' constitutes an odd 11th horn, falsely purporting to represent God.

REMAINING SEVEN HORNS:

Horn 5 (deadly wound healed, Rev 17:5. 1st head of the beast	IMPERIAL RESTORATION JUSTINIAN, 554 A.D. Pope supreme in West. Beast ridden by a Scarlet Woman/Church)

Horn 5 (deadly wound healed, Rev 17:5. 1st head of the beast — IMPERIAL RESTORATION JUSTINIAN, 554 A.D. Pope supreme in West. Beast ridden by a Scarlet Woman/Church)

Horn 6 (The 2nd head, etc) — FRANKISH KINGDOM CHARLEMAGNE 774 A.D. crowned by Pope 800 A.D.

Horn 7 — HOLY ROMAN EMPIRE OTTO THE GREAT Crowned by Pope 962 A.D.

Horn 8 — HAPSBURGS CHARLES THE GREAT Crowned by Pope 1520 A.D.

Horn 9 — NAPOLEON'S KINGDOM Crowned by Pope 1805 A.D.

Horn 10 — ITALY – GARIBALDI to MUSSOLINI who declared the Roman Empire

Horn 11 — E.E.C., E.C., EUROPEAN UNION which is to become the UNITED STATES OF EUROPE

This chart shows a perfect match between the prophetic framework of the Bible and actual events. Of the last horn (the EU) it is said that:

*"...ten kings...receive power as kings **one hour** with the beast."*
Revelation 17:12

The same expression of an 'hour' is used in chapter three:

*"...the **hour** of temptation which shall come upon all the world, to try them that dwell upon the earth." Revelation 3:10*

The final top political leadership of the EU will consist of eleven leaders, the ten 'kings,' or leaders of principal nations. Not only are there *eleven* leaders in the final European empire, but there are in all *eleven* prophetic horns that span world history since the Babylonish Empire. These elevens all falsely purport to represent God!

The *hour* of temptation (testing) to come upon the world mentioned in Revelation chapter three is obviously not a literal sixty minutes. The actual period is enumerated in several prophecies. For example, the time period is explained twice in Revelation chapter eleven:

Verse 2 "...the holy city [Jerusalem] shall they tread under foot <u>forty and two months</u>. v3 And I shall give power unto my two witnesses, and they shall prophesy a thousand two hundred and threescore days...v8 And their dead bodies shall lie in the street of the great city...where also our Lord was crucified...v15 And the seventh angel sounded and there were great voices in heaven saying, The kingdoms of this world have become the kingdoms of our Lord and of his Christ and he shall reign for ever and ever."

What is referred to here is a crisis period of 42 months (three and one half years) which immediately precedes the second coming of Christ. This is the same end-time crisis referred to by both Daniel and Christ in similar words:

"And at that time [the time when the King of the South pushes]...there shall be a time of trouble, such as never was since there was a nation even to that same time; and at that time thy people shall be delivered..." Daniel 12:1

"And there shall be signs in the sun, and in the moon, and in the stars; and upon the earth distress of nations, with perplexity; the sea and the waves roaring; Men's hears failing them for fear, and for looking after those things which are coming on the

earth: for the powers of the heavens shall be shaken." Luke 21:25-6

"For then shall be great tribulation, such as was not since the beginning of the world to this time, no, nor ever shall be." Matthew 24:21

The final crisis period – the 42 month Great Tribulation[3] referred to by Jesus Christ - is precipitated by one specific event described in Daniel the eleventh chapter, where the 'King of the South' pushes against the 'King of the North.' That Northern King (relative to the Middle East) will be a *United States of Europe*. It will occupy approximately the same territory as that of the original Roman Empire.

But who is the biblical King of the South? The obvious candidate always was the nation of Iran. Iranians are the ancient Persians of the Medo-Persian Empire; they are not of the Arab stock of surrounding nations. Is it rational to believe in coincidence to the extent that the following, though it harmonises with the above chart of Bible prophecies, can be dismissed:

Horn 6   ITALY – GARIBALDI to MUSSOLINI
Axis powers lasted **444 weeks** - Mussolini declared the Roman E.

Horn 7   THE UNITED STATES OF EUROPE
Tehran siege lasted **444 days** - emergent Iran, the catalyst for war

The 444 pattern is the clue that Iran will be the catalyst that plunges the world into the next great crisis. Iran is likely to gain control over much of Iraq after American forces leave.

Most Iranians share the same Shiite religion as the masses of Iraq. Shortly following the 2003 invasion, the two countries linked their electricity grids to enable Iran to supply the electricity the Americans could not. The power of Iran is growing and its ability to humiliate America and its allies is increasing. Iran supported the Hezbollah proxy army that so successfully resisted Israel in the month long war between Israel and Lebanon in the summer of 2006. It has defied the U.N. and the White House over the issue of uranium enrichment. The USA appears powerless to stop Iran developing nuclear weapons. It can only watch from the sidelines as a new empire rises in the Middle East. Iran is likely to take Iraq into protective custody. In August 2007 Iraq's foreign minister Hoshyar Zebari forecasted catastrophe if the Allies were to withdraw prematurely, a *"real disaster"* which would see Iranian intervention sucking in neighbouring Sunni Arab countries and Turkey. He predicted an oil disaster; that *"oil and terrorism would combine forces...I am worried, absolutely worried."* Any progress for the Americans in Iraq has always been described by their own generals as fragile and reversible.

America has squandered its bank of international trust and goodwill over Iraq. Its leadership of the world lacks credibility. The world has moved into a new and dangerous phase since 2003. Efforts to block Iran must ultimately fail, just as the bid by President Carter to release the Tehran hostages failed so ignominiously.

The above framework of Bible prophecies indicates that the present phase of world history will end at the time of the emergence of a fully united Europe. The 444 pattern confirms that Iran is the King of the South because it links that nation to the prophetic 'horns' of the sixth and seventh resurrections of the Holy Roman Empire. Europe is well advanced in its long march towards unity. By the time of the 'push' by the King of the South it is likely that the USA will have withdrawn into isolationism. The resultant power vacuum could then be filled by an emergent European Union: a *United States of Europe.*

The book of Daniel adds further detail in respect of this United States of Europe. What is described in the second chapter of Daniel is clearly a kingdom with five characteristics:

1. It is composed of two halves (East and West)
2. It lacks natural cohesion; some nations are stronger than others
3. It occupies the territory of the original Roman Empire
4. It has ten final leaders or 'kings' and an eleventh as dictator
5. It exists immediately prior to the second coming of Christ

These five points offer a strong match with the EU of today, the last of the eleven prophetic horns. The series of eleven horns had started with Babylon. The following detailed prophecy harks back to an era before the seven resections of the Holy Roman Empire. It is given in Daniel chapter two, where the prophet explains the dream of King Nebuchadnezzar of Babylon:

> *"But there is a God in heaven that reveals secrets, and makes known to the king Nebuchadnezzar what shall be **in the latter days**. Thy dream...v31 Thou, O king, saw, and beheld a great image...v32 This image's head was of fine gold, his breast and his arms of silver, his belly and his thighs of brass, v33 His legs of iron, his feet part iron and part clay. v34 Thou sawest until a stone was cut out without hands, which smote the image upon his feet that were of iron and clay, and brake them to pieces. v35...the stone that smote the image became a great mountain, and filled the whole earth." Daniel 2*

Daniel then relates his interpretation of the dream:

> *verse 38 "...thou art the head of gold (Babylon). v39 And after thee shall arise another kingdom inferior to thee (silver, the Medo Persians), and another third kingdom of brass (Greco-Macedonian Empire), which shall bear rule over all the earth. v40 And the fourth kingdom shall be strong as iron (Rome)...v42 And as the toes of the feet were part of iron and part of clay, so the kingdom shall be partly strong, and partly broken. v43 ...they shall not cleave one to another, even as iron is not mixed with clay. v44 And in the days of these kings shall the God of heaven set up a kingdom which shall never be destroyed...it shall stand forever."*

These four kingdoms and their and symbols are as follows:

THE CHALDEAN EMPIRE (625 – 538 BC, gold)   HEAD

THE PERSIAN EMPIRE (558-330 BC, silver)   CHEST

GREECE (beginning 333 BC Alexander, brass)   BELLY

THE ROMAN EMPIRE (31 BC to 476 AD, iron)   LEGS

The character of the successive empires is accurately reflected in the symbols given. Rome was in two divisions or 'legs', East and West, with an eastern capital in Constantinople. The statue bypasses the issue of revivals for Rome and jumps right down to the final (seventh) manifestation of the Holy Roman Empire we see today. The symbols of iron and clay succinctly describe the states of the present day European entity whose existence is 'smashed by a stone carved from a mountain.' Mountains are a symbol of government and this stone (Christ) sent back by the Father grows into a great mountain filling the earth. Before that epoch-changing event the EU will 'revert to form' to begin to display many of the evil characteristics of previous European revivals.

The affairs of the EU are increasingly dominated by a newly unified Germany, the largest exporter in the world. It is unusual that Germany now finds itself with a female Chancellor. According to Sir Bernard Ingham, press secretary to Baroness Thatcher throughout her eleven year premiership, women in leadership can operate somewhat differently to men. In his autobiography he argues that as a woman she was able to get away with certain initiatives a man would not dare to take.

It can be argued that a woman, by virtue of her maternal instinct, is trusted to a greater degree than a man. A male is assumed to be more aggressive and in need of restraint. Therefore the female leader may introduce radical measures or ideas unchallenged, where a male leader might not. For example, in the 1970's the conflicts between the government and the miners carried many risks for the government, as Edward Heath discovered. It was only Margaret Thatcher who was later able to defeat them. It is less likely that a male Prime Minister would have so forcefully confronted the miners; or seized back the Falkland Islands from Argentina.

The electorates of Europe were always more likely to acquiesce to a woman reviving the stalled Constitution for a United Europe. Germany looks a lot less menacing *with a woman at the helm.* Mrs. Merkel's ally Hans-Gert Pöttering, the President of the European Parliament, said at the March 2007 celebration of the 50th anniversary of the Treaty of Rome that the Constitution *"can have a different name but the content must be preserved."* In a celebratory summit of all 27 EU leaders, Merkel then set December 2007 as the deadline for drawing up a fast-track replacement of the Constitution rejected by voters. Writing in *The Times* of April 23rd 2007, in an article entitled *Referendum double-talk in London, and danger in Berlin*, Sir William Rees-Mogg referred to:

> "...an increasingly bureaucratic Europe, in which power is still moving towards non-elected bodies. The proposed constitutional treaty, which will be renegotiated at the Berlin summit in June, would centralize power even further...it is clear...that Ms. Merkel contemplates rewrapping the constitutional treaty without altering the contents of the parcel."

The worst fears of Rees-Mogg were realised in July of that year, when the EU 'Reform Treaty' based on 240 of the original 250 clauses of the 2005 Constitution was unanimously adopted. Any changes were *"more cosmetic than real"* thought ex-French President Valery Gisgard d'Estaing, who had written the original document.

There is little chance of ever procuring any future referenda on the new measures, which included a euphemistically named 'High representative' to coordinate the Union's *"competence in matters of common foreign and security policy."* As the EU juggernaut rolled forward, the British Parliament was nigh on becoming little more than a regional assembly. In the same month, British war veterans in Harwich were told they would have to pay £18,000 for 'security' to hold their annual Remembrance Day parade: the event was cancelled. As fast as UK sovereignty was eroded those who fought to preserve it were marginalised or forgotten.

In a modern and enlarged Germany of the 21st century, now the world's largest exporter, we see an anomaly. This is the largest nation of Europe, boasting the fourth largest economy in the world, yet it has a military policy still largely controlled by 'war guilt.' Perhaps this is not surprising in view of the recent global calamities the world has suffered due to a nationalistic Germany. The victors of World War Two, Roosevelt, Stalin and Churchill, had resolved to prevent German rearmament.

Today, nuclear weapons remain the most immediate threat to the world's future. It is striking that whilst lesser countries such as Pakistan and India possess such weapons, as do Germany's neighbours France and the UK, it is still unthinkable that Germany would be permitted to arm in this way. For how much longer can the unspoken embargo against a nuclear Germany hold?

Today's world has been defined by the struggle against Nazism. This is seldom discussed. The older generation is tired of war memories, while the younger generation - fascinated as they are by all things Nazi - take today's world for granted. Yet all major post-war institutions such as NATO, the U.N. the IMF and the World Bank were set up in the wake of World War Two. The Manhattan Project to develop the atomic bomb was a key plank in America's response to the scourge of German nationalism. The Nazis had also been racing to develop such a weapon.

Fears over the misuse of such weapons, which is the real human threat to the planet rather than global warming, often dominate news headlines today. But through the shifting power structures of the EU it appears inevitable that German politicians will one day gain sway over existing nuclear stockpiles, or be permitted to make their own.

Is it inconceivable that *"the savage furies latent in the most numerous, most serviceable, ruthless, contradictory and ill-starred race in Europe"* to quote Winston Churchill's description of the Germans, could surface for the third time in little more than a century? Following the July 2007 EU treaty agreement *The Times* reported on comments made by the President of the European Commission:

> *"...we are unique in the history of mankind. Sometimes I like to compare the EU as a creation to the organisation of empires. We have the dimension of empire but there is a difference. Empires were made with force with a centre imposing diktat. Now what we have is the first non-imperial empire."*

José Manuel Barosso's claim that the EU is non-imperial could be debated. Imperialism is, by definition, the 'acquisition of dependencies.' Tens of millions of EU citizens are dependant on grants from Brussels yet most feel they have no adequate representation. What Barosso means by his non-imperial claim is 'without violence.' However, the prophecies of the Bible show that Europe will revert to type. Churchill would have been alarmed at the

186

speed with which Germany is gaining the upper hand in today's Europe:

> "Twice within our lifetime, and three times counting that of our fathers, they have plunged the world into their wars of expansion and aggression. They combine in the most deadly manner the qualities of the warrior and the slave. They do not value freedom themselves, and the spectacle of it in others is hateful to them. Whenever they become strong they seek their prey, and they will follow with an iron discipline anyone who will lead them to it. The core of Germany is Prussia. There is the source of the recurring pestilence." Winston S. Churchill, page 139, book 9, The Second World War

Today's world has literally been born out of the struggle against Germany and her Axis allies in an alliance that lasted 444 weeks. That era was the sixth resurrection of the Roman Empire. *Daniel* chapter eleven, the longest prophecy in the Bible, reveals that there will be a seventh and final revival in Europe, a final manifestation of Churchill's *"recurring pestilence."* This nascent revival exists already today in the form of the EU. It is, *in embryo*, the 'beast' of Revelation chapter thirteen:

> "And I stood upon the sand of the sea, and saw a beast rise up out of the sea, having seven heads (revivals) and ten horns, and upon his horns ten crowns, and upon his heads the name of blasphemy. v2 "...and the dragon (Satan) gave him power and great authority. v4 "And they worshipped the dragon which gave power unto the beast: and they worshipped the beast, saying, Who is like unto the beast? Who is able to make war with him? v5...power was given unto him to continue forty and two months. v6 And he opened his mouth in blasphemy against God..." Revelation 13:1-6

Notice the time period specified: 42 months. This political combine is assisted by *another prophetic beast* described in the same chapter of Revelation. It is, as will be shown, an *institution* of immense influence in today's world. It is described in the prophecies of *Isaiah* as 'The Lady of Kingdoms.' This is the horn 'stouter than its fellows' that, supplementing the ten horns of the beast, completes the series of eleven. This eleventh prophetic horn falsely claims to represent God: it is none other than the Church of Rome.

Man has invented his own religions under the sway of the Adversary. But God reserves the right of intervention, while allowing us here below to reap the bitter fruits of our own way. God still controls human affairs and has marked history with the defining numbers of the Grid: the seven resurrections of Europe from Justinian to the EU, plus the eleven prophetic horns. This gives 7 and 11 a whole new meaning.

1. Hitler was appointed Chancellor of Germany on January 30th 1933 by President Hindenburg. An election held on 5th March to *validate* Hitler's appointment won 44% of the popular vote. Hitler's first electoral breakthrough had occurred in the election of September 1930 at which time he dreamed of winning 100 seats; in the event he won 107.

2. Pope John Paul 2 lived 31,000 days exactly according to the British Israel Society bulletin reporting on the papal funeral. One should earnestly enquire as to whether his organisation truly represents the Deities of the Bible.

3. The number 42 carries both positive and negative connotations. In Ezra 2:64 the Jews returning from Babylon to build the second temple numbered 42,360. In 2 Kings 2:24 forty-two youths mocked Elisha and were killed by bears: a bear possesses forty-two teeth.

32

DECLINE AND FALL

*"In that same hour came forth fingers of a man's hand
and wrote...upon...the wall..." Daniel 5:5*

The USA has been the greatest military and economic power in history. At the start of the 21st century it is undoubtedly suffering from imperial overreach.

Due to its hundreds of overseas military bases, record trade imbalances and unaffordable tax cuts for the wealthy, the US government was in financial jeopardy before the crisis of 2008-9. America possessed the largest trade deficit in the world by 2008 but fewer dollar reserves than India. Its economy had been sustained by an asset bubble, beneath which government deficits and liabilities piled up. Even the basics were being neglected, with $1.6 trillion required merely to make good existing road infrastructure. The country was spending just 0.65% of GDP on such works compared to 9% in China.

With astronomical unfunded pension liabilities in the tens of trillions, and the ever-escalating costs of sustaining military adventurism, it was certain that US government finances would move into a crisis phase. The US banking collapse of autumn 2008 duly brought it. The election of President Barack Obama then guaranteed the continuation of deficit spending, the very cause of the crisis in the first place. No great power in history has ever survived long under such policies.

Collapse

Another world power, Russia, has hugely influenced the contemporary scene. The Soviet economic and political implosion of the 1990's was dramatic. It changed completely the world's perception of military risks. For a while, the cold war became a distant memory. During the halcyon days of 1990 – 2003 the USA was considered the only superpower. People perceived the world to be a safer place, but then came 9/11 and the invasion of Afghanistan and Iraq.

The USSR had been a clean and tidy adversary in comparison to Islamic fundamentalism. The logic of Mutually Assured Destruction (MAD for short) was that each side had an overwhelming self-interest in maintaining a state of masterful inactivity. This had the eventual effect of producing a settled state of mind in the citizenry of each side, even optimism. This turned to joy when in November 1989 the Berlin wall fell.

The fall of the Wall was an event completely misinterpreted by the media and the public. The general view was that an oppressed group of East European nations had been delivered from the mailed fist of the Russians into glorious Western-style democracy and liberty. Rumanian and East German dictators were quickly deposed. To most citizens, lacking the perspective of Bible prophecy, these events appeared encouraging. Few people realised that the Berlin Wall had been a dam holding back the scourge of ages: a renewed, enlarged and unified Germany. Churchill's 'Hun' was to begin his long march back to military greatness.

Sclerosis?

Is it possible that a resurgent Europe could eclipse the USA? Many of the same sociological tell-tale signs of imminent collapse seen in Soviet society in the 1970's and 80's are now characteristic of US society. Most notably these are a decline in life expectancy accompanied by a rise in infant mortality rates. Moreover, twice as many hours are worked by the average US household to pay bills than in the 1960's. It is a system with one foot in the grave; a once great society that can now be compared unfavourably on almost every count to an increasingly prosperous and confident Europe.

America's decline can be plotted without recourse to Bible prophecy. Some religious groups have mistakenly branded the USA as the biblical beast power, but the beast is a European entity as amply demonstrated in the earlier chapter *A brief history of History*. In Revelation 13:4 a question is posed:

190

"Who is like unto the beast? Who is able to make war with him?"

This is not referring to the USA, whose innings as world policeman is nearly over. The massive economic slump that began unfolding in autumn 2008, or even some other yet unforeseen catalyst, could trigger *another soviet-style implosion* this time involving America. The USA in such circumstances might well draw back into a protectionist shell. The mantle of the USA's world leadership would then fall not on China, but onto the shoulders of the seventh and final resurrection of the 'Holy Roman Empire', as described in the books of Daniel and Revelation. Europe will rush in to fill a power vacuum.

The emergence of Europe as a political power-bloc has, against all the odds, and much gainsaying media chatter concerning 'Euro-sclerosis', stagnation and low growth, won out over the skeptics. In 2006 the economic performance in terms of its growth rate of the EU exceeded that of the USA, whose growth rates vis-à-vis Europe have always been exaggerated by extraneous activities such as prison building and anti-pollution programs. On true annual GDP, the two rivals have been neck and neck. Reuters reported on 14th March 2008 that:

> *"The US economy lost the title of 'world's biggest' to the Euro zone this week as the value of the dollar slumped in currency markets. Taking the gross domestic product of both economies in 2007, the combined GDP of the 15 countries which use the Euro overtook that of the United States..."*

Citizens of the USA are incarcerated at twenty-five times the EU rate. When such non-productive elements are stripped out of the GDP figures, the USA has not enjoyed a growth rate any greater than the EU for decades. Although China offers serious long term competition, the EU is the world's single present day empire in the making. In an article entitled *It is the West that's starting this new Cold War,* Anatole Kaletsky of *The Times* asked on June 7th 2007:

> *"Is it so very unreasonable to view this EU-Nato juggernaut as the world's last remaining expansionist empire, or even the natural successor to previous German and French expansions that were considerably less benign?"*

191

Russia is right to worry about the new Europe. Most of Hitler's aspirations for the Continent are already a reality in a Europe a lot less 'benign' than *The Times* assumes. Kaletsky's vision of an emergent Cold War took a leap forwards in August 2008 when Prime Minister Putin ordered Russian troops into Georgia.

The Mother of Harlots

The ascendancy of the EU has confounded the pundits because it has been 'fated' to happen. It is the seventh of a series of seven prophetic horns - actual resurrections - depicting successive revivals of the Holy Roman Empire of Europe (see the chart in the chapter *A brief history of History*). This is the time spoken of by the prophet Daniel when he was told:

> *"But thou O Daniel, shut up the words and seal the book, even to the time of the end: many shall run to and fro and knowledge shall be increased." Daniel 12:4*

It is the time spoken of by Jesus in the Olivet prophecy in Matthew 24:20-22:

> *"But pray that your flight be not in the winter, neither on the Sabbath day: for then shall be tribulation, such as was not since the beginning of the world to this time, no, nor ever shall be. And except those days should be shortened, there would no flesh be saved..."*

Revelation 13:13 reveals that a great miracle-working leader will emerge to galvanise the world behind him, bringing 'peace' to the Middle East. His idea of peace will be an invasion of the Holy Land. This leader is mentioned in Daniel chapter eleven, the longest prophecy of the Bible. The 'time of the end' is the time setting:

> *Verse 40 "And at the time of the end shall the King of the South [Iran] push at him [King of the North, EU]: and the King of the North shall come against him like a whirlwind...41 He shall enter also into the glorious land...42 He shall stretch forth his hand unto the countries: and the land of Egypt shall not escape...44 But tidings out of the east [China] and out of the north [Russia] shall trouble him: therefore he shall go forth with great fury to make away many. 45 And he shall plant the*

192

tabernacles of his palace between the seas [Jerusalem]; yet he shall come to his end, and none shall help him." Daniel 11

This closing section of Daniel chapter eleven is one part of the huge jigsaw puzzle of end-time prophecy. In this passage a great military power is seen invading the Holy Land. The prophecy refers to that land between the Mediterranean and the Sea of Galilee, *"between the seas"* and a religious leader *"planting the tabernacles of his palace"* – a place of *religion.* Matthew chapter twenty-four predicts the rise of false prophets:

"And many false prophets shall arise and shall deceive many... For there shall arise false Christs and false prophets, and shall show great signs [Gk. semeion – indication, supernatural miracle] and wonders..." Matthew 24:11 and v24

In Revelation chapter thirteen one specific great leader is depicted as a beast with two horns:

"And I beheld another beast coming up out if the earth; and he had two horns as a lamb [falsely representing Christ], and he spoke as a dragon [like Satan]. v12 And he exerciseth all the power of the first beast [Holy Roman Empire] before him, and causeth the earth and all them that dwell therein to worship the first beast, whose deadly wound was healed [The imperial restoration through Justinian 554 A.D.]. v13 And he doeth great wonders, so that he maketh fire to come down from heaven on the earth in the sight of men: v14...saying to them that dwelt on earth that they should make an image [the Vatican is an 'image' of the government of ancient Rome] to the beast...v15...that the image of the beast should both speak ["great things", the little horn of Dan 7:20-21 that "made war with the saints" is a great persecuting church], and cause that as many as would not worship the image of the beast should be killed." Revelation 13:11-15

This leader is going to make fire come down from heaven in the sight of men! How much searching of history is required to find a government or institution resembling the above? There is only one credible explanation. Once the Bible had been translated it was the realisation that Rome fulfilled the description of a whore in Revelation chapter seventeen that added impetus to the Reformation:

"When considering the phenomenal growth of the Reformation from its birth in A.D.1517 to the year A.D.1585, the Protestant movement was swiftly gaining converts all across Europe. The Reformers revealed...The Church of Rome is the "MYSTERY, BABYLON THE GREAT, THE MOTHER OF HARLOTS AND ABOMINATIONS OF THE EARTH" Rev 17:5" Revelation in History, Daniel R Wilhelm.

Two Beasts

So in Revelation thirteen two 'beasts' are shown to be contemporaneous. One of them, a human being in league with occult forces, performs great miracles as mentioned above:

"...he exerciseth all the power of the first beast [EU]...v13 And he doeth great wonders so that he maketh fire come down from heaven on the earth in the sight of men." Revelation 13:12

People who do not know and believe the scriptures will be deceived. The wondrous miracles of this religious leader will convince them that he is God's servant, but he will be the tool of Satan. Meanwhile the other 'first beast' is a great political power. This description tallies with an account of two leaders in Revelation chapter nineteen:

Verse 20 "And the beast was taken, and with him the false prophet that wrought miracles before him, with which he deceived them that had received the mark of the beast, and them that worshipped his image. These both were cast alive into a lake of fire..."

The prophecy concerns two human leaders, the beast and the false prophet, representing the political and religious dimensions. This false prophet will be electrifying. His miracles will exceed those of the ancient Pharaoh's sorcerers who opposed Moses. Anyone who criticises him or disagrees with his church will be pronounced 'anathema', just as in Europe's Dark Age.

This is the condition Western society is about to revert to; a dark age of Catholic dominance. In their groping blindness men will turn to

194

this leader in their hundreds of millions, a man who will fit Paul's description of false leaders in the second epistle to the Corinthians:

"For such are false apostles, deceitful workers, transforming themselves into the apostles of Christ. v14 And no marvel: for Satan himself is transformed into an angel of light. v15 Therefore it is no great thing if his ministers also be transformed as the ministers of righteousness; whose end shall be according to their works." Corinthians 11:13

This great religious leader will work hand in glove with a secular European leader able to eclipse Hitler's deeds. No one can understand today's world without knowledge of the events of the twentieth century as they relate to German nationalism. Today's power structures and international institutions are the outgrowth of the postwar settlement. The present world has been defined by the defeat of Nazism.

Hitler had joined the German Workers' Party, the embryonic National Socialists that became the Nazi Party, in January 1920. He always claimed that his membership number was 7. The number on his membership card was actually 555, a picture of which is reproduced in the book *Hitler: hubris* by Professor Ian Kershaw. If the universe runs according to numbers, that may mean something. Prophecy shows that there is a successor to come whose number is 666. Alternatively, the recorded number universally quoted in all modern Bibles may be wrong. In a 2005 discovery, fragments of the earliest known version of the Greek New Testament clearly show the number of the beast as 616, not 666. If there is such an error it is not a unique occurrence. There are two other significant (and many more insignificant) errors of this kind in today's New Testament. Both involve fraudulent attempts to introduce trinitarianism, these being seen in Math 28:19 and 1 John 5:3.

The next Hitler

In the *Introduction* to the present book, Harvard law professor Alan Dershowitz was quoted as describing President Ahmadi-Nejad of Iran as *"the Hitler of the 21ˢᵗ century."* While Ahmadi-Nejad is highly significant, in the light of Bible prophecy one could question the professor's diagnosis. There is another claimant waiting in the wings. There is to come, according to Revelation chapter 13, a leader in Europe who will *"cause that as many as would not worship the image* [Vatican] *of the Beast* [a revitalized German-led EU] *should be killed."* In this regard the words of German philosopher Karl Jaspers concerning Nazism are worth repeating:

> *"That which has happened is a warning. To forget it is guilt. It must be continually remembered. It was possible for this to happen and it remains possible for it to happen again at any minute. Only in knowledge can it be prevented."*

There will be a leader who heads up the European combine. He will work in conjunction with a great religious figurehead. These two will head the 'beast power' through which will be channeled a resurgent German nationalism. This is already beginning behind the cloak of the almost fully fledged United States of Europe.

The Empire Number

At the end of the first world empire of Babylon, the head of Gold in Daniel chapter two (see the chapter *A brief history of History*), there was a feast held by king Belshazzar during which a mysterious hand appeared and began *writing on the wall*. It wrote the words MENE, MENE, TEKEL, UPHARSIN. Taking the number of letters in the original Chaldean language the pattern 444 emerges:

KJV translation	MENE	MENE	TEKEL	UPHARSIN
from original:	mêne	mêne	t^{e}qal	(p^{e}rac) [U-] pharsin
Letter characters	**4**	**4**	**4**	8
(Strong's ref. no.	4484	4484	8625	6537)

Meaning: 'numbered' 'numbered' 'to balance' 'to split up, divide'

The above words from Daniel chapter five were written in the Chaldean language. The message is in *two* parts:

1.   <u>444 (Empire number)</u>
   God has numbered (twice, 4 and 4, *mêne*) your kingdom and
   finished it. It is weighed (once, 4, *t ªqal*) and found wanting
   (lacking).

And therefore:

2.   <u>8 (New beginning)</u>
   Your kingdom is divided and given to the Medes and the
   Persians.

Chaldean was the language of the Persians who now comprise
modern day Iran. The number 444 in respect of the Tehran siege
regularly appears in the British press. On April 5th 2007, comparing
the thirteen day Iranian 'hostage' crisis involving British sailors and
Marines to the 1979 Tehran siege, *The Times* reported:

> *"This row was never like the 444-day US hostage crisis…"*

Again, on May 25th 2007 *The Times*, in an article discussing the era
of President Carter, recalled that:

> *"America watched helpless as its diplomats were held hostage
> by Iranian revolutionaries for 444 days."*

One of those revolutionaries is now, by all accounts, President
Ahmadi-Nejad of Iran. The number 444 can be linked to all of the
following:

Fall of Babylon in 539 BC, mêne, mêne, t ªqal, (letters) 444
Axis alliance between Hitler and Mussolini (weeks)   444
Tehran siege that released Iran from US control (days) 444

These events were connected to major changes in the fortunes of
great world powers. The original fall of Babylon was the precursor
to its second and final overthrow in the end-time. Just as the ancient
Persians brought about the fall of Babylon, today's nascent Persian
Empire will be instrumental in triggering the destruction of its latter
day equivalent, a United States of Europe. The first was a kingdom;
the head of Gold in Daniel's prophecy discussed in the chapter

A brief history of History. The second Babylon is to be a kingdom of nation states under the sway of a false religious system:

> *Revelation 17: 5 "...MYSTERY BABYLON THE GREAT, THE MOTHER OF HARLOTS AND ABOMINATIONS OF THE EARTH. V6 And I saw the woman drunk with the blood of the saints...v9...the seven heads are seven mountains [successive governments, specifically resurrection of the Holy Roman Empire] on which the woman sitteth."*

Babylon is therefore the *first* and the *eleventh* human kingdom. These incorporate the seven resurrections of the Holy Roman Empire as can be seen from this summary:

1. Babylon
2. Medo-Persia
3. Greece
4. Rome 31BC to 476AD
5. Imperial Restoration – Justinian
6. Frankish Kingdom – Charlemagne
7. Holy Roman Empire (German) - Otto the Great, crowned by the pope 962AD
8. Habsburg (Austrian) – Charles the Great, crowned by the Pope 1520AD
9. Napoleon's kingdom – crowned by the pope 1805
10. Italy (and later Germany), Garibaldi to Mussolini (with Hitler)
11. The EU (with the Catholic Church, the final fulfillment of BABYLON)

This is a different list of eleven to that of the horns in the chapter *A brief history of History*. In that list the first three horns were the three Germanic tribes that overran Rome. The fourth 'little horn' (stouter than its fellows) was the Roman church. In its *eleventh* manifestation this modern tyrannical Babylon the EU is, in effect, claiming to represent God the Father through his number 11.

In a June 2007 editorial comment on the then impending revival of the stalled Constitution for Europe, the London *Daily Mail* lamented:

"And doesn't bitter experience teach us that the EU juggernaut, with Germany at the wheel, is all but unstoppable?...80 per cent of our laws are [already] imposed by unelected bureaucrats in Brussels ... [Mr. Blair] is prepared to accept a change in the voting system which will reduce Britain's influence by 30 per cent (while leaving Germany's intact naturally)."

In the coming decades a religious entity, riding a 'beast', will incite religious totalitarianism in Europe. This 'King of the North' will then retaliate against a 'King of the South' in an invasion of the Holy Land, subduing in the process a modern Iranian empire. The King of the North prophecy does not, as some have assumed, apply to Russia. A schedule of prophesied *resurrections* of the Roman Empire is given in the chapter *A brief history of History*. It makes a watertight case for Europe as the territory of the coming 'beast power.' Russian history cannot provide the detailed events to correspond with the Bible's predicted pattern of revivals, neither has it had a powerful religious figure able to

"persecute the saints of the most High and intend to change times and laws." Daniel 7:25

The clinching argument is seen in Daniel chapter two, where the modern Europe is depicted as the feet of iron mixed with clay on a great statue. The feet are smashed by a rock hurtling from the sky, which is the returning Christ – a veritable Northern Rock! The descriptions given of a territory with two halves both mixed with strong and weak elements perfectly depict modern Europe. The picture bears no resemblance to Russia, past or present. Another confirmation is the smashing of the chief centre of false worship (Satan's seat, Rev 2:12) at the time of Christ's return, which can find its fulfillment only in Rome.

A New World Order

Once the death throes of this present evil world are over, something altogether different will replace it. That has been the message of Christ all along, that the Kingdom of God has as yet only been planted as a *seed*. It will gradually grow to fill the whole Earth during his millennial rule.

Freedom of choice will not be removed, but that malevolent spirit now holding sway over the affairs of man will be bound for 1000 years (Revelation 20:2). His greatest propaganda coup has been to convince the 'educated' of the world that he does not exist. His next greatest coup has been to convince religions that this is God's world while all the time Satan himself remains as Paul told to the Corinthians *"...the god of this world..."* (2Corinthians 4:4).

But at this future time of restitution the religions of this present world will be abolished by decree. The reader may draw hope and encouragement from the proof given in this book of the existence of real Creators, who move clearly into our view once the fetid cobwebs of mans' pagan traditions have been pushed aside to reveal the simple truth. We have been making our own religions all along.

Rome is the one religious power centre of the Earth that has held sway over nations for the greater part of 2000 years. The eight power cuts of 2003 drew a giant figure of 8 across the Earth, centered on Rome. The number 8 denotes a new beginning. Noah, the gematria of whose name is 64, or 8^2, was also marked by this number. It is a sign of the end of this age. Regarding Noah, Christ made this statement:

> *"For as in the days that were before the flood... Noah entered into the Ark...so shall also the coming of the son of man be."*
> *Matthew 24:27-28*

This age will be brought to a close not by a flood *but by fire*. The prophet Isaiah refers to this time:

> *Isaiah 24:6"...therefore the inhabitants of the earth are burned and few men left."*

Iniquity not yet full

Why should such a cataclysm as this happen in the present age? There has always been sin in the world. Why visit this terrible ordeal on today's society? The same chapter four of Isaiah quoted above gives some possible reasons:

1. Haughtiness – v4 *"...the haughty people of the earth..."*
2. Defiling the Earth – v5 *"The earth is also defiled under the inhabitants..."*

3. Commandment breaking – v5 *"...because the have transgressed the laws..."*
4. Calendar changes – v5 *"...changed the ordinance..."*
5. Covenant breaking – v5 *"...broken the everlasting covenant."*

Evidently these conditions will reach a climax before punishment is meted out. God had told Abraham that his descendents would one day return to their land. Before then its inhabitants would reach a peak of wickedness, at which point God would eliminate them:

> *"But in the fourth generation they shall come hither again: for the iniquity of the Amorites is not yet full." Genesis 15:16*

Has sin reached a pinnacle in the 21st century? Can internet pornography be brought under control? But the above list of five sins includes something not normally thought of as sin - an alteration to the calendar; they *"changed the ordinance."* Today's western calendar is called the Julian Calendar after the Roman Emperor who instituted it in collusion with the Catholic Church, the 'little horn' of Daniel chapter seven. The description given is an accurate description of this 'Lady of Kingdoms':

> *"And of the ten horns that were on his [HRE/EU] head, and of the other [horn] that came up, and before whom three fell; [Vandals, Ostrogoths, Heruli, see v24 "shall subdue three kings"] even of that horn that had eyes, and a mouth that spoke very great things, whose look was more stout than his fellows, v21 I beheld, and the same horn made war with the saints, and prevailed against them, v22 Until...the saints possessed the kingdom...v25 And he [the little stout horn] shall speak great things against the most High, and shall wear out the saints of the most High, and think to change times and laws..." Daniel 7:20*

Here would be another marker of the end-times: an alteration to the Calendar. Never has the seven day cycle been broken by a calendar adjustment. The normal working week has comprised Monday to Friday in the western world. But the 'little horn' who appears *"stouter than his fellows"* will seek to *"wear out the saints of the most high."* It is possible that the Sabbath (Saturday) will be moved to a mid-week position making it extremely testing to observe the Sabbath day, a time period made holy from creation[1].

Savage furies

The *god* of this present age is preparing a grand new deception – a miracle working leader in Europe. By this means an iron band of delusion will be wrapped around the minds of men and women the world over. The coming warfare will be a religiously inspired blitzkrieg which, to quote again Winston Churchill's description of the Germans, will be the final end-product of

> *"the savage furies latent in the in the most numerous, most serviceable, ruthless, contradictory and ill-starred race in Europe."*

Is it so difficult to believe that this nation would once more, as Churchill put it:

> *"...combine in the most deadly manner the qualities of the warrior and the slave"*

and visit the *"recurring pestilence"* of German nationalism on the world all over again.

Might a pacifist UK government hand over Britain's nuclear deterrent to the European Council, only for it to be used against them?

---

1. The Sabbath was created at the time of the Earth's refurbishment (see the chapter *The Great Breach*). The late Cardinal Gibbons wrote: *"You may search the scriptures from Genesis to Revelation but find no basis for Sunday observance. The Bible enjoins Saturday, a day we have never reverenced."* Faith of our Fathers, page 76, 1980 edition.

33

THE PROPHETIC GRID
"O Daniel...seal up the book, even to the time of the end." Daniel 12:4

The Bible, particularly in respect of its many prophecies, is a puzzle to most people. When Christ spoke in parables to the crowds his disciples asked the reason why. He responded:

> *"To you it has been given to know the mysteries...but to them it has not..." Matthew13:11*

If that was the case, one might ask, why bother preaching to them at all? And if Christ was the Messenger of the New Covenant why deliberately obscure that message?

In the twelfth chapter of the highly prophetic book of *Daniel* that author is told, *"Seal up the book until the time of the end..."* It was not to be understood in his time, although today nearly all of it can be. There are also prophecies about today's world in *Revelation, Daniel, Hosea, Isaiah, Matthew* and many other sections of the Bible where momentous future events are predicted in detail.

About one third of the Bible is prophecy, much of which is yet to be fulfilled. Churchill, addressing the Unites States Congress in May 19th 1943, said *"There is a purpose being worked out here below."* He didn't know what that purpose was but the Bible reveals it and the Grid affirms it. Four patterns of prophetic significance can be shown:

PATTERN ONE

In the chapter *A brief history of History* the following points were established:

1. Bible symbols match closely 4 world-ruling kingdoms

2. A powerful religious entity was matched to one particular prophetic horn
3. Seven horns were matched to 7 resurrections of the Holy Roman Empire, the fourth kingdom
4. There were 11 prophetic horns in the entire framework of prophecy
5. The four kingdoms and the seven resurrections (revivals) of total eleven in all, from ancient Babylon to the EU Babylon.
6. Key turning points: the original fall of Babylon, the Axis Powers and the release of modern Iran from US control were all marked by the number 444

Bullinger overlooked completely the pattern of 7 and 11 in the above prophetic horns. All of the numbers of the Grid seem to align well with key factors in history:

1	**7**	**11**
The little horn	Seven resurrections	Eleven prophetic kingdoms
(Catholic Church)	(European kingdoms)	(Babylon to Babylon)

44	**40**	**4**
	The kingdom	Four world ruling kingdoms:
	number 444	Babylon, Medo-Persia,
		Greco-Macedonia and Rome

In this, everything is anti-God. All things stand for the malevolent forces of the world. The number of God's spirit is taken by the little horn, its antithesis.

PATTERN TWO

As previously shown the top line of the Grid, 1 7 11, is very similar to the 1711171 pattern of 19 days and is a shorthand version of it. It was first suggested in the earlier chapter *A 19 year Middle East Cycle* that the pattern 1711171 is the 'fullest expression' of 1711. Conversely, one could say that 1711 is the abbreviated form of

1711171. The 1711171 pattern is in itself highly prophetic. There are many future events depicted by the 7 holy days contained within it:

Passover	1	The memorial of Christ's death
Days of unleavened bread	7	holy days, repentance pictured
Day of Pentecost	1	Holy day; Church as the firstfruits
Day of Trumpets	1	Depicting the return of Christ
Day of Atonement	1	Depicting the removal of Satan
Feast of Tabernacles	7	1000 year rule of Christ on Earth
Last Great Day	1	Depicting the general resurrection

From the Day of Trumpets through to the Last Great Day, four major epoch-making developments are predicted by the 1711171 pattern. The top line of the Grid 1 7 11 therefore not only identifies the two Deities and their spirit, but in its fullest expression 1711171 it codifies their plan for the human race. In this way the top line of the Grid is also prophetic. Certain prophecies predict the worldwide observance of these holy days following the return of Christ. An example is seen in the book of *Zechariah*:

"For I will gather all nations against Jerusalem for battle...v3 Then shall the Lord go forth and fight against those nations...v4 And his feet shall stand in that day upon the Mount of Olives...v5...the Lord my God shall come, and all the saints with thee...v8...living waters shall go out from Jerusalem...v9 And the Lord [Jesus Christ] shall be king over all the earth...v16 And it shall come to pass, that everyone that is left of the nations that came against Jerusalem shall even go up to keep the Feast of Tabernacles...v18...the Lord will smite the heathen that come not up to keep the Feast of Tabernacles..." Zechariah 14:2

Unless the Old Testament is dismissed as poetry it is plain that these days will be observed universally during the millennial rule of Christ on the Earth. The early Church kept these days. One particular passage in Colossians 2:16-17 affirms that the seven holy days were current at the time of Paul. This is explained in detail in the chapter *That They May Be Taken*. These vital days are all but forgotten by today's apostate churches.

PATTERN THREE

There is a vertical pattern suggestive of the number 144:

1	7	11
44	40	4

Firstly, this 144 pattern appears to correspond to the Sun and Moon as each heavenly body occupies $1/720^{th}$ of the sky (half a degree). In respect of one of the few things they have in common – their size as viewed from Earth – they are therefore marked with the same number, 720. Simplified to 72 this suggests for the two heavenly bodies together the total 144. (The eleventh Fibonacci number as discussed in *Appendix VIII*). The above insight also squares with another critical historical fact, the three days and three nights Christ spent in the tomb - 72 hours (again discussed in the chapter *The Mark*).

The number 144 is an important number in the book of Revelation. It is the number of the first resurrection, or God's *first fruits*, the 144,000. Yet all those who have ever lived on planet Earth can be said to have been in on a ground floor opportunity. The universe appears to be uninhabited. It will not be left in this empty and futile state indefinitely. The Bible speaks of a time following the millennial rule of Christ when 'sons of God' will release his vast creation from its empty condition. The book of *Romans* depicts the creation as groaning in travail, waiting for this future time:

"For the earnest expectation of the creation [Greek ktisis – original formation, building, creation] waits for the manifestation of the sons of God. v22 For we know that the whole creation groans and travails...v23... waiting for the adoption [or sonship¹]...v29...that he [Christ] might be the firstborn [Gk. prototokos - firstborn or first begotten] among many brethren." Romans 8:19

PATTERN FOUR

The '88' pattern within the Grid is prophetic. This pattern is derived from the bottom line (44+40+4) and the sum of the outside columns (1×44, 11×4), a line and a circle implying pi:

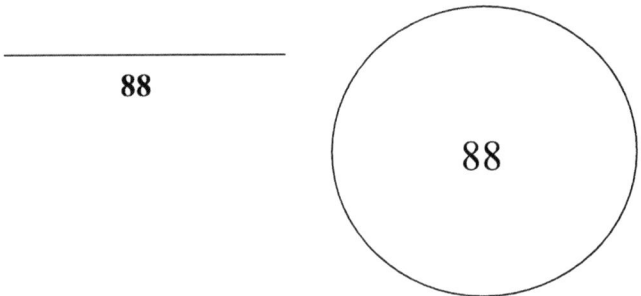

The number 88 matches by multiplication the 2 sex chromosomes and 44 autosomes that record human life. Also encoded in the Grid are the relative sizes of the human cells of reproduction 1:44. Both languages of the Bible, Hebrew and Greek, have been used to encode 44 as the *number of reproduction*. This is the pivotal theme of the Grid. The number 88 is the greater expression of 44. It encapsulates the greatest prophecy of God, that He will put his creation (the universe) under the control of a family of literal sons and daughters, composed of spirit. That would be the answer to Churchill's statement concerning a *"... purpose being worked out here below."*

It has been shown that the Grid encodes the 1711171 and 444 patterns, both of which are highly prophetic. But the greatest encoded message within the Grid, namely that of 1-44, is its most potent signal. That pattern is to do with reproduction and refers to a yet future event: God is going to self-replicate, by way of a resurrection, through the human race.

1. Translation is ever an issue in dealing the Bible. The Greek is *huiothesia,* of which Strong's Exhaustive Concordance states (ref.5206) "From a presumed compound of 5207 and a derivative of 5087; the *placing* as a son, i.e. *adoption* (figuratively Christian sonship in respect of God):- adoption (of children, of sons). It should be noted however that this explanation is an *interpretation* by Bullinger, or others, because reference is made to "a presumed compound." It is illuminating to examine in full the two other words, references 5207 and 5087, that make up this *presumed* compound:

5087 tithemi – a prolonged form of a prime *theo,* (which is used only as an alterative in certain   tenses); to *place* (in the widest application, literally and figuratively; properly in a passive or horizontal posture, and thus different from 2476, which properly denotes an active and upright position, while 2749 is properly reflexive and utterly prostrate):- + advise, appoint, bow, commit, conceive, give, × kneel down, lay (aside, down, up), make ordain, purpose, put, set (forth), settle, sink down.5207 *huios* – apparently a prime word, a "son" (sometimes of animals), used very widely of immediate, remote or figurative kinship:- child, foal, son.

There appears to be ample latitude for a literal sonship, as indicated elsewhere, within this range of meanings. Who is to dictate that the meaning of *place* necessarily eliminates all possibility of a begettal by God through his spirit. Numerous Bible passages indicate an act of begettal. Besides, the idea of *placing* is in keeping with the statement by Christ in John 14:2 that *"In my Father's house are many mansions (positions)...I go to prepare a place for you."*

208

34

A STRANGE 2003

"...ye can discern the face of the sky...
can ye not discern the signs of the times?" Jesus Christ

The first rumblings of a global phenomenon started in Manhattan. Cash machines stopped, mobile phone networks collapsed and thousands of commuters were forced to sleep in the open. Dazed New Yorkers wandered through the streets next morning as if to check that the city was still standing. In the deserted silence of the previous evening, Manhattan had been a blackened silhouette against a dramatic orange sunset.

That Thursday afternoon on the 14th of August 2003, the largest power cut in history, historic for its astonishing reach, plunged cities in America and Canada more than a century back in time. A woman in Detroit was rescued after eighteen hours trapped in a lift. In New York one commuter slept on top of a bus shelter while nearby a lady collapsed and died after climbing several flights of stairs. Residents were warned to boil water, airports closed down and thousands were trapped on subway trains in temperatures of 90f. A hundred miners in Ontario escaped after a night underground while politicians in Canada fell over themselves to blame the US for the disaster. But nobody could agree on what caused it.

In London 22 days later at 9.30pm on Friday the 5th September the magician David Blaine climbed into a Plexiglas box by Tower Bridge. He was to fast 44 days to finishing at 9.30pm on Sunday 19th October. In view of Blaine's several Christ-impersonating stunts it was no great mystery that he'd decided on a fast. This would not be for 40 days - the period Christ had fasted in the wilderness - but for 44 days. Blaine was born on April 4th 1973 (the 4th day of the 4th month) and conducted his fast, as did Jesus Christ, at the age of thirty.

Eight days earlier, on Thursday the 28th August, London had been struck by an enormous and unprecedented power cut. Two hundred and fifty thousand commuters were trapped in the tunnels of the London Underground for more than an hour only two weeks after their counterparts in New York had suffered the same fate. Mayor Ken Livingstone called it *"an absolute disgrace."* Above their heads the rain-sodden public sought shelter in cafeterias and bars lit by candle light. Traffic lights and CCTV cameras failed, creating grid-lock on the roads of the "world's coolest city."

Some astonishing events occurred in 2003 that could have led many people, had they been alerted to the facts, to wonder if the world was going mad. Eight unprecedented and highly unusual power cuts involving electricity grids on different continents had occurred in quick succession, over 44 days, overlapping symmetrically with Blaine's 44 day fast.

The fast overlapped at exactly the half-way point - after 22 days - within the 44 day period of power cuts. The fast and power cuts therefore encompassed a period of 66 days. In view of the all these facts should David Blaine be credited with the power to cause eight worldwide power cuts?

Both the first and the last of these power cuts were billed in the press as the largest ever. The first began in New York and the last in Rome. The black-out centered on Rome disrupted the ordination ceremony for 31 new cardinals, in preparation for the succession of Pope John Paul. In the event the pontiff used a public address system powered by an emergency generator. Here is a summary of the power cuts as reported by various media:

Power cut 1: NEW YORK

"Not an electric light was to be seen, with Manhattan silhouetted against the sunset after Thursday's huge power failure…"
"America and Canada were fighting last night to restore power to huge sections of their most densely populated areas, after the largest power cut in history…"

"America and Canada blamed each other for the crash, which began on Thursday afternoon (14th August) when a single fault appears to have sent uncontrollable surges racing across high-voltage lines, knocking out twenty-one power stations and leaving fifty million people without power."

"But the blackout was historic for its astonishing reach, denying power to residents of six US states and millions of Canadians." Daily Telegraph. August16, 2003

Power cut 2: REPUBLIC OF GEORGIA

This next power cut followed on almost immediately on 18th August. Even in such a small country there were 4.5 million people affected. The metro was halted and sabotage blamed. (www.english.people.com)

Power cut 3: HELSINKI, FINLAND

A power cut occurred on 23rd August and the airport had to be shut. Authorities were at a lost to explain the problem with what was supposedly a first rate electricity grid. (www2.hs.fi/english/archive.news)

Power cut 4: LONDON

"Blackout Thursday – A massive power cut plunged London and many parts of the South East into chaos last night…tube passengers stuck…about 250,000 people were trapped. London Mayor Ken Livingstone said 'We have never faced a crisis like this before. It's an absolute disgrace…It seems the national grid has simply collapsed.'" Daily Mirror. Aug. 29th 2003

Power cut 5: KUALA LUMPUR, MALAYSIA

On 1st September the capital and five other states were struck by a massive blackout. Workers in the Petronas Towers were trapped in the elevators, traffic lights were out and traffic halted. "It's perplexing, blackouts are very rare...." (www.archives.tcm.ie.breakingnews.com)

Power cut 6: CANCUN, MEXICO

Preparations for the World Trade Organisation meeting were disrupted by a six hour power cut affecting 3 million people. (www.solarstorms.org/sblackout)

Power cut 7: SWEDEN & DENMARK

Four million people were affected while trains, airport and factories stopped. It was described as "highly unusual in Scandinavia" by Yahoo/Reuters, 23rd September 2003 (www.news.bbc.co.uk/1/hi/world/Europe)

Power cut 8: ROME

The Daily Mail and Daily Telegraphic on Monday 29th September variously reported of the previous day headlined as "Rome's Darkest Day" and described "Fifty-seven million Italians blacked out as another grid crashes" and "the world's biggest power cut" and "This dramatic picture - taken shortly after 3.30am [on Sunday 30th September] - shows the usually floodlit dome of St Peter's looking down on a city in darkness. All of mainland Italy's power went down in the night due to fierce storms. Rome was holding a *Sleepless Night* event with shops, museums and other attractions open until dawn. Tens of thousands were on the capital's streets when the power failed. Hundreds were trapped in lifts and 30,000 stranded on trains which juddered to a halt." Later that day, in a ceremony delayed by the power cut, Pope John Paul ordained a further 31 cardinals in preparation for an eventual succession, using a microphone powered by emergency generators.

So it is seen that in electricity grids separated by whole continents, 8 rapid-fire power cuts coincided with the worldwide spectacle of

David Blaine and his 44 day fast. How often does someone make global news with a stunt lasting 44 days? And when they do, how often will that overlap symmetrically with an inexplicable global phenomenon also lasting 44 days?

These facts take a great deal of explaining. What are the chances of events like these happening randomly? Could David Blaine have arranged the power cuts, perhaps using witchcraft? It's just as easy to speculate that outside powers caused these phenomena. If so, what are they trying to tell us? Surely they were saying to New York, "Do you think *you* are the great financial power?" And to Rome, "Do you think *you* are the great spiritual power? *We will show you who the great powers are.*"

Soon after the power cuts of 2003 and David Blaine's fast another highly unusual and perhaps unique event took place between October 19th and November 4th of that year, as mentioned in the *Introduction*. Eleven giant solar flares[1] had erupted from dark earth-sized spots moving across the face of the Sun. Enormous damage to satellites and electrical equipment around the world would have resulted had these been directed towards the Earth. Yet 99% of the inhabitants of our planet had no idea that the Sun had, according to numerous scientific websites, experienced the most violent storm of activity in 144 years. Nothing like it had been seen since 1859 when the Western world had witnessed a single solar flare so fierce that it melted telegraph wires in North America. That was the year of the first great solar storm in modern times.

In 2003 the number 11 was signalled to the whole world through these solar flares. The number 44 was signalled worldwide by the power cuts and David Blaine's fast, an event universally broadcast. That same year, American author Dan Brown began a meteoric rise to fame with *The Da Vinci Code*, the sixth best-selling book of all time and translated into 44 languages. Interest in the history of the Catholic Church has been greatly stimulated by the book which generated a tsunami of religious controversy. *The Da Vinci Code* first became available in the UK on July 1st 2003, the beginning of the seventh[2] month.

Due to the activities of these two influential Americans in the UK, David Blaine and Dan Brown, the numbers 44 and 7 were marked on the year 2003. The unprecedented burst of solar activity in that year signalled the number 11 from a source already marked by that number, the Sun. Thus the Grid's numbers feature in the year 2003:

7, 11 and 44

The first of these - a power cut of historic proportions - burst on the world scene from Manhattan, home town of David Blaine. The Manhattan project was the name under which the first Atomic Bomb was developed. The name has a meaning: drunkenness. That place name fits into an unfolding picture of the future. In subsequent chapters it is shown that 'Manhattan' will come to Rome.

1. Eleven giant X-class solar flares erupted from the surface of the Sun over a 387 hour period. This was unprecedented. The 11th flare was the largest ever recorded at a magnitude of X45, over twice the previous record. Remarkably this occurred when the 11-year sunspot cycle was already in decline. See www.weathermaine.com for the list of eleven incidents and www.spaceref.com . The BBC website www.bbc.co.uk reported that "Solar scientists have confirmed that Tuesday's explosion on the Sun was, by far, the biggest flare ever recorded, capping an energetic solar period." The event 144 years earlier on September 2nd 1859 is discussed on www.science.nasa.gov . Also www.firstscience.com relates that "Even 144 years ago, many of Earth's inhabitants realised something momentous had just occurred. Within hours, telegraph wires in both the United States and Europe spontaneously shorted out..."

2. Dan Brown's fourth book *The Da Vinci Code* was first published in the USA on March 18th 2003. That American author's exploits match those of New Yorker David Blaine's activities in London, where the later London release date of his book (on the 1st of the seventh month) would be applicable.

35

THE CRISIS

"Madness is a rare thing in an individual- but in groups, parties, peoples and ages it is the rule." Friedrich Nietzsche

Writing in 1946, in the foreword to a series of twelve books entitled *The Second World War*, Winston S. Churchill referred to *"...the awful unfolding scene of the future."*

Churchill was in trepidation for the human race. But since his death in 1965 the world has experienced many crises, and the things he dreaded have not come to pass. The Russian Invasion of Afghanistan, the Falklands War, the Gulf War, the 9/11 attack – none of these changed the world's way of doing business. But the 'Meteor' – the financial panic of 2007 – might just.

Then one might reasonably compare that crisis to the emergence of Iranian President Mahmoud Ahmadi-Nejad in 2003. That certainly signaled the start of a perilous new phase in history. There can be no doubt that, in respect of what he has said repeatedly of the Jews, he is the first elected leader since the thirties *to employ language of a character comparable to that used by Adolf Hitler.* He is a holocaust denier, having organised an international conference for that very purpose. In view of Iran's geographical location, its volatile president is, so far, the most high-risk personality since World War Two. The US Tehran Embassy staff can recall Ahmadi-Nejad as one of the student coordinators of the November 1979 siege; it lasted 444 days. That pivotal empire-changing event brought about the ousting of the unpopular Shah, widely considered a stooge of the CIA. In its wake we have seen the release of Islamic fundamentalism into the wider world and a huge decline of US influence in the Middle East.

Iran has waged war by proxy on America's one true Middle Eastern ally Israel, now all but spurned by President Obama. As the Obama administration will inevitably withdraw from Iraq, Iran will duly move in. A new kingdom and a deadly foe of the West is in the making. The world faces two crises: one economic and the other military. It is the second that is immediately species-threatening, to a degree global warming could never be.

A maverick outsider, Ahmadi-Nejad's democratic ascendancy in Iranian politics was initially via the mayoral elections in Tehran in March 2003. His success was only made possible by the White House's greatest folly of all follies, the Iraq invasion. At that time 77% of the American people, according to ex-presidential candidate Al Gore, believed Saddam Hussein was behind the 9/11 disaster. Clearly, the American public had been deceived.

Churchill likened governing Iraq to being perched on the edge of *"an ungrateful volcano."* At the end of the First World War the British maintained 410,000 troops in Iraq to control a population a fraction of its present size. In 2003 Donald Rumsfeld sent a mere 140,000 troops in defiance of the advice of his generals. The invasion of Iraq commenced in March 2003. Due to a refusal to heed the lessons of history, the US has committed the greatest foreign policy blunder since World War Two[1]. Deception was used to justify that war, an escapade dogged by incompetence throughout. At the same time the US concocted a *new military doctrine*. It claimed the right to invade any sovereign nation on the mere suspicion they may be harboring forbidden weapons. The War on Terror has been used as the justification for this although terror tactics are hardly new. As a justification for this new aggression, the World Trade Centre debacle proved to be an all too convenient 'New Pearl Harbour.'

The dangerous preemptive US doctrine sprang from an evangelical zeal that gazed in disapproval over a world it thought it could control, but did not adequately understand. Its *modus operandi* was to deceive its own people and the world with a smoke screen of propaganda and half truths about terrorism. This it did in order to promote the neoconservatives' agenda for Full Spectrum Dominance - the total control of all land, sea, air and outer space. This was a curious agenda for a nation printing IN GOD WE TRUST on its bank notes. It was implemented by 'born again' Christians in the White House and it was madness.

By July 2008 that agenda - and the crippling deficit financing it demanded - looked foolish when the second man on the Moon, Buzz Aldrin, reproached NASA with the following observation:

> *"To me, it's abysmal that it has come to this: after fifty years of NASA, and after putting in about $100 billion into the space station, we can't get our own astronauts to our space station without relying on the Russians."*

Such feebleness in the 'world's only superpower' is part and parcel of the prophetic trend highlighted by the present discovery. The appearance of the Grid in 2003 and its pattern 444 confirms what was postulated in an earlier chapter *The Signal*: upon the invasion of Iraq in 2003 *the world turned a corner it never meant to turn*. Iran responded to the Iraq invasion by voting the radical Ahmadi-Nejad into the presidency. He is a man who, at first, inspired mirthful disdain from political peers due to his mixing of Shakespearian-style verbiage with street slang. But Ahmadi-Nejad has nonetheless proved a cunning survivor. His assertion of Iran's right to a nuclear capability unshackled by foreign supervision was, at first, popular across all classes of Iranian society. As the main force in Iranian politics he appears unstoppable. In June 2006, Professor of political science Nasser Hadian-Jazy affirmed, *"He's more popular than a year ago. He is on the rise."* In March 2007 *The Sunday Times* reported that the speeches of Iran's supreme leader Ayatollah Sayyid Ali Khameini *"...tended to back his president's belligerent approach."*

Ahmadi-Nejad has made many widely reported pronouncements about the State of Israel. The precise wordings are sometimes in dispute but his comments amount to saying that 'Israel should be wiped off the face of the Earth.' That is certainly the perception he has deliberately created. If there was a Richter-scale of military risk he would, by comparison, put Saddam Hussein in his latter years on a par with comic characters from the British TV sitcom *Dad's Army*.

Saddam Hussein's government had been secular in character, whereas Ahmadi-Nejad believes we are in the 'end time' and would

217

like to accelerate that time-table by attacking the enemies of his Shiite Moslem Faith. By doing this he hopes to clear the way for Islam's saviour, the Mahdi[2]. He expects this Mahdi, also called the hidden Imam, to soon appear on the world scene. His arrival is to be heralded by warfare. In his infamous 2005 speech to the U.N., Ahmadi-Nejad invoked the coming Mahdi, at which point a colleague claimed to have seen a green light around his body. This is what Ahmadi-Nejad said of the event:

> *"I felt it too. I felt all of a sudden the atmosphere changed there. And for 27-28 minutes the leaders did not blink...It's not an exaggeration, because I was looking. They were astonished, as if a hand held them there and made them sit."*

Delegates were certainly astonished that someone as reckless as this could gain such a powerful office. The usually soft-spoken human rights activist Elie Wiesel describes Ahmadi-Nejad as *"pathologically sick."* Harvard law Professor Alan Dershowitz, speaking at a December 2006 rally in Toronto, called Ahmadi-Nejad *"The Hitler of the 21^(st) Century."*

But what many will wish to know is whether this bombastic leader is a man of destiny or a man of straw. According to Tehran Embassy staff, at the time of the 1979 siege Ahmadi-Nejad's pedigree as a radical, at least, was second to none. It was that humiliating episode, involving the siege lasting 444 days, which paved the way for the rout of Jimmy Carter by Ronald Reagan in the 1980 US elections. The effect of the siege was the loss of US control over Iran and its ill-fated switch towards an alliance with Saddam Hussein, whom they backed against Iran in the Iran/Iraq war. The USA still felt it had to have an ally in the region in addition to Israel.

Frequent and unprincipled collusion with foreign dictators has probably created more problems for the American government than it has ever solved. By spurning the advice of its military leaders and brushing aside the fears of its western allies, America's politicians have brought the world to a calamitous position over Iraq. The *Sunday Times* in December 2006 likened the position of the Middle East to that of Europe at the time of the Thirty Years' War because now the *"...sectarian vision has begun to create the real possibility of a wider war."* *The Times* in January 2007 commented on the parlous position of the ruling Sunni's of Saudi Arabia who were *"...more worried than Israel about an Iranian A-bomb..."* Both Israel and Saudi Arabia were urging President Bush to attack Iran, but by then it was too late. The goodwill of allies had been squandered over Iraq.

While the world's attention is easily turned by other matters – the Credit Crunch and flu epidemics – the greatest threat to human survival is still the nuclear one. By 2011 a man whose pronouncements bear the same hallmark of racist bile as Hitler's will have the atom bomb.

The conditional Bible prophecy of Matthew chapter 24 that, *"...no flesh would be saved alive..."* could never be completely fulfilled through famine, disease or conventional war alone but only by a most intense nuclear radiation.

After watching us for thousands of years, are powers greater than those of New York and Rome finally preparing to save us from ourselves?

1. It now appears that a new era began with the controversial March 2003 invasion of Iraq. This badly judged attack has altered military doctrine, subverted the U.N. and gravely undermined the Atlantic Alliance cemented by Churchill and Roosevelt. The pretext for the Iraq war, the alleged Weapons of Mass Destruction, has been shown to be untrue and the perception that trickery was used in its deployment cannot be shaken off. A British expert on Iraq's weapons who disagreed over the evidence died in highly suspicious circumstances. He had pointed out that Saddam's military teeth had already been efficiently extracted by weapons inspectors. The largely spurious war on terror has, as many predicted, only spawned a greater terror. It is now certain that before the March invasion, the White House never properly anticipated or planned for any serious 'blowback' from neighboring Iran. They naively bought into the notion that democratic voting in Iraq would cause the masses of the Middle East to turn moderate. Instead they are electing hard-liners.

2. The same Mahdi is said to have originally vanished at the revered Golden Mosque in Samarra, bombed by Sunni Moslems in February 2006 in an incident described as Iraq's 9/11. That explosion in particular set off a violent chain reaction of sectarian conflict.

36

THE LADY

"I will show unto thee the judgment of the great whore…" Revelation 17:1

Can it really be shown that powers from outside planet earth have controlled the affairs of man for thousands of years? How extensive is the evidence? And what of the possibility that sinister unseen powers *on planet earth* could be behind much of the misery and horror experienced by its dominant species?

It has been shown that four successive world empires, culminating in the Holy Roman Empire, have matched detailed prophecies in *Daniel* and *Revelation*. The last of the four world-ruling kingdoms was centred in Europe and has seen many revivals. One religious entity in the region has been a constant energising force. But since the Reformation, that institution, the Roman Catholic Church, has appeared to take something of a back seat in world affairs. The spread of democracy in Europe and the New World did much to neutralise its influence.

In the chapter *A brief history of History* the evidence was shown that the Catholic Church is described in Daniel chapter eleven as *"The little horn that speaks great things."* But a church is pictured as a woman in Bible symbolism, not a horn. However, the explanation in this case is simple. The horns were political powers and in addition to heading a religion the Vatican has been a *great power*. In the earlier chapter *The Little Horn* it was stated that the first great leader in Rome to promote a form of Christianity was Simon Magus, a magician with the claimed power of levitation. Such powers are not natural to man. The fruits of this egregious organisation suggest that its power can come from only one place, the occult.

Revelation chapter seventeen describes a tyrannical and persecuting religious entity, as a *mother of harlots,* described as *abominable*:

"...with whom the kings of the earth have committed fornication"
Revelation 17:2

This is an entity that is both Church and State, a power able to muster the armies of nations. *Isaiah* speaks of a certain *Lady of Kingdoms* who boasts:

"I shall be a lady for ever" Isaiah 47:7

Isaiah foretells her destruction in verse 11 of that chapter. What institution other than the Catholic Church could possibly fulfill such numerous prophecies of a great false church?

During the Second World War, the science fiction author H.G. Wells occupied the post of Minister of Allied Propaganda under Churchill. In that post he came to perceive the extent of the power of the Catholic Church and its corruption. After his resignation Wells wrote *Crux Ansata – an Indictment of the Roman Catholic Church.* In that book, long shunned by publishers for fear of reprisal, he recalled that the bones of the great English scholar John Wycliffe, who had rendered the world the service of translating the Bible, were dug up and burned upon the edict of Pope Martin the fifth. H.G. Wells asked why we did not bomb Rome:

> *"Not only is Rome the source and centre of fascism, but it has been the seat of a pope, who, as we shall show, has been an open ally of the Nazi-Fascist-Shinto Axis since his enthronement. He has never raised his voice against that Axis, he has never denounced the abominable aggressions, murder and cruelties they have inflicted upon mankind, and the pleas he is now making for peace and forgiveness are manifestly designed to assist the escape of these criminals, so that they may presently launch a fresh assault upon all that is decent in humanity...No other capital has been spared the brunt of this war. Why do we not bomb Rome?"*

In a 1944 interview Wells said of the Catholic Church:

> *"I think that it stands for everything most hostile to the mental emancipation and stimulation of mankind. It is the completest, most highly organized system of prejudices and antagonisms in existence...It presents many faces towards the world, but everywhere it is systematic in its fight against freedom..."*

The Catholic Church has been the sponsor and ally of many European dictators. It is not commonly realised that the wartime pope, Pius XII, was Papal Nuncio in Germany in January 1933, at the time von Papen maneuvered Hitler into executive office:

> *"The Third Reich, like the EU, was an attempt to revive the Roman Empire. The higher strategy of the Vatican and the acquiescence of the Catholic Central Party had brought Hitler to power. Instrumental in this strategy were Reich Chancellor Franz von Papen and Papal Nuncio, Monsignor Pacelli, the future Pope Pius XII." Page 12, Papal Rome and the European Union, by Richard Bennett and Michael de Semlyen.*

But the master propagandists controlling this ancient institution know how to mould opinion. The *Daily Mail* reported in an article *"Hitler's Pope is close to sainthood"* that:

> *"Pope Pius XII has been approved for beatification...Pius, Pope from 1939-1958, was described in a Foreign Office wartime memo as 'the greatest moral coward of our age.'" The Daily Mail, 14th May 2007*

Rome has, for the greater part of two thousand years, been that constant and unchanging power centre to which almost all western governments and empires have given their obeisance. An organisation for which the end has long justified the means, their propagandists have never been idle for long. In June 2007 the London *Daily Telegraph* reported in an article entitled *A fairy Tale from Brussels* that school teaching material involving a narrative by a priestly sounding 'Good Father Houpette' was informing pupils in November 2005 that, *"You will see that the EU is a necessity."* Children are then described in the story reading the rules and regulations of an indoor sports club. Father Houpette informs them that, *"Not long ago the EU was given rules such as these. With this new constitution everything will go like clockwork, just like in your club."*

Two hundred thousand children were issued with this propaganda at a time when the Constitution has been rejected by voters in France and Holland. The largely unaccountable elite of Brussels use as a logo the twelve stars, a symbol of Mary worship. For sixty years most of the prime movers within the European movement for integration have been Catholics. For Rome, the end frequently justifies the means. History records how it has eradicated those who disagree with it:

> *"No computation can reach the numbers who have been put to death...opposing the corruptions of the Church of Rome. A million poor Waldenses perished in France, 900,000 Orthodox Christians were slain in less than thirty years after the institution of the order of the Jesuits...the inquisition destroyed, by various tortures, 150,000 within thirty years. These are a few specimens...the total amount will never be known till the earth shall disclose her blood..."* John Dowling, History of Rome in Scott's Church History; book eight.

What other organisation so exactly matches the Bible description of a persecuting Church, the agents of whom *"say they are Jews* [Christian 'converts', see Romans 2:9], *and are not, but do lie"* (Revelation 3:9) and then shed blood to enforce dogma:

> Revelation 17:5 *"...the mother of harlots...v6 And I saw the woman drunken with the blood of the saints, and with the blood of the martyrs of Jesus..."* 18:23 *"...for by thy sorceries all nations were deceived. v24 And in her was found the blood of prophets, and of saints, and of all that were slain on the earth."*

This describes a religious entity in league with world powers:

> Revelation 17:7 *"...I will tell thee the mystery of the woman, and of the beast that carrieth her, which hath the seven heads and the ten horns."*

Churches and religious movements are pictured in Bible symbolism as women. Whilst the true Church is the affianced Bride of Christ, the great false Church is described in Revelation as a prostitute:

> Revelation17:2 *"...with whom the kings of the earth have committed fornication and the inhabitants of the earth have been made drunk with the wine of her fornication"*

The prophecy shows that at the end of the present age political leaders will turn against this powerful institution:

> *v16 "And the ten horns [leaders, see Rev17:12] which thou saw, where the whore sits [Europe], these shall hate the whore, and shall make her desolate and naked, and shall eat her flesh, and burn her with fire."*

The description of a religious movement given in Revelation chapter seventeen fits perfectly the Catholic Church. It can be none other than the Great Whore of Revelation 17:1. This is a church that has circulated Bibles with the book of Revelation missing because it identifies her; a church that changed the Ten Commandments by removing the second commandment forbidding graven images, freeing itself to bow down to statues. It uses weasel words to deny that it teaches Mary-worship, claiming instead that its followers merely 'reverence' her. This idolatrous false church prays to Mary and other dead mortals. The Catholic Church has perpetrated gross fraud. It constitutes a giant conspiracy against humankind. This is the same entity John had seen in vision:

> *Revelation 17:3 "So he carried me away in the spirit into the wilderness: and I saw a woman sit upon a scarlet coloured beast, full of names of blasphemy, having seven heads and ten horns."*

The psychological techniques used by this false church are powerful. It has swayed the minds of the masses with impressive architecture and rousing music; splendid clothing, colourful Christmas trees, mystical ceremonies and the burning of candles and incense. All of this pomp and paraphernalia – mere form over substance - has proved an effective drug. People cannot see God but they can see the world around them. The truth, by comparison, is a very abstract pursuit when one's life may be at stake. It has always been easier to conform and accept the delicious *opiate of the masses* handed out by the priests.

A distinct style is adopted by the agents of counterfeit religion. Their manner of doing things can be easily recognised from ages past, as human nature does not change. False teachers are rebuked twenty eight times in the four gospels. Of such Christ told his disciples that *"all their works they do to be seen of men"*; that they *"love the uppermost rooms at feasts"* and to be called *"Master, master."* Today that could be 'holy father', 'reverend' or 'right reverend.' Such titles properly belong to God.

These are those of whom Christ said:

> "...for a pretence make long prayer... (and would) compass sea
> and land to make one proselyte (convert) and when he is made
> you make him twofold more the child of hell than yourselves"
> Matt 23: 14-15

Acting is an ancient profession. Nowhere has it been more subtly
practiced than by ministers of religion, of which Christ said:

> "In vain do they worship me, teaching as doctrine the
> commandments of men" Matt 15:9

Today the arguments deployed by false teachers may differ, but the
basic strategy for the maintenance of power remains unaltered. It
rests on two pillars: the selective use of scripture (or its complete
suppression) and the maintenance of political links with civil
powers. In this way control of the masses can be maintained. The
price is that anti-Christian ideas and practices are perpetuated under
a false Christian badge. The flagship concept for all Catholic dogma
is the Trinity doctrine. One scripture superficially appears to give
direct support for this idea. It is a good example of how false
teachers operate:

> Matthew 28:19 "Go therefore, and teach all nations, baptizing
> them in the name of the Father, and the Son and the Holy
> Ghost."

Even if accurate, this would not amount to a proof of anything.
Scholarly sources are consistent in saying that the idea of the Trinity
is not taught explicitly in the Bible. *Sacramentum Mundi, an
Encyclopedia of Theology* states:

> "TRINITY, DIVINE. A. The scriptural doctrine of the Trinity
>
> 1. *The Old Testament: since revelation and salvation come in
>   historical form, it cannot be expected that the Trinity of God
>   should have been explicitly revealed in the Old Testament...*
>
> 2. *The New Testament: there is no systematic doctrine of the
>   'immanent' Trinity in the N.T. The nearest to such a
>   proposition is the baptismal formula of Matthew28:19, though*

it must be noted that modern exegesis does not count this saying among the ipsissima verba of Jesus."

This is not the only source to suggest that Matthew 28:19 has been tampered with. Many other sources suggest the same thing. The *Scaff-Herzog Encyclopedia of Religious Knowledge* states in its article on the Trinity:

"...the formal authenticity of Matt 28:19 must be disputed..."

The only passage in the King James Version of the Bible *known* to have been falsified is I John 5:7 which purports to support the Trinity. In this case words were added[1] to the Catholic's Latin *Vulgate* in the fourth century. This was about time that the Trinity appeared in 'Christendom.' Suspiciously, ministers and theologians almost never refer to this scripture! Clearly, fraud has been used to try and establish an idea that already looked highly suspect. The International Standard Bible Encyclopedia says the Trinity teaching is:

"...contrary to the facts of early Church history."

Husting's Encyclopedia of Religion states:

"The Trinitarian form is not suggested in early Church history...until the time of Justin Martyr."

Many trinities appear in paganism. There can be no doubt at all that it is a man made idea that does not originate from scripture. Almost every greeting in the epistles of Paul is given on behalf of 'God the Father and his son Jesus Christ.' There is never any reference to a Holy Spirit Person. David Kemball-Cook records in *Is God a Trinity?*

"There can be considerable doubt as to whether the first century church had a belief in God as a Trinity at all...It was not until the late fourth century AD that the Holy Spirit was admitted to full status as the 'Third Person'. Early formulations were 'binitarian', admitting only two divine entities..." page 21

The flagship doctrine of Catholicism, the 'holy' Trinity, can justly be described as nothing more than a pagan sham. The Bible talks about two personalities that have the status of the God *type*, never

227

three. There has never been anything that could justifiably be called *evidence* of a 'Third Person.' Again, *Is God a Trinity* states:

> "...*texts that supposedly show a Trinity are outnumbered by far by texts in which the two 'Persons' God ('the Father') and Jesus Christ (or 'the son') are described together without any reference to the spirit...Taking the example of the salutations...not a single salutation is from the Holy Spirit as well." page 84*

The Anglo-Saxon world has enjoyed much respite from popery since the rift between Henry VIII and Rome in 1533, though not all of the religious persecution by 'Christians' has emanated from that source. Much of the emigration to the New World was a reaction against religious persecution in England. Protestants had also been burning to death critics of the Trinity doctrine, as recorded *in Is God a Trinity?*

> "*Michael Servetus, who wrote De Trinitaris erroribus (1531) and Dialogues on the Trinity (1532), was condemned by the Inquisition as a heretic. However he was condemned to death and burnt at the stake, not by the Catholic Church, but by the authorities in Protestant Geneva with the approval of Calvin.*" *(Owen Chadwick, The Reformation, pp198f) Page 163*

The Catholic Church is described in the Bible as a whore, and like a whore it has spawned daughter churches as the conduct of Calvin attests. These 'churches' may have jettisoned such obvious heresies as the sale of indulgences and Mary worship, but they remain apostate in character. Like the Catholics, their authority for doctrine remains, in effect, the Emperor Constantine - not the Bible. These churches also believe in a Trinity, going to heaven or burning in hell (in perpetuity, as a torture), the error of the immortality of the human soul, Christmas, the myth of a Sunday resurrection (see the later chapter *The Mark*) and the use of crucifixes - another non-biblical innovation of Constantine's.

The instrument of torture on which Christ suffered is described as a tree, a stake - but never a cross. Christ died on a stake, an item that resembled the pole on which the brass serpent was mounted by Moses in the wilderness. The snake on a pole was a *type* of Jesus

Christ, the antidote for sin. Some may wonder if using the Sun and the Moon as secret symbols for deities would be idolatrous, but the account of the snake on the pole offers clarification on that point. This strange object pointed to Christ in a way that was prophetic and appropriate to the situation being addressed (A fuller explanation is given in Appendix XII).

In the 21st century a revival in Catholic power is underway. The funeral of John Paul II was a *tour de force* of international diplomacy from which no major President or Prime Minister was absent. The status of the Vatican as power broker is as unique in the modern world as it was in ancient times. Its power and prestige is immense. Some of its recent actions indicate that its wartime collusion with Germany has continued without a break.

A present day partnership between the German political elite and the Vatican would certainly be consistent with links in former ages. The Catholic/German relationship, though like many marriages often adversarial, has for a millennium been a dominant theme in European politics. Recent evidence of present day collusion between Germany and the Vatican could be seen at the start of the Balkan war of the 1990's. A dormant conflict was fanned to life through the recognition, by Germany and the Vatican acting in concert, of Croatia and Slovenia as separate nation states in December 1991. This step was taken against the wishes of the entire international community. It was an indication of the growing power of Germany that the same states that had once fought Nazism soon fell into line with her views.

Two world wars still cast a long shadow over the Continent. It is the fear of renewed warfare that has, more than anything else, been the driving force behind the unification of Europe. The Times, reporting on the fiftieth anniversary party of the EU in Berlin, recorded March 26th 2007 that:

> *"Angela Merkel, the Chancellor, said at the weekend that Europe was still a matter of war and peace."*

To a eurosceptic this could sound like blackmail, as it implies that everyone must either support the unification of Europe or suffer war. It is also testimony to the perceived untrustworthiness of Europe's leaders and nations, especially her own, that a German Chancellor should publicly state this. It is as if these supposedly civilized nations are like wild animals, bent on violence; nations that can only be constrained by being chained tightly together, cheek by jowl, in a rather unnatural union. But if German politicians have a vision for Europe so do the diplomats of the Vatican. Catholic power-politics was in evidence during the same weekend of the EU 50th anniversary when EWT News reported the comments of Archbishop Dominique Mamberti, the Vatican's top foreign diplomat, speaking to the Committee of the European Bishops:

> *"The Holy See would not support any expansion policy of the EU that would threaten the principles and values forged by Christianity that have made Europe a beacon of civilisation for the world."*

This swelling statement relates to the issue of Turkey's proposed membership of the EU. It would be an anathema to the Catholic Church to admit Moslem Turkey. One might wonder if the archbishop had the Spanish Inquisition in mind when he described Europe as *"...a beacon of civilisation for the world."* This is yet another propagandist statement about a region that had recently plunged the world into the greatest conflict in history. With the Vatican's blessing it has been a beacon of barbarity, not civilisation!

Proponents of the European project would have their subjects believe that the dangerous nationalisms of the past can be subsumed within a supra-national identity. Not everyone has agreed with this optimistic scenario:

> *"You have not anchored Germany to Europe; you have anchored Europe to a newly dominant, unified Germany. In the end, my friends, you'll find that it will not work."* Margaret Thatcher, Colorado Springs, October 1995

During the 1988 European summit held in Bruges, Belgium, the then Prime Minister Margaret Thatcher reminded fellow European Leaders of an uncomfortable fact:

> *"Europe has only been united under tyranny."*

Shortly after the November 1989 fall of the Berlin Wall, Britain's

Prime Minister Margaret Thatcher was asking regarding the Germans *"But can we trust them?"* She was not alone in her doubts. The former French President François Mitterrand once said *"I like Germany so much I would prefer to have two of them."*

Margaret Thatcher's real opinion of the Germans emerged from a remarkable cache of official Kremlin records smuggled out of Moscow. Reported by the *Times* in September 2009, the records detail how the Russians reacted to the tumultuous events of 1989 and Britain and France's secret opposition to German unification, despite official pronouncements to the contrary.

"We do not want a united Germany" Margaret Thatcher had said in a frank exchange with President Gorbachev:

> *"This would lead to a change to postwar borders, and we cannot allow that because such a development would undermine the stability of the whole international situation and could endanger our security."*

She urged Gorbachev to do what he could to stop it. Even after the Wall was down in 1990 she told Gorbachev:

> *"All Europe is watching this not without a degree of fear, remembering very well who started the two world wars."*

Jacques Attali, the personal advisor to President Mitterrand, said he would *"fly off to live on Mars"* if unification went ahead.

In March 1990 Thatcher called a gaggle of historians to Chequers to discuss the German issue. Professor Norman Stone told her *"There's no danger of a fourth Reich."* Yet twenty years after the fall of the Wall we virtually have a fourth Reich. Europe's currency is controlled from Frankfurt and its most powerful politician is German. Germany has led Europe out of recession, and its military influence in flash points such as Afghanistan and the Horn of Africa increases month by month; it has become unstoppable.

While Thatcher and Mitterrand had good reason to fear a united Germany, the longest lasting tyranny in the world has been that of the Catholic Church. They have campaigned for a United Europe continually throughout the ages. This was also Hitler's dream according to Winston Churchill in *The Second World War*, book 6, page 155:

> *"Hitler can no doubt force his way through Spain, just as he can dominate Italy. His deterrent is found in the political sphere. His aim is to establish a United States of Europe under the German hegemony and the New Order. This depends not only upon the conquest, but even more upon the collaboration of the peoples."*

A satanically inspired resurrection of the Holy Roman Empire led by Germany will be responsible for the next world crisis. The book of Revelation says of this military colossus, described as a beast with a 'whore' astride it:

> *Revelation 13:4 "...and they worshipped the beast saying, Who is like unto the beast, who is able to make war with him?"*

There is an abundance of evidence that the Germans have a unique history of war-making going back thousands of years. They have never been content with peace for very long. In concert with the Catholic Church, the Great Whore of Revelation chapter seventeen, they have terrorised the peoples of Europe down through the ages. *The Germans are of the same stock as the warlike ancient Assyrians:* this will be demonstrated in the next chapter.

Bible prophecy is quite specific: it will be the modern descendents of the Assyrians who will plunge the world into its next great crisis. In this seventh revival of the Holy Roman Empire, the Vatican has yet to take centre stage. As will shortly be shown, the manner in which they eventually do so is described in the Book of Revelation.

---

1. Of I John 5:7 the book *Is God a Trinity* by David Kemball-Cook says: "'For there are three that bear record in heaven, the Father, the Word, and the Holy Ghost, and these three are one' These words are widely acknowledged to be a medieval addition, as they occur in no Greek manuscript until the fourteenth century, except for one manuscript in the eleventh and one in the twelfth century where they have been added in another hand."

37

GERMANI

"O Assyrian, the rod of mine anger…" Isaiah 10:5

The invading tribes that warred with the Roman Empire gained the collective name *Germani*: it has the meaning of *war-men*.

These were ancient tribes. Could they have come into contact with the Israelites of the Holy Land? Before acquiring from the Romans the name 'Germani' these peoples had been known by other names. Centuries before Rome became a world empire, the Israelites had been conquered and taken captive by a great military power, the *Assyrians*. Prophecies concerning this powerful ancient race abound in scripture. It is evident that the writers of the Bible envisaged that the Assyrians would be involved in major events at a time they called the 'close of the age.'

The British museum houses a priceless black stone cylinder covered with writings commissioned by King Sennacherib of ancient Assyria. He had inscribed details of a campaign against the Israelites' southern kingdom (Judah), with its capital at Jerusalem, during the reign of King Hezekiah (2 Kings 32:1 – 33). The kingdom was not conquered at that time although captives had been taken. Sennacherib boasts of his accomplishments in this description of the siege inscribed on the cylinder:

> *"I fixed upon him…And of Hezekiah [king of the] Jews, who had not submitted to my yoke…I captured 200,150 people small and great, male and female…I reckoned [them] as spoil. [Hezekiah] himself [was] like a caged bird within Jerusalem."*

These ancient persecutors of the Israelites appear in several 'end-time' prophecies[1] of the Bible. Is it likely that such a mighty empire-building people would vanish from history, or are they still a great nation today? The historian Josephus in *Antiquities*, I, ix, 1, records four ancient kings that formed a military alliance. These kings were Assyrians:

> *"At this time, when the Assyrians had dominion over Asia, the people of Sodom were in a flourishing condition…the Assyrians made war upon them…every part of the army had its own commander…Amraphel, Arioch, Chodorlaomer and Tidal…"*

These same four kings are also mentioned by name in the Bible, in Genesis chapter fourteen, where Abraham makes war with them circa 1900 B.C. The Assyrians had been the first race to establish a world empire:

> *"According to Ktesias and others, the Assyrians were the first to establish a world-dominion." Lange's Commentary volume 1, page 403*

This ancient warlike race was known by different names at different times. Much later than the time of Abraham they were known as Chatti, a name derived from the Hebrew for Hittite: *Chitti* (meaning terror, break down by violence). They shared the name Chatti, or Hittite, with a second ethic group. Of the Hittites, James Hasting's Dictionary of the Bible (1899), volume 2, article "Hittites" says:

> *"Besides the northern Hittites, other Hittites, or 'sons of Heth,' are mentioned in the Old Testament as inhabiting the south of Palestine…The Assyrians…caused the name of 'Hittite' in the Assyrian period to be applied to all the nations west of the Euphrates."*

Evidently, more than one tribe was once known by the name Hittite (or Chatti). The *Encyclopedia Britannica* (11th edition, volume 13, article *"Hittites"*) acknowledges a problem in tying down the identity of the ancient Hittites:

> *"The identification of the northern and southern Hittites, however, presents certain difficulties not yet fully explained; and it seems we must assume Heth to have been the name of both a country…and of a tribal population not confined to that country."*

That there were two different tribes of the Hittites (or Chatti) is again born out by *Lange's Commentary* volume 3, page 123, analysing I Kings 10 verse 29:

> *"The Hittites are the same as those mentioned in chapter 9:20, but were an independent tribe, probably in the neighbourhood of Syria, [biblical Syria was just north of modern Syria] as II Kings 7:6 mentions them as in alliance with the Syrians."*

These were the Germanic tribes that fought Rome. The *Encyclopedia Britannica* says in volume 6, article *"Chatti."*

> *"(Chatti)...an ancient German tribe (which) frequently came into conflict with the Romans during the early years of the first century."*

The same source in another article *"Hesse"* (volume 13) gives further confirmation of the identity of the 'ancient Germans':

> *"The earliest known inhabitants of the country [Germany] were the Chatti, who lived here during the first century A.D...Alike in both race and language, the Chatti and the Hessi are identical."*

In this way the Germanic tribe *the Hesse* received their name, as the Old High German spelling for Hesse is Hatti. In several ways it is shown that one of the earliest Germanic tribes, known as the Chatti, have descended from the Assyrian Chatti that had migrated from Asia Minor. The Black Sea region was the Assyrian's place of origin:

> *"Now consider what Sylax, the author of the 'Periplus' who lived about 550 B.C., writes of this region: 'The coast of the Black Sea...is called Assyria.' (from page 261 of Perrot and Chipiez's History of Art in Sardinia, Judea, Syria and Asia Minor, Vol. II)"*

Both Germanic tribes and the Assyrians had originally lived north of the Black Sea before migrating across Europe; they are one and the same race according to the historians Jerome and Pliney the Elder:

> *"Jerome, who lived at the time when the Indo-Germanic tribes were invading Europe, gives the answer: 'For Assur (the Assyrian) also is joined with them.' (Letter 123, chapter 16, Nicene and Post-Nicene Fathers)...Only 300 years before*

Jerome, the Roman naturalist Pliny the Elder declared the Assyriani - the Assyrians - were dwelling north of the Black Sea (Natural History, IV, chapter 12, page 183). But the Assyrians did not remain there! They are not there today! Of course not – they migrated into Central Europe – where the Germans live today!" Dr. Herman L. Hoeh, The Plain Truth Magazine December 1962

Further confirmation as to the identity of the Hittites (the Chatti) is seen in James Hasting's *Dictionary of the Bible*, volume 2, the article *"Hittites:"*

"The Hittites seem to have had a special fancy for combining the parts of different animals into strangely composite and sometimes grotesque forms."

The Germans have long had the two-headed eagle as a *symbol of the German empire*. One other ancient symbol of the Chatti (or Hatti) resonates in the modern world:

*"The ancient kings of Assyria called themselves Khatti-sars – meaning the '**Kaisers of Hatti**' or 'Kings of Hatti'. The chief people of Hatti regarded themselves as Assyrians. The Assyrian kings wrote of the tribes of Hatti – the ancestors of the Hessians: 'As Assyrians I [ac]counted them' (D. D. Luckenbill, 'Ancient Records of the Assyria and Babylonia', volume II, chapter 29). The ancient capital of the land of Hatti was popularly known among the Romans as 'Ninus Vetus – the old Nineveh' ('History of Art in Sardinia, Judea, Syria and Asia Minor' by G. Perrot and C. Chipiez, volume II, page 272). Nineveh was Assyria's capital!...[ILLUSTRATION: under which there is a reference to the standard of the Hatti recovered through excavation in ancient Anatolia.] Notice the swastikas." Dr. Herman L. Hoeh, The Plain Truth Magazine, January 1963. (Emphasis added)*

Could Hitler have been a modern day Hittite? Once the Assyrians are understood to be modern Germans, the meaning of many Bible prophecies concerning the 21st century become clear. Leonard Catrell in *Anvil of Civilisations* wrote:

"In all the annals of human conquest, it is difficult to find any people more dedicated to bloodshed and slaughter than the

*Assyrians. Their ferocity and cruelty have few parallels **save in modern times**." (Emphasis added)*

It is the nation of Germany that has demonstrated the greatest dedication to bloodshed, slaughter, ferocity and cruelty *in modern times*. Germany, under the Nazis, embarked on a quest for world domination in what historian Richard Holmes (in the BBC Radio Collection *Battlefields*) calls *"Quite simply the greatest event in History."*

Is it conceivable that this 'greatest event in history' would not have been anticipated in Bible prophecy? It has already been shown in the chapter *A brief history of History* that the German/Italian Axis power was the sixth of the seven revivals of the Holy Roman Empire predicted in *Daniel* and *Revelation*. The modern Assyrians fought a titanic war for world domination against which, for many months, Great Britain stood as the only bulwark.

Are the British mentioned in the prophecies of the Bible? They are, and the identity of the English speaking world in Bible prophecy is the key to understanding contemporary events.

---

1. Most Bible prophecy is focused on, or finds its ultimate fulfillment in 'end-time' events. Many such prophecies are prefixed with the expression *"in that day."*

38

SAMARIA ON THAMES

*"There is a generation pure in their own eyes,
and yet is not washed from their filthiness." Proverbs 30:12*

At the close of his resignation speech of 10th May 2007 British
Prime Minister Tony Blair stated:

> *"This country is a blessed nation. The British are special, the
> world knows it, in our innermost thoughts, we know it. This is the
> greatest nation on earth. It has been an honour to serve it."*

Is the nation of Britain, proclaimed by an outgoing Prime Minister to
be the world's greatest nation, mentioned in the Bible? This is the
nation who at the time of the Reformation threw off Rome's yoke to
establish democratically accountable government. Britain then went
on to develop the greatest empire in the history of the world, on
which the Sun never set. It seeded the greatest military power ever
seen, the USA. Anglo-Saxon values, entertainments, aviation,
fashions, business methods and ultimately their war waging
capability completely dominated the 20th century.

Britain once stood alone against a world-imperiling Nazi tyranny.
Later, in partnership with its Allies, it went on to defeat what was
the sixth revival of the 'holy' Roman Empire (see the earlier chapter
A brief history of History). Is it conceivable that Bible prophecies
concerning the *end times* do not record the nation of Great Britain?

The Old Testament book of *Hosea* is almost entirely prophetic. It
appears immediately after the highly prophetic book of *Daniel*.
Assyria (Germany) is repeatedly mentioned in *Hosea* and almost
always in connection with a nation called *Ephraim*. It is apparent
that Ephraim is a special nation to God but has done much to
displease him. According to Hosea, that nation is to be brought

239

down by means of the Assyrians who, in the chapter *Germani,* are shown to be the modern Germans. Does the nation of Ephraim exist today and by what evidence can it be identified?

The name Ephraim first appears in the book of *Genesis* as one of the grandsons of Jacob. (By that time Jacob's name had been changed to *Israel*). Jacob (Israel) was the grandson of Abraham and the rightful inheritor of some transcendent promises. A special nation was to be brought into being, a servant nation that was to play the pivotal role in the working out of the divine purpose in a world yet under the sway of the invisible adversary. Close to death, the patriarch Israel made a prophetic pronouncement over the heads of his two grandsons:

> *"And Israel stretched out his right hand, and laid it upon Ephraim's head, who was the younger, and his left hand upon Manasseh's head, guiding his hands wittingly, for Manasseh was the firstborn. v16... [He said] bless the lads; and let my name be named on them...v18 And Joseph [the boys' father] said unto his father, Not so, my father: for this is the firstborn; put thy right hand upon his head. v19 And his father refused, and said, I know it, my son, I know it: he also shall become a great people, and he also shall be great: but truly his younger brother shall be greater than he, and his seed shall become a multitude of nations." Genesis 48:14*

What *"great nation"* has ever appeared that was closely linked to a *"multitude of nations"* in the annals of history? The British Empire encompassed the largest territories of any empire, dwarfing that of Rome. This vast dominion was accumulated by Napoleon's 'nation of shop-keepers' with lightning speed and minimal force. Consider the following selection of better known territories that came under the control of the British Empire:

India, Ceylon, Hong Kong, Cyprus, North Borneo, Brunei, Rhodesia, Nigeria, Gold Coast, Sierra Leone, Gambia, German West Africa, Uganda, Egypt, Sudan, Ontario, Quebec, Nova Scotia, Australia, new Zealand, Fiji, the Bahamas, Jamaica, Barbados, Bermuda, Trinidad, Tobago and the Falklands.

Seventy-three countries and territories are listed in the 1921 book *The Destiny of the British Empire & the USA* by Lt-Col. W. G. Mackendrick. Never was there an empire remotely like this. Then, in history's climactic war, Britain fought side by side with history's greatest and wealthiest nation, the USA. Could there be a connection between the stupendous promises made to the two grandsons of the patriarch Israel and the two great English-speaking powers?

Such a suggestion brings immediate denials from historians, principally on linguistic grounds. In an article *The 'Jewish' Conspiracy is British Imperialism* Henry Makow PhD states:

> *"Hebrew is a Semitic* [of the sons of Shem, Noah's son] *language, while English is of Aryan* [Germanic] *origin."*

Here is seen the single greatest objection to the notion of what has been called British Israelism. Historians can see that the English language appears to have common roots with the German language. There is however an explanation for this factor. It is seen in the evidence of migrations from the area of the Middle East to Western Europe, as will be shown.

The movement called British Israelism mixes truth with error. It promotes the erroneous idea that the British Empire is, or was, the Kingdom of God on Earth. (If this were true, according to the prophecy of Christ's Kingdom in Isaiah chapter eleven, one might expect to see lions led about by small children in the streets of London!). A further fallacy attached to British Israelism is the assumption that the twelve tribes of Israel, having migrated across Europe from an area to which they had been taken as captives near the Caspian Sea, *all* arrived on one small group of Islands, the British Isles. These errors notwithstanding, there remains substantial evidence that the British Isles were colonized by certain Israelite tribes.

It has been shown that the Assyrians are the Germans of today. Could 'Germans' have deported the Israelites after the fall of the ancient Israelite capital Samaria? History records that both Germans and Israelites originated from the same geographical area:

241

*"The Germans, then, can be traced in historical records to the
regions surrounding the Black and Caspian seas, which border
on the ancient biblical land of Mesopotamia. This is the region
where civilisation commenced and from where the Bible
patriarchs came." Smith's Classical Dictionary, article,
Germania, page 361.*

If the Assyrians (Germans) did take captive the northern kingdom of
Israel, with Ephraim at its head, it could account for the fact that
"English is of Aryan [Germanic] *origin."* Is it possible that
'Hebrews' could, after centuries of migration, come to be using a
language of Aryan character? This is a vital question: their Germanic
language is the principal obstacle to the proposition that Anglo-
Saxons have Hebrew blood. Yet it can be shown that the only
hypothesis that fits all of the facts is that certain tribes of the ancient
kingdom of Israel (its capital in *Samaria*, not Jerusalem) were taken
as captives by the Assyrians (Germans) across Europe, where some
settled in the British Isles. Although their language was transformed,
many striking phonetic similarities have endured. Changed firstly by
exposure to that of their captors the Assyrians, and later due to
transplantation across Europe, it can nonetheless be shown that
many contemporary words, place names and customs have an
Israelite origin.

Where then is the evidence that ancient tribes migrated, or were
forcibly removed, across land and sea two and a half millennia ago
to settle in Western Europe? This evidence is of several types and
involves in the first case legends, traditions and ancient poems. More
substantive are the numerous place names seen across Western
Europe, the enduring linguistic features of the language of the
British Isles, certain ancient customs, and not least of all the well
chronicled history of an ancient stone – the Stone of Scone, or
'Jacob's pillar stone' – kept until recently in Westminster Abbey.
That stone was returned to Scotland by Prime Minister John Major
on 15th November 1996 after 700 years exactly. If divine providence
be a factor in that it does not bode well for the future of the Union or
our monarchy. Most monarchs since antiquity have been anointed
sitting on this stone, including Queen Elizabeth II.

To all of this must be added the remarkable congruence between the extravagant promises made to Abraham in *Genesis* and the unique place of the English Speaking Peoples in the annals of world history. Only the existence of these nations can come close to a fulfillment of the promises to Abraham.

If the Bible is divinely inspired, the promises must at some point be fulfilled. They were to involve real land, actual physical descendents (*"as the stars of heaven"* Genesis 26:4) and the wielding of political and military power with the possession of strategic outposts (*"the gates of your enemies"* Genesis 22:17 & 24:60) such as Gibraltar, Panama, Hong Kong, the Straits of Hormuz and the numerous naval outposts of the British and latterly the American Empires. The prophecies describe a nation and a company (or commonwealth) of nations. Perhaps mostly tellingly, these inheritors were to contend with the Assyrians, today's Germans.

The evidence of a linguistic trail from the Middle East to the British Isles is much stronger than supposed by most historians. The facts have been ignored, and thereby effectively suppressed, with a veil of secrecy drawn across the ancient history of the British Isles not unlike that which obscures that of the early Church.

Did some of the ancient Hebrews settle in these islands? The widow of the clergyman G.A. Rogers MA, in 1904 amassed evidence for this in her book *The Coronation Stone*. Relating the Irish legend concerning a voyage of the prophet Jeremiah to Northern Ireland, she mentions a letter from another clergyman F.R. Glover regarding the origin of the grave of a claimed Israelite princess Queen Tephi. He made this comment regarding her burial mound, the 'Sacred Hill of Tara':

> *"I don't know if you are aware that the tomb is called the great Mergech, which is not a Celtic word or name, but it is a Hebrew one."* F.R. Glover, author of *England the remnant of Judah*.

According to Irish legend, this was the daughter of the last king of Israel, King Zedekiah, whose eyes were put out by the invading Nebuchadnezzar. Jeremiah had carried her to safety, intending to transport her to Denmark but due to a storm at sea was deposited, by divine providence, on the north-western shores of Ireland. In *The*

243

Destiny of the British Empire the already quoted Mackendrick relates the following:

> *"Ten different Irish histories record the fact that at this time a wonderful Hebrew Prophet arrived in Ireland who caused Baal worship to cease, married his ward (Zedekiah's daughter) to the then ruling king of Ulster...today, in the Court of Justice, Dublin, in the position of honour, is the medallion of Jeremiah..."*

Jeremiah had ensured the continuation of a dynasty. Israel's kingly line was never to be broken, according to promise made to King David:

> *"I will set up thy seed after thee...I will establish the throne of his [Solomon's] kingdom forever...Thy throne shall be established forever."* II Samuel 7:12-16

Where is that throne today? Where has been the longest, the greatest and most prosperous and *illustrious royal lineage in history?* It resides in London – the centre of today's world. In 1887 Queen Victoria was presented with a chart showing her descent from King David. This knowledge is not lost to the royal family. To those who would deny the obvious one must ask *why*: why the British Empire, why the USA, why the two world wars and the phenomenon of Germany; defeated in the greatest war in history, but now again at the political helm of Europe and one of the world's great economic powerhouses.

Once the big picture is seen it is evident that the development of democracy and the industrial revolution only became possible after the break with Rome in 1533. Now, after several centuries of unparalleled greatness for the English, the well camouflaged tentacles of Rome have spread back into our body politic and are slowly strangling it. The Treaty of Rome and its successive mutations have become a cancer to modern Britain, starving the nation's life blood and power away from Westminster.

The sons of Joseph, Ephraim and Manasseh, Britain and America, have been the inheritors of the promises of national greatness given to Abraham. These promises were delayed 2,520 years because of disobedience. Moses had prophesied in Leviticus chapter twenty-six:

"I will punish you seven times more for your sins." The prophetic year consists of 360 days which when multiplied by 7 is 2,520 days. Bible prophecy employs a 'day for a year principle.' One example of this was Ezekiel's task in portraying the sins of Israel over a span of 390 years by lying on his left side for 390 days:

> *"For I have laid upon thee the years of their iniquity, according to the number of the days..." Ezekiel 4:5*

According to this day-for-a-year principle the blessings promised to Abraham were delayed 2,520 years. From the time of the captivity and transportation of the Israelites from 721 – 718 B.C. by the Assyrians, that period of delay ended around 1800 - 1803 A.D.

England had been through stormy times since breaking with Rome. The Catholic Queen Mary, Bloody Mary, introduced a reign of terror against those who would not 'bow the knee to Baal.' Later the Catholic Guy Fawkes tried to destroy parliament in 1605 by a gunpowder terrorist attack. The civil war, the Jacobite rebellions that climaxed in 1715 and the Spanish Armada of 1588 were all about restoring the tyranny of the Pope.

By the end of the eighteenth century, a stable foundation for freedom had been laid. This was precisely the time that the English-speaking peoples on both sides of the Atlantic leaped to national greatness. The English slave trade was being closed down just as the great railroads of North America were being laid. These two nations, the sons of Joseph, were to come into possession of over two thirds of all cultivated resources and wealth on the globe.

The period that the promises had been delayed, a span of 2,520 years, actually makes a dual appearance. The same span of years controlled the destiny not only of the kingdom in Samaria, but the original kingdom of Judah. A prophecy concerning the release of Jerusalem from gentile (non-Israelite) powers was fulfilled when General Allenby's troops accepted the surrender of the Turks from the Palestinian Mayor of Jerusalem, Haj Amin Nashashibi on 9th December 1917. This was 2,520 years after the southern kingdom of Judah (capital at Jerusalem) was conquered and taken into captivity in 604 B.C. The prophecy of Isaiah 31:4 *"...as birds flying, so will the Lord of Hosts defend Jerusalem"* fits well that occasion, as military bi-planes were in action over Jerusalem at the time.

Thus the fortunes of the modern Middle-Eastern state of Israel are tied to the Birthright nations, the sons of Joseph. This is confirmed in a vital statement in Hosea regarding Israel ("let my name [Israel] be named on them"; *Ephraim* and *Manasseh*) which denotes the English speaking peoples:

> *"And the pride of Israel doth testify to his face: therefore shall Israel and Ephraim fall in their iniquity; Judah shall also fall with them." Hosea 5:5*

Both history and current events bear testimony that the fortunes of Britain, USA and the Middle-East state of Israel are closely entwined. The following chart illustrates the historical fulfillment of key prophecies regarding the two Israelite kingdoms and the remarkable fit of the 2,520 year span in each case:

	NORTHERN KINGDOM (Capital Samaria)	SOUTHERN KINGDOM (Capital Jerusalem)
Comprised of:	9 or 10 breakaway tribes led by Ephraim	Judah (Jews), Benjamin, Levi, led by Judah
Taken captive:	718 – 721 B.C. by Shalmanesser of Assyria	604 B.C. by Nebuchadnezzar
Transported to:	Assyria	Babylon
Blessings delayed:	2,520 years	2,520 years
Fulfillment:	1800 – 1803 British Empire & USA begin to acquire vast territories	1917 Gen. Allenby captures Jerusalem from the Turks

The sons of Joseph, Ephraim and Manasseh, constitute the Birthright Nations of Britain and America. All along they have been destined to become the greatest of nations. As the patriarch had said:

246

"...he [Manasseh] also shall be great: but truly his younger brother [Ephraim] shall be greater than he, and his seed shall become a multitude of nations." Genesis 49

That is why Tony Blair could say with little fear of ridicule:

"The British are special, the world knows it, in our innermost thoughts, we know it. This is the greatest nation on earth."

Evidence of the true history of the world has been suppressed, just as the true religion has been suppressed. The 1865 book *Our British Ancestors: who and what were they* by the clergyman Samuel Lysons MA, FSA, shows an example:

"The Anglo-Saxon Chronicle distinctly states that "The first inhabitants of this land were Britons, and that they came from Armenia [near the Black Sea] ...Some later transcriber of the Chronicle, actuated by some fancy of his own...takes a pen and interpolates Armorica, i.e. the present Brittany in France. This reading, however, while admitted into the margin, is not admitted into the text of the best editions of the Chronicle." page 22, Our British Ancestors.

Lysons' book revealed more than a glimmer of a secret - and at times suppressed - history of Europe. Lysons gives over 3,000 examples illustrating similarities between British words and names, and Hebrew words of the same or similar meaning. There is no positive information concerning our forbears by which such a powerful indicator of ethnicity - that of language - could be dismissed. Our *ancient* history is for the most part shrouded in mystery. The resources with which to *disprove* a link between Palestine and the British Isles are simply not available. Lysons makes that all important point:

"...before we are justified in rejecting histories and traditions which have the sanction of a respectable antiquity, their impugners are called upon to show that they are either impossible or improbable, or that they are refuted by more trustworthy authorities..." Preface xiii, Our British Ancestors.

The strategic centre of the world landmasses, London and the island on which it is situated, were barely known of by historians 2,300 years ago as Lysons recounts:

> *"We do not find Britain positively spoken of in any written work before the time of Herodotus, who flourished in 445 B.C...he only mentions the Cassiterides, now called the Scilly Isles, from whence he says that tin was exported...Aristotle, who lived a hundred years later, mentions the British Isles by name, as Albion and Ierne, that is, what we now call England and Ireland." page 1, ibid.*

Then around 1,900 years ago the first century historian Josephus, quoted by Lyons, makes a most illuminating observation concerning the Welsh:

> *"Gomer, the eldest son of Japheth, was father of the Gomerites, called by the Greeks Galatians,"..."Of this opinion, too, is the learned Bocholt: and if he be right, those who derive the Cimmerians or Cimbri (Cymri) from Gomer have some grounds for their derivation, the Cimmerians seeming to be the same people with the Gauls or Celts under a different name; and it is observable that the Welch [sic], who are descended from the Gauls, still call themselves **Cymri** or Kymry (Universal Histories,. vol. i. p140)." page 23, ibid.*

Here is evidence that the Welsh - who up to the present day use the name **Plaid Cymri** - are of Israelite origin. Lysons is able to call on the support, *with many quotations*, of ancient historians: Theophilus Antiochenus, Eustachius of Antioch, Apian, Diodorus, St Jerome and Isidore. He also uses the Chronicle of Alexandria. All of these sources support the Gomer-Galatian -Gaul link:

> *"It has been a universal tradition, and an unbiased and independent one, and therefore worthy of respect." Samuel Lyons, page 25 of Our British Ancestors.*

From a completely different angle, Lysons then brings in another piece of evidence about the origins of the Welsh. Citing Rawlinson's Essay on the fourth book of Herodotus:

> *"Mr. Rawlinson observes, 'The very closest possible resemblance between the Greek name Κιμμέριοι and the Celtic **Cymry**, and the presumption is in perfect harmony with all that*

enlightened research teaches of the movements of the races which gradually peopled Europe...the Celts had an unvarying tradition that they came from the East." page 27, ibid

Lyons concludes on the matter of the link between Armenia and Britain:

"The way in which Mr. Rawlinson, in the Essay from which I have quoted, brings the Cymric Celts from Armenia to Britain is most masterly; it confirms all the traditions of the Welch [sic], the views of Nennius, and the Anglo-Saxon Chronicle, and all our earliest histories...if a variety of independent and undersigned testimony is requisite...we have it as fully as it is possible to expect." p 28, ibid

Lysons' glossary of over three thousand English words is, according to him: *"apparently – judging by the sound and sense – derived from the Hebrew, Chaldee, or Syriac."* One example is the English word *naughty*, for which the equivalent Hebrew is, suggests Lysons, כ מ ת. This sounds like 'Nateh' with a meaning of 'turning away from the right way.' Here are some further examples:

Beryl	ב ו ר ל א	pron.	"byrila"	A beryl, a precious stone
Castle	ב מ ל	norp.	"casel"	Strength, support, comfort.
Duck	ד ה ה	pron.	"duchen"	To dive down, to plunge into water
Exempt	ע מ ת	pron.	"zempt"	Cut off, excluded
Fist	פ ז	pron.	"fist or fix"	Solid, compact

(See Appendix XIII for a longer list)

Compelling evidence of Israelite origins can be seen in place names across Europe. In the forty-ninth chapter of *Genesis* a detailed prophecy is given by the patriarch Israel concerning the twelve tribes that were to be his progeny. There are actually thirteen tribes when the sons of Joseph, Ephraim and Manasseh, are taken into account; the heraldry of the USA is replete with the number 13.

One of the tribes Dan was, according to Genesis chapter forty-nine, to be a *"serpents trail."* This idea is echoed later in the book of

249

Judges chapter eighteen, where Danites took a town called Laish and renamed it Dan. They also renamed another place *Mananeh-Dan* in remembrance of their father Dan. Clearly, this tribe had a strong predisposition to name places after themselves. Their migrations can be seen in the names Danube and Danzig; the rivers Don, Dnieper and Dniester; Denmark (Dan's mark); Donegal, Dundee and Dunraven. In Western Europe the prefix Dan, Dun or Don is seen in dozen of places. Irish annals record settlers called "Tuatha de Danaans" prior to 700 B.C. The Irish name Dunn means judge and Dan was to "judge his people" (Genesis 49:16).

A map prepared by the Rev. L.G.A. Roberts, Commander R.N., secretary of the Imperial British-Israel Association, purports to trace the peregrinations of the British race from Palestine into Britain. This map is reproduced on page 25 of Mackendrick's *Destiny of the British Empire* and gives place names that are *"found on very ancient maps [and] can be verified in the British Museum."* In Northern Ireland the place name Dan-Rans is seen; the name of Sweden is rendered Swe-Dan, the names DAN-E-MARK, Angli, Co-Dan and Danzig all appear close together at the top of what is now Germany.

Until quite recently, a belief in an ancient migration from Palestine to the British Isles amongst the educated classes of the English was not uncommon. In *Destiny* Mackendrick quotes the British Admiral who had built up the British Navy in anticipation of the Great War. Lord Fisher, writing on page 223 of his *Memoirs*, wrote the following:

> *"Jerusalem, the Capital of the Lost Ten Tribes of Israel, whom we are without doubt..."*

From even a superficial investigation it is obvious that much true knowledge concerning this subject is accessible. Although space only allows for limited examples here, it can be seen that the evidence is extensive (if fragmented by time and neglect) and that substantive migrations from the Middle East across Europe have been recorded. The belief that these began more than two millennia ago, and involved Israelite tribes transported by the Assyrians, persists to the present day. The evidence that the Assyrians of old substantially comprise today's Germans is equally compelling.

Might the nation of Nietzsche and Goethe, Mozart and Wagner, Brahms and Beethoven take up arms against their European neighbours again?

Incoming Prime Minister Gordon Brown, on the occasion of his first official visit to Camp David for discussions with President Bush in July 2007, invoked Churchill by referring to *"the joint inheritance of the English-speaking world"* and the *"shared destiny"* of Britain and America. These two nations are blood brothers. They are destined to experience a re-run of the violent transplanting suffered by their ancient forebears at the hands of *The Assyrians*.

Further information is given concerning the Israelite migrations in Appendix XIII.

39

KELVEDON HATCH
"That God's most dreaded instrument in working out a pure intent
is man - arrayed for mutual slaughter, Yea, carnage is his daughter."
William Wordsworth

What really happened to the ancient kingdom Israel with its capital
at Samaria?

The Jews of Jerusalem had gone into captivity over a century before
Samaria was besieged. Their origin is known to this day. Therefore,
how could the larger northern kingdom of the Israelites, containing
three times as many Israelite tribes, disappear completely from the
face of the earth when the Jews did not? Might it be that those tribes
still exist but have *lost their identity?*

What is not disputed is that there were two kingdoms of Israelites.
The original unified kingdom had split in two after the death of King
Solomon. A smaller southern kingdom with its capital in Jerusalem
remained, being composed of the tribe of Judah (Jews), plus the
tribes of Simeon and Levi.

A larger northern kingdom with Samaria as its capital ruled all other
Israelite tribes including Ephraim. This larger northern kingdom was
never comprised of 'Jews' but merely Israelites. Those of the tribe of
Judah were originally *the Jews*, though later the tribes that settled
with them at Jerusalem, Simeon and Levi, also became known by
that name. Today's Jews are a greater mixture than those of the
ancient kingdom, other races having intermarried with them. The
swarthiness sometimes seen in their features is not characteristically
Israelite.

The larger northern kingdom, with *Ephraim* as its leading tribe, was
taken captive by Assyria - taken prisoner by Germans! They were

removed hundreds of miles westwards, away from Samaria on the
Mediterranean coast, through Assyria, and into the cities of the
Medes towards Armenia. If Assyrians have survived down to the
present day, why not Israelites? No promises of national greatness
were given to the Assyrians, although they were to be *"The rod of
mine anger"* (Isaiah 10:5). Yet staggering promises were made to
the Israelites, the descendents of Abraham:

> *"...thou shalt be a father of many nations...kings shall come out
> of thee...I will establish my covenant...in their generations for
> an everlasting covenant...Genesis 17:4-8 "...I will multiply thy
> seed as the stars of the heaven...thy seed shall possess the gates
> of his enemies. And in Thy seed shall all the nations of the Earth
> be blessed; because thou hast obeyed my voice." Gen22: 17-18*

The British Empire, and later the Commonwealth, was a blessing to
the world. Was not the defeat of Nazism also a blessing to the
world? The Israelites have indeed survived to the present day *but
their identity has been lost*. The Jews are known as Abraham's
descendents because they keep the *sign* of identity. It is the sign of
the seventh day Sabbath:

> *"Israel shall keep the Sabbath...for a perpetual covenant v17 It is
> a sign between me and the children of Israel for ever; for in six
> days the Lord made heaven (Heb. shameh – lofty, the sky) and
> earth [a renovation, see the earlier chapter The Great Breach]
> and on the seventh day he rested..."Exodus 31:16-17*

The Sabbath is both a memorial of what was a restoration, a
renovation, of the surface of the planet and a sign of 'God's people.'
But the wandering Israelites were to lose this sign:

> *"I said, I would scatter them into corners, I would make the
> remembrance of them to cease from among men." Deuteronomy
> 32:26*

Notwithstanding the above, evidence of their Israelite identity is
available. Events in the modern world strongly support the scenario
of an English speaking nation and company of nations in ongoing
conflict with Assyrians (Germans). That conflict is now in recess
although many tensions lurk beneath the surface. The book of Hosea

offers much detail concerning the dealings of Ephraim, the younger brother of Manasseh, with their ancient foes:

"When Ephraim saw his sickness…then went Ephraim to the Assyrian, to king Jareb (meaning: contention)…" Hosea 5:13

Economic sickness is the most common concern of national leaders. This future diplomatic move must occur at a time when the delusions of pacifism, disarmament and the trusting of former enemies take hold in Ephraim (Great Britain):

"Ephraim is like a silly dove [pacifist] without heart…they go to Assyria. When they go I will spread my net upon them; I will bring them down as the fowls of the heaven; I will chastise then as their congregation hath heard." Hosea 7:11

The key nation Ephraim is here seen interacting with the Assyrians, the modern Germans. Who therefore is this nation Ephraim? God says in the above prophecy of this key nation that he will *"bring them down"* by means of the Assyrians. Throughout history God has used the Assyrians to chastise those he wishes to punish:

"O Assyrian, the rod of mine anger, and the staff in their hand is mine indignation. v6 I will send him against an hypocritical nation…to tread them down like the mire of the streets." Isaiah 10:5

The final world crisis will involve the nations of a United States of Europe. It will invade the Holy Land (see the chapter *A brief history of History*). It is evident from the congruence of the many end-time prophecies that this mystery nation Ephraim will also suffer invasion at that time. What modern nation might fulfill the above scenario concerning Assyria (Germany) and *"a hypocritical nation?"* Which nation have the Germans primarily contended with for world power? Put that way, it is obviously Great Britain.

Might Germany go to war against the British for a third time? Britain today is placing its faith and hope for security in foreign alliances more than in God:

"For they are gone up to Assyria, a wild ass alone by himself [Germany is a law unto itself]: Ephraim hath hired lovers…v12 I

have written unto him the great things of my law, but they were counted as a strange thing." Hosea 8:9

Whoever this mystery nation of Ephraim is, if it is Britain then its church system is derived from the religion of *Rome*, which is largely the religion of Constantine. Although more liberal than Rome, the Anglican Church carries the same false doctrines. It rejects the only Bible-ordained system of worship, proved here to be the 7 holy days of Leviticus chapter twenty-three. These are encapsulated within the 19 day cycle as explained in the later chapter *The 1711171 System*. These days are contemptuously dismissed as 'Jewish' together with the true Sabbath, but they remain the only authentic system of worship given in the scriptures. They have never been repealed.

God holds Britain's religious leaders responsible, to some degree, for the sins of the people. That is evident from the principle given in Ezekiel:

"...if thou warn...thou hast delivered thy soul." Ezekiel 3:21

But the Ecclesiastical leaders of Britain have been dumb dogs that cannot bark:

"His watchmen are blind: they are all ignorant, they are all dumb dogs, they cannot bark; sleeping, lying down, loving to slumber. v11 yea, they are greedy dogs which can never have enough, and they are shepherds that cannot understand: they all look to their own way, everyone for his gain..." Isaiah 56:10

Of the moral conditions of their nation these leaders have little to say. In a June 2007 article by the London *Daily Mail* entitled *"Like the Bolsheviks..."* the moral condition of the nation is laid bare:

"It is as if the Government, like the Bolsheviks after the Russian Revolution, actively hates the idea of the bourgeois family and seeks to destroy it by making people totally dependent on the state. It no longer makes financial sense for a young man and a young woman of average means to raise a family together. Under the current welfare system, it makes far more economic sense for a woman to have a series of children by different fathers, and for a young man to drift about without embarking on any strong family commitment."

Successive Governments in modern Britain have crafted a complex web of financial inducements that destroy the fundamental institution created for the happiness and welfare of human beings, that of the family.

British society is becoming anti-God and anti-Christ. But not all religious leaders in the U.K. remain silent about moral conditions. Commenting on the June 2007 floods in Yorkshire the Bishop of Carlisle, Dr. Graham Dow, offered the following as reported in the July 3rd London *Guardian*:

> *"This is a strong and definite judgment...The sexual orientation regulations are part of a general scene of permissiveness. We are in a situation where we are liable for God's judgment which is intended to call us to repentance."*

The prophet Isaiah's end-time prophecy concerning 'Jerusalem' could describe modern Britain with devastating accuracy:

> *"...I will give children to be their princes, and babes shall rule over them. v5 And the people shall be oppressed, everyone by another...the child shall behave himself proudly against the ancient...v9 They declare their sin as Sodom, they hide it not...v12 As for my people, children are their oppressors, and women rule over them. O my people, they which lead thee cause thee to err..." Isaiah 3:4*

Those sixty-six words from Isaiah describe modern Britain as nothing else could: a matriarchal society in which down is up and up is down. The natural order has been overthrown and ecclesiastical leaders rendered 'dumb dogs that cannot bark.'

Family life in corrupt Britain is being torn asunder, with millions of children living a daily hell of emotional anguish and loneliness while their estranged parents fill out paperwork for the state. Unusually for a City Editor, Alex Brummer in the Daily Mail of December 24th 2008 invoked the Bible prophet Amos in discussing the UK's social malaise. He cited four cardinal sins: idolatry, adultery, murder and the persecution of the poor. Brummer's exposition of the last point was along the lines of 'the greed that caused the rich and powerful to take advantage of the helpless to pervert justice in order to get their way.'

In 2007, voices in the British press and parliament expressed increasing concern at the nation's obvious decline. Male life expectancy in the drug drenched estates of Catholic East Glasgow stood at 54. That was lower than The Gambia and nearly a decade less than Bangladesh, considered the poorest country in the world.

The Daily Mail of July 9th commented on *"Britain's deepening social catastrophe"* noting that *"On just about every measure of social dislocation, Britain is top of the international league"* and has *"the worst rate of family breakdown"* and is committing *"social suicide."*

The actual act of suicide is becoming an increasingly popular opt-out for the embittered youth of the UK. Youth violence against the police is also increasing. In July 2008 two policemen were attacked by school children in the prosperous London suburb of Croydon. A teenage girl had dropped litter and was asked by the police officers to pick it up. As they walked off she picked it up and threw it at them. A crowd of youths quickly gathered. The witnesses said:

"All of a sudden punches were thrown and the yobs got involved. The girls were biting and kicking. It was disgusting." Another recounted: "The crowd had made a semi-circle...like a gladiatorial ring. How often do you see a police officer being punched in the face?"

One labour MP revealed that a single parent with two children working 16 hours per week will receive, due to the welfare system, the same money as a father working 116 hours per week (which would be impossible) on the minimum wage to support a wife and two children. In modern *Ephraim* the traditional family, with the disciplines and mores that go with it, is almost dead in the water.

Anciently, the capital of the twelve tribes of Israel was transferred from Jerusalem northwards to Samaria. Today's sumptuously rich *Samaria on Thames,* overflowing with wealthy (and often criminal) tax exiles who feed themselves at the expense of the more highly taxed poor, with its 1000's of public phone booths perennially smothered in lewd advertisements by prostitutes, has led the world in decadence. Its feral youth slaughter one another with knives. Twenty teenagers died this way in the first half of 2008.

London's financial Square Mile and a housing bubble are all that has propelled an otherwise sluggardly Britain to the fore in the economic growth stakes. While the Germans manufacture things people want to buy, the UK's up and coming mathematicians work for hedge funds. The country has had most of its eggs in one roulette-wheel-shaped basket. Its judgment slumbers not:

> Hosea 4:1 "...there is no truth, nor mercy, nor knowledge of God in the land...v6 My people are destroyed for lack of knowledge...v9 Ephraim will be desolate in the day of rebuke...v13 When Ephraim saw his sickness, and Judah saw his wound, then went Ephraim to the Assyrian...v14 For I will be unto Ephraim as a lion...I will take away and none shall rescue...v15...in their affliction they shall seek me early...7:11 Ephraim also is like a silly dove without heart...they go to Assyria. v12 When they shall go I will spread my net upon them; I will bring them down as the fowls of the heaven, as their congregation hath heard...9:3 Ephraim shall return to Egypt, and they shall eat unclean things in Assyria...v17...they shall be wanderers amongst the nations."

The Anglo-Saxon nations have led the world into disastrous wars and are now, with Northern Rock, Bear Sterns et al, possibly leading it into a depression. Writing in the *Times*, February 2009, Sir William Rees-Mogg considered a sobering possibility:

> "...a full blown depression...of a kind that appears only once or twice a century...we are moving though a critical period in global politics as well as in global economics."

Increasingly, blame for the woes of the world will be laid at Anglo-Saxon doors. Bible prophecy is inexorable: attack, enslavement and the deportation of a remnant population at the hands of Germany is the fate of modern Britain. The monuments of its great cities will be reduced to blackened, irradiated wreckage.

The possibility of nuclear attack is something the nation has been anticipating for six decades. A few miles north of Brentwood, driving through the sleepy town of Ongar there is a curious brown coloured road sign, the oxymoron *Secret Nuclear Bunker*. Upon following this and other signs, one is eventually directed around what resembles a winding farm track and then into a car park. More signs indicate a short walk down a narrow path past an elephantine green army lorry of World War Two vintage. Some steps lead up to a rather common looking and strangely misplaced bungalow. Its walls are almost half a metre thick.

The underground government bunker and its entrance thus disguised were created in the fifties. This structure was decommissioned and opened to the public in 1994. That decision was as a result, the literature suggests, of the end of the Cold War. No mention is made of any further bunkers being built. The impression left hanging is that the government no longer needs a bunker. This warren of steel staircases and concrete corridors was, one is led to believe, a Cold War phenomenon that is now, courtesy of the kindly Mr. Gorbachev (the one that Mrs. Thatcher could do business with), relegated forever to the status of a bad dream.

There is plenty of time to contemplate, on a quiet drizzly mid-week visit, what might have been. Wandering around the labyrinth of landings and conference rooms, passing rows of bunk beds and an operating theatre, walking through room after silent room of aged telephony equipment and obsolete computer cabinets, there is time to imagine the electrifying atmosphere had a real war driven six hundred caterers, telephonists, map readers and secretaries below ground to live, cheek by jowl, with the Prime Minister of the day and his or her cabinet.

Aged T.V. sets positioned at various points run continuous loops of old films instructing householders on how to build shelters under kitchen tables and wipe fallout from the tops of their food tins. In other rooms there are documentaries about nuclear explosions and discussions on nuclear disarmament from the 1980's.

Huge steel doors, each the weight of a car, lead out of one side of the bunker to the underground air-conditioning units, otherwise a weak point in the bunker's defenses from what would have been a desperate, maddened population. In the centre of the bunker there is a planning room containing several huge maps. A Formica board with the word 'Tote' at the top is sectioned for score keeping of a macabre kind, with the headings *One megaton," "Two megatons,"*

"Three megatons," for the purpose of keeping track of air-born radiation.

Nearby a rather fetching life-sized manikin of a Navy wren stands to attention. Behind her are the steel doors of an exclusive Cabinet safe room: Winston Churchill would have once stood at the spot, scrutinising the hinges for any sign of weakness. A few feet along from that, a mock-up of Margaret Thatcher stands behind a glass panel in a purpose built studio full of old fashioned communications equipment, addressing the nation. Nearby there is a corridor from which the bedrooms of cabinet ministers can be viewed. A grinning John Major lies in a cheap metal framed bed with an army blanket over him, but it is a fantasy. The actual prime ministerial bed will be in the real bunker, because none of the equipment at Kelvedon Hatch has been renewed since the 1960's and no Prime Minister in 1990 would have ever gone there.

Naturally there is another bunker somewhere. It will serve to delay the capture and deportation of a few senior people but no more than that.

Just as the ancient Assyrians moved captive populations, the Germans of today will make full use of the Channel Tunnel. The Jewish holocaust had been facilitated by the most efficient mode of mass transportation ever devised - the train. That orgy of murder was only the forerunner of what is to befall sinful Britain: God's principal birthright nation, together with brother America, at the time of the seventh and final resurrection of the Holy Roman Empire.

Meanwhile, in today's Europe, pleas for fair play are frequently heard but seldom heeded. After Prime Minister Blair's last Intergovernmental Conference Sir William Rees-Mogg in a June 2007 article in *The Times* warned:

> *"The British will not forgive another betrayal. In all honour, give us a vote on Europe...the reform treaty [just agreed] is the same in substance as the 2004 constitutional treaty [that was voted out by France and Holland] ...the government has promised it [a referendum]. [The treaty] creates a new president, a new foreign and defence minister...and an extension*

of majority voting in 70 or more areas...it will result in a subordinate Britain...it takes us closer to a European superstate...The British do not want to belong to such a single European state; if we are forced to go in, we shall immediately start to look for the exit."

As the leading Birthright nation, joint inheritor with America of the mighty promises to Abraham, Britain should have been a beacon to the whole world and perhaps for a time it was. For this present day the Bible predicts that God will bring the Assyrians against a hypocritical nation who, in a third and final war, will be *"...trod down as the mire of the streets."*

Who else in today's world, other than Britain, could the nation spoken of so frequently in Hosea be? Genesis describes a nation that has given rise to a *"multitude* (or commonwealth) *of nations."* Hosea describes a nation that has contended with the Assyrians, both in ancient and modern times; it is Britain that has twice fought and defeated modern Germany. It was England that defied the *"Lady of Kingdoms"* to shake off her yoke of tyranny, splitting from Rome in 1533 to establish new freedoms for the ordinary citizen. *Ephraim* can be none other than Great Britain, the nation whose fortitude saved the world from Nazism.

In a reaction to the Irish 'No vote' of June 2008, the German journalist Wolfgang Münchau reflected, unwittingly perhaps, the true character of the European juggernaut. He found the result of the referendum *"shocking."* In his Financial Times article of June 16th 2008 entitled *Europe's hardball plan B for the Lisbon Treaty* he stated:

"...Europe's leaders have a plan B. It is not a pretty plan...what he [the German Foreign Minister Frank-Walter Steinmeier] is saying in effect is that Ireland should quit the EU...I know this appears to be in contravention of European law...the strategy most likely to be successful...is to play hardball. This is plan B."

Having lived for centuries under a parliamentary democracy where elected individuals spoke for their constituencies, the British Isles are now almost entirely ruled by unelected bureaucrats allied to the

Prussian fist. Displaying a bias familiar to anyone who has followed the European debate, this German writer further ruminated:

> *"It is difficult to formulate any specific concessions, since nobody knows what the Irish electorate wants."*

Perhaps what they want is to not be ruled by Germans. The 20th century's eruption of German nationalism in two wars spanning twenty seven years succeeded in dislodging Britain from its position as leader of the free world. It was a satanically inspired movement.

Nazism is marked with certain number patterns that counterfeit the truth. One such number is 66, the number of books in the Bible. Hitler's first bid for the chancellorship was in the *Putsch* (uprising) of November 9th 1923; this marked the beginning of a 66 year sequence of events.

It was 66 years *to the day* from the Munich Putsch to the fall of the Berlin Wall on November 9th 1989. Seven years *to the day* before that Wall fell, Pope John Paul II made the most important public pronouncement of his pontificate when on November 9th 1982 he exhorted Europe to *"rediscover her roots."*

That process of rediscovery referred to by Pope John Paul II is now well under way in a Europe which is *rooted in counterfeit Christianity*, inquisitions, war, tyranny and genocide. It will culminate in a Church/State combine, a *United States of Europe* empowered by a persecuting religious entity. Today Great Britain has taken the nascent German-dominated European beast power as a 'lover', but this will in no way secure her future:

> *"Wherefore I have delivered her [Israel[1], or Ephraim and Manasseh] into the hand of her lovers, into the hands of the Assyrians..."* Ezekiel 23:9

Prophetic utterances concerning 'Israel' therefore primarily refer to *Ephraim* (the younger but greater brother) and those who have come out of him, *not* the Middle-Eastern nation called 'Israel' today. Many Bible prophecies have been obscured through this missing knowledge. When the Bible speaks of 'Israel' in end-time prophecy, it is speaking of the English-speaking peoples, as in this description of their decline and fall:

"Son of man, the house of Israel is to me become dross...[as] the dross of silver...v20...as they gather silver...into the midst of the furnace, to blow the fire upon it, to melt it, so I will gather you in mine anger and in my fury, and I will leave you there, and melt you." Ezekiel 22:18

The Berlin Wall was erected in August 1961. For 28 years it divided the 'modern Assyrians,' effectively thwarting any overt revival of German nationalism. Since the Berlin wall came down, eastern Germany has provided a source of cheap labour, propelling Germany to the position of the world's number one exporter with the fourth largest economy. Their humiliation of 1945 is long forgotten.

For over a thousand years, since the time of Otto the Great, Germanic peoples have dominated the continent of Europe. Now they increasingly control the institutions of the EU. The new Constitution for Europe, cunningly repackaged by Angela Merkel as a treaty and waved through by outgoing Prime Minister Tony Blair, contains a little noticed 'ratchet clause' allowing the EU to abolish almost all remaining vetoes simply by 'notifying' British MPs. British Shadow Foreign Secretary William Hague lamented in August 2007:

"There is a declaration attached to the treaty that is supposed to protect our independence in foreign policy - but now it turns out that it is not legally binding."

Hitler would not have needed to go to war in today's Europe. All of his aims are being achieved today by stealthy takeover. In the same month, a leading European think tank, the Skeptika Group, pronounced that:

"...everything that was said about the new constitution [treaty] has turned out to be a lie."

Skeptika uncovered a paragraph, signed by Tony Blair as one of his last acts in office, in which Britain agrees to eventually hand over its seat on the UN Security Council to the EU. While the Foreign Office call this a 'myth', Skeptika say there would be a 'creeping agenda' where, beginning with less controversial issues where there was a common view, the EU would speak collectively to cover issues such as Iran, Iraq and Afghanistan.

Also in August 2007 Christopher Booker in the *Sunday Times* described the new 'treaty' as a coup d'etat:

> *"The attempt to ram through this treaty in 10 weeks is an immense new EU power grab...The new treaty formally sets up the body known as the European Council as the government of Europe...the new treaty makes clear [that] when the heads of government meet in council they are no longer to represent their own countries... their first loyalty will be to the EU...the Union will have power to shape and decide policy in almost every field...to take any powers...under a new version of Article 308...its new wording amounts to a blank cheque...powers to levy its own taxes...a government we cannot dismiss and which is unaccountable. It is nothing less than a complete coup d'etat."*

Is it coincidental that this takeover is happening under a German presidency of Europe? Churchill's harsh view of the Germans may be understood in the light of the times in which he lived. Yet he was a historian and had, uniquely, held high executive office during both world wars. Were Churchill still alive he would have been sounding warnings about German influence in the EU. The record of his discussions with Stalin and Roosevelt at the 1943 Tehran Conference, as described in book ten of his series *The Second World War* is instructive.

Although the three parties at that juncture had felt assured of eventual victory, there was still a considerable fear of subsequent German rearmament. Churchill felt that their duty was to make the world safe for at least fifty years through the foreign supervision of German factories, plus a prohibition of all German aviation. He anticipated a century of chaos if the Allies did not maintain a post-war military dominance. Stalin feared that Germany would foment a third world war within fifteen to twenty years, unless German factories were constrained.

The postwar German Constitution forbade the manufacture of armaments for export, inhibiting the development of a military-industrial complex. In 2009 the German government scandalously took steps to circumvent this restriction by relocating armaments production to the United Arab Emirates. The website German-foreign-policy.com reported in October 2009:

"The German ThyssenKrupp Corporation is cooperating in the production of warships with the Abu Dhabi state enterprise and has buried plans for a common European shipbuilding company. ThyssenKrupp announced it would renounce on civilian ship production to concentrate its dockyards solely on arms production. It is entering a "strategic partnership" with the Abu Dhabi Mar Co. from the United Arab Emirates (UAE). Their deal seals the military alliance between Germany and the Emirates, possibly creating the opportunity for circumventing German arms exports regulations and ending efforts aimed at forging a German/French ship production - under German leadership. Paris successfully repudiated German hegemony in warship production and therefore will now be confronted with the fait accompli. If a German dominated "European solution" cannot be accomplished, Berlin will do without "Europe" and go it alone."

In November 1934 Winston Churchill, speaking during his 'wilderness years' had given warning of the war-like Germans:

"Only a few hours away by air there dwells a nation of seventy million of the most educated, industrious, scientific, disciplined people in the world; who are being taught from childhood to think of war as a glorious exercise, and death in battle as the noblest fate for man. There is a nation that abandoned all its liberties in order to augment its collective strength. There is a nation that for all its strength and virtue is in the grip of ruthless men preaching a gospel of intolerance and racial pride, unrestrained by law, by parliament or by public opinion. It is only twenty years since these neighbours of ours fought almost the whole world, and almost defeated them. Now they are rearming with the utmost speed..."

On becoming Prime Minister in 1940 Churchill made the following resolution:

"...to wage war against a monstrous tyranny never surpassed in the dark and lamentable catalogue of human crime..."

Has there ever been any other nation like Germany? Today's legacy of their lust and ferocity is an unstable world menaced by nuclear weapons. But the Germans remain a paradox. They were the only nation ever to repent at the word of a prophet of God. The king of Nineveh, Assyria's ancient capital, commanded a nationwide fast when the prophet Jonah warned him of God's coming chastisement. The nation was spared. But Britain will not follow that unusual pattern, and the existence of Weapons of Mass Destruction today makes possible the fulfillment of many fiery prophecies concerning the nation of Ephraim.

Few people are aware of the scale of the destruction that could be wreaked by the most modern weaponry. When on June 27th 2007 Gordon Brown replaced Tony Blair as Prime Minister, his first task, after seeing the Queen and delivering a short inaugural speech outside 10 Downing Street, was the writing of a handwritten message to be read only by the commander of a Trident nuclear submarine. All incoming Prime Ministers must write such a letter following a briefing by the chief of the defence staff on the damage that can be caused by a single Trident missile. According to Peter Hennessy, Professor of Contemporary History at Queen Mary London University, Tony Blair *"went white"* at his briefing.

The writing is on the wall for the Anglo-Saxons and the letters are radioactive. The best laid plans of politicians will not determine Britain's fate because it has already been written. As mentioned in the earlier chapter *Decline and Fall*, Jesus Christ made this statement regarding Noah:

> *"For as in the days that were before the flood... Noah entered into the Ark...so shall also the coming of the son of man be."* Matthew 24:27-28

The appearance of a *giant figure of eight* over the Earth, the *Sign of Noah* centered on Rome, should alert all thinking people to the fact that the time table for the present age is nearing its end.

---

1. Again, prophecies regarding 'Israel' can be understood in the light of the following statement by the patriarch Israel, who said of his grandsons *Ephraim* and *Manasseh*: Genesis 48:16 *"...bless the lads...let my name [Israel] be named on them."*

40

ROME IN THE CROSSHAIRS

"The priest and prophet have erred through strong drink…
…all tables are full of vomit." Isaiah 28:7-8

An explanation is needed concerning the eight power cuts of 2003 described in the *Introduction*. What is the claim of a giant figure of 8 based on? Here is the explanation of the power cuts which occurred in the following order:

1. New York
2. Georgia
3. Helsinki
4. London
5. Kuala Lumpur
6. Cancun
7. Sweden & Denmark
8. Rome

A first attempt to join these by lines on a map, starting from New York, in the order in which they occurred reveals no clear pattern but only the shape below:

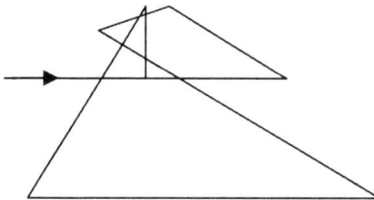

A better pattern might emerge via a clue, rule or principle in the manner seen in crossword puzzles. The data themselves could possibly provide ideas for this. It can be observed that if power cut number 1 is connected to power cut number 2 by a straight line, it passes through Rome, which was power cut number 8. This suggests the following method:

"Link points by ascending number order, passing through any other higher numbered points on the way such that each is passed through at least once."

Using this procedure of passing through points on the way, in combination with the use of a curve, a figure of 8 is formed. Of all eight points, only Rome is intersected twice, as if in the cross-hairs of a rifle's telescopic sights!

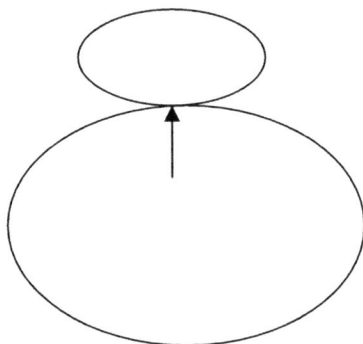

Because Rome is passed through twice to form a pattern of 8, that gives by inference three 8's or 888. But then consider the place numbers passed:

New York	1
Rome	8
Georgia	2
Helsinki	3
Sweden/Denmark	7
London	4
Rome	8
Kuala Lumpur	5
Cancun	6
Place numbers total	*44*

Each place is passed over once, except Rome which is counted twice, resulting in a total of 44. This finding reinforces the existing pattern, that of the 44 day fast of David Blaine overlapping with the 44 days of power cuts. What could a giant figure of 8 possibly mean? Eight is the number of redemption and new beginnings. There were eight people on the Ark including Noah:

> "...wherein few, that is, eight souls were saved by water." I Peter 3:20

And also:

> "But God saved Noah the eighth person..." 2 Peter 2:5

Regarding Noah, Christ made this statement:

> "For as in the days that were before the flood... Noah entered into the Ark...so shall also the coming of the son of man be." Matthew 24:27-28

As in the days of Noah, today's world is brimming over with wickedness. The name of Noah is marked by the number 8. In a book first published in 1863 entitled *Palmoni: or, the Numerals of Scripture* (the book that had inspired Bullinger to write *Number in Scripture*) its author Dr. M. Mahan stated on page 73:

> "Taking, then, the letters of Noah's name...the longer spelling, namely, the word that means "comfort", we get sixty-four, that is, eight times eight."

Thus the name of Noah has a gematria of 64 (or 8^2) which can be understood as 8 *concentrated*. Therefore it is not surprising that the time of the second coming of Christ is marked in some way by the number 8. This number also has significance within pi as its first 8 digits total 31, the deity number.

Furthermore, the number 31 featured in an event that was underway in Rome during the final power cut. By switching off the power in Rome and disrupting a ceremony in which the Pope[1] was ordaining 31 cardinals, God demonstrated who the real Holy Father is. *He did it with the eighth power cut: 8 disrupted 31.* The true Father trumped the false by plunging him and his cardinals into darkness, reminiscent of the plagues of Egypt, in the following manner:

TRUE FATHER SENDS:	FALSE FATHER SUFFERS:
8th and final power cut	**Counterfeit church targeted**
True Father is number 11 (31 is the 11th prime)	**The Pope's 31 new cardinals are plunged into darkness**
A worldwide figure of 8 (The number of Noah)	**Rome is in the cross hairs, a new beginning is imminent**

The Pharaoh of Egypt had been plunged into darkness in the penultimate ninth plague (Exodus 10:22) on the land. To the Catholic Church in 2003 it was as if God was saying:

"You are preparing your government to carry on your vain traditions in perpetuity, but I am preparing a government to displace you and the starting gun has fired."

That future God-ordained government is referred to by the prophet Daniel:

"And many of them that sleep in the dust of the earth shall awake, some to everlasting life, and some to shame and everlasting contempt. 3 And they that be wise shall shine as the brightness of the firmament: and they that turn many to righteousness as the stars for ever and ever." Dan. 12:2

An aspect of this future government was further revealed by Christ in the *vision* (as he called it) of Moses and Elijah described in Matthew chapter seventeen. Evidently Christ will have his own 'cardinals' and none will have been trained by Rome.

It should be obvious that God loathes the system of idolatry perpetuated by this Church, to be destroyed at Christ's return:

"For as in the days that were before the flood they were eating and drinking, marrying and giving in marriage, until the day that Noah entered into the Ark, 39 And knew not until the flood came and took them all away: so shall also the coming of the Son of man be...42 Watch therefore: for you know not what hour your Lord doth come." Matt 24:38-42

It should be obvious that the European combine now possessed of a currency, an anthem, a flag and a largely integrated legal and governmental system is the seventh and final resurrection of the 'Holy Roman Empire.' This European colossus even displays a highly meaningful Bible symbol at its seat of government. It is a statue of the woman *Europa* riding a bull. This is a symbol of a powerful church, the "Lady of Kingdoms" referred to by the prophet Isaiah:

> *"...O virgin daughter of Babylon...v3 Thy nakedness shall be uncovered, yea, thy shame shall be seen: I will take vengeance...v5...thou shalt no longer be called the Lady of Kingdoms...v7 And thou hast said, I will be a lady forever...v8...thou that art given to pleasures...v9...for the multitude of thy sorceries ...v10 For thou hast trusted in thy wickedness...thy knowledge, it hath perverted thee... v11...desolation shall come upon thee suddenly..." Isaiah 47:1*

Which church could possibly match all of the characteristics ascribed by scripture to this great false church, a woman described as a whore?

> *"...she shall be utterly burned with fire, for strong is the Lord who judgeth her." Revelation 17:8*

Who could deny in the light of the evidence that the power cuts of 2003 signalled something of great importance? But there is one seeming anomaly regarding the 44 day fast of David Blaine and the overlapping 44 days of power cuts: why were they not simultaneous? The overlap of the fast with the power cuts was exactly 50:50 as shown below:

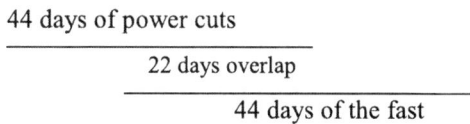

44 days of power cuts
_____

22 days overlap
_____

Figure 1.   44 days of the fast

There was a reason for this overlap: it defined a total of 66 days. The first 22 days can be counted from the evening of Thursday 14th August (New York) to Friday the 5th of September at 9.30 pm (David Blaine begins). The next 22 day period ends on Saturday

273

evening the 27th of September at 9.30 pm. That Saturday was the Day of Trumpets – the centre of the 1711171 pattern:

Day of Trumpets

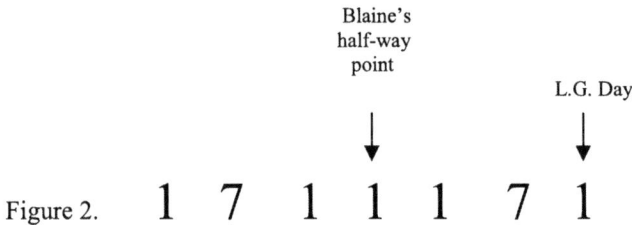

Blaine's
half-way
point

L.G. Day

Figure 2.   1 7 1 1 1 7 1

The end of the pattern was Blaine's finishing point. He emerged on Sunday the 19th of October at 9.30 pm. The Last Great Day was Saturday 18th, which was the last '**1**' of the pattern.

Rome was smitten with a power cut beginning at 3.30 am on Sunday morning, just about six hours after the sunset close of the *Day of Trumpets*, picturing the day of Christ's return. This was the halfway point for David Blaine. *Fasting can be a form of petitioning God!* Unwittingly, Blaine, the 30 year old Jew so physically similar to many people's idea of Jesus Christ, and the same age, acted out a powerful message in Figure 2. His second 22 day period encapsulated the second half of the 1711171 pattern therefore, by implication: *his whole fast represented the complete pattern.* It was as if he was unwittingly saying: *"Shut down Rome, come back to Earth and reinstitute the true religion, 1711171."*

God has plainly marked out with the number 8 the baleful institution of the Catholic Church - as if in the cross hairs of a sniper's rifle - for total destruction in the coming world crisis.

1. The magazine *Crown and Commonwealth*, summer 2005 edition, calculated that the life span of Pope John Paul had lasted exactly 31,000 days. He was ordaining 31 cardinals when the lights went out in 2003 during 'Rome's Darkest Day,' the last of eight inexplicable power cuts spread across the globe.

41

THE SICKNESS
"...a month shall devour them and their portion..." Hosea 5:7

The 2003 attack by the US on Baghdad was billed as *"Shock and Awe."* TV viewers around the world were treated to a fireworks show of 1,000lb bombs detonating as desperate Iraqi mothers cowered in pantries and under staircases, clutching their children to them.

Those events seemed quite distant as General Petraeus delivered his September 2007 evaluation of the Iraq war to Congress, but that month brought an event that bridged the four year gap: the spectacle of depositors queuing in the streets of Britain's major cities, a phenomenon not seen since 1866. The Northern Rock was a building society that 'borrowed short and lent long.' It was permitted, by a complacent Financial Services Authority, to jeopardize the entire UK banking system, resulting in sending reverberations around the world.

Queues were in place as the Sun rose on Friday 14th September. Repeated assurances from the Chancellor of the Exchequer failed to calm investors' fears over the weekend. That weekend the *Mail on Sunday* headlined: *"Police called to break-up Northern Rock panic queues as customers withdraw Millions."* In one incident, police went to a branch in Cheltenham, when two joint account holders had barricaded the bank manager in her office after she refused to let them withdraw £1 million from their account.

On Monday evening a fateful decision was made: the Government was to guarantee all twenty-four billion pounds of depositors' money, superseding the limited safeguards of the regulators. This desperate volte-face opened a bottomless pit of moral hazard and

275

uncertainty. The tax payer's cheque book had become the principal tool of bank regulation.

The following day the queuing increased. Finally, after further clarity was added to the authorities' position that evening, calm prevailed on Wednesday 19th. This hitherto unprecedented historic 'run on the banks' began in the night hours following the *Feast of Trumpets 2007*. The UK Government was shown up to be as impotent as the Pope had been during the night hours after the *Feast of Trumpets 2003*, when the ordination ceremony for 31 new cardinals was plunged into chaos by the biggest power cut in history - the last in the worldwide series of eight. Consider again how this relates to the 1711171 pattern:

Feast of Trumpets (picturing the Return of Christ)

$$\downarrow$$
$$1 \ 7 \ 1 \ 1 \ 1 \ 7 \ 1$$

These developments bring *The Signal* of 2003 forward by four years. This new manifestation of *The Signal* makes it a 2003/7 phenomenon. The time gap is, by the reckoning of the holy day system 1711171, four years precisely. Four is the 'number of power.' This event affecting the financial system of the world is another display of power 'from someplace.' It was the 'power cut' later referred to by Prime Minister Gordon Brown in March 2009 when meeting with the newly elected President Obama.

In 2003, the financial and religious power centres of the world New York and Rome had in effect been told: *"You think you are in control, but you are arrogant, ignorant and wrong."*

There is actually a precedent for the way God deals with the proud and boastful of the Earth. Thousands of years ago Nebuchadnezzar, king of Babylon and the mightiest ruler of his day, was on what today could be termed an ego-trip:

> *"Is not this great Babylon...built...by the might of my power and for the honour of my majesty?" Daniel 3:30*

Nebuchadnezzar had been warned in a dream that something very humiliating would befall him if he didn't change his approach:

"I saw in the visions of my head upon my bed, and, behold, a watcher and an holy one came down from heaven; he cried aloud and said thus, Hew down the tree, and cut off his branches, shake off his leaves, and scatter his fruit: let the beasts get away from under it, and the fowls from his branches: Nevertheless, leave the stump of his roots in the earth, even with a band of iron and brass...let a beast's heart be given unto him..."

The proverbial madness of Nebuchadnezzar lasted seven years during which he crawled naked on all fours eating grass, his hair and nails hideously overgrown. At the end of this period his senses returned to him, when he said:

"...I Nebuchadnezzar lifted up mine eyes unto heaven, and mine understanding returned unto me...all the inhabitants of the earth are reputed as nothing: and he [God] doeth according to his will...all those that walk in pride he is able to abase."

Modern Britain and America have been proud nations. In pride they marched together in 2003 into a self-righteous battle to 'depose a dictator' when they have been, and remain, the friends of many dictators. They proclaim their righteousness in prayers and songs, with politically correct mantras and socialist feel-good handouts. But Britain appeared bottom out of twenty-one advanced nations in a U.N. sponsored survey to determine the happiness and wellbeing of its children. This depraved country's senior citizens are robbed and humiliated by street hoodlums, the care home system and the government.

Britain's leaders assured their electorate that they had abolished 'boom and bust' while selling off gold reserves to Frankfurt. They inflated the economy on 'stealth' taxes and money from home equity release, unleashing the 'mother of all property bubbles.' They of course *knew* that weapons of mass destruction existed in Iraq. Then everything in the UK, from schools to dentists' surgeries had to be 'reformed' causing costs to rocket and standards to decline. They intervened in Afghanistan, only to start a war that cannot conceivably be won. And Britain's 'New Labour' politicians have created regional assemblies that seem able to destroy 300 years of The Union between Scotland, England and Wales. They then signed away what little is left of the UK's sovereignty to the EU without a promised referendum.

Yet politicians were mute when speculators on both sides of the Atlantic hoarded homes they didn't need, dispossessing the next generation of their own countrymen. If there is anything needing government oversight it is the nation's housing needs. In the UK, 400,000 new landlords were created under New Labour in a nation of only 61 million. Over the same period in the USA, the proportion of homes bought for 'investment' shot up from 10% to 28%. Ludicrous sums were lent to buyers with little job security. Strings of homes were snapped up by speculators without proof of income. But even as the transatlantic property bubble inflated, the fortunes of the English speaking peoples had changed decisively.

In 2003, politicians on both sides of the Atlantic had been hyperventilating in their eagerness to sell an illegal war. They were to find that the *"pride of their power"* (Leviticus 26:19) was to be broken. As explained in the chapter *Samaria on Thames,* these nations are mentioned in prophecy. The name of the patriarch Israel, where his grandsons *Ephraim* and *Manasseh* are collectively labeled 'Israel' (*"Let my name be named on them"* Genesis 48:16), is prominent in prophetic passages regarding the end times:

> *"The pride of Israel [Ephraim and Manasseh] testifieth to his face: and they do not return to the Lord their God...Ephraim also is like a silly dove without heart...they go to Assyria [Germany]. When they shall go I shall bring them down as the fowls of the heaven; I will chastise them as their congregation hath heard."* Hosea 7:10-12

This prophecy indicates that, at a time soon approaching, the British will look to Germany to save them from economic ruin. When Britain has once again become the sick man of Europe they will implore the wealthy Germans to rescue them - but to no avail:

> *"...now shall a month [a new moon] devour them with their portions [wealth]...When Ephraim saw his sickness and Judah saw his wound then went Ephraim to the Assyrian, and sent to king Jareb..."* Hosea 5:10 -13

278

In the above passage, 'portions' refers to wealth. Both the UK and the State of Israel (Judah) - a country created by Great Britain - are to suffer together at the end of the present age.

The Unfolding Plan

Tensions in the international financial system had been building up during 2007, but the Northern Rock was indisputably the *catalyst* setting off a state of panic. Economics is only one part of the picture. In prophecy, the fate of the English speaking people is also tied to that of the modern State of Israel. These are the Jews, identifiable in the prophecy of Hosea (above) as *Judah*. Britain (Ephraim) is to suffer a sickness and Judah (in modern times the State of Israel) a 'wound.' The State of Israel and Britain will fall together:

> "...Israel [Ephraim and Manasseh] hath forgotten his Maker...Judah hath multiplied fenced cities: but I will send a fire on his cities, and it shall devour the palaces thereof." Hosea 8:14

In 2007 we saw the completion of the wall – a fence - separating the State of Israel from its enemies, just as *Hosea*, above, describes. The world's attention, when not distracted by economic catastrophes, is frequently riveted on a tiny patch of land by the Mediterranean Sea. It is increasingly clear that the whole world will be affected by events and conflicts centered on the State of Israel. The UK is drawing closer to that nation. In his July 2008 visit to Israel Prime Minister Gordon Brown described the Iranian President's call for *"Israel to be wiped from the face of the earth"* as totally abhorrent. He pledged an *"unbreakable partnership"* with Tel Aviv.

At this critical time, by way of the proofs shown in this book, it is fitting to restore the Bible to its rightful place as the premier book in all human history. It has no peers; it is quite simply incomparable to any other writings. It foretells the rise and fall of kingdoms, records the history of the world - principally from the perspective of God's chosen nation Israel (all of the twelve Israelite tribes) - and predicts the eventual emergence of an immortal spiritual nation derived from all human tribes.

The plan encoded within the Bible by which that purpose will be accomplished continues to unfold regardless of human schemes and machinations, and irrespective of the satanic counterfeits in false religion that have blinded the whole world. The events of 2003, and four years later in 2007, have shown that the one whose *power is numbered as 4* is firmly in control. After four years and with perfect timing this power has jabbed a cosmic screwdriver into the wheels of the world economy. This he is easily able to do:

"Behold, the nations are as a drop of a bucket, and are counted as the small dust of the balance...all nations before him are as nothing...he that sitteth on the circle of the earth, and the inhabitants thereof are as grasshoppers...that stretcheth out the heavens as a curtain...as a tent to dwell in...behold who has created these things...he calleth them all [stars] by names by the greatness of his might..." Isaiah 40:15

Planet Earth is a mere speck of cosmic dust. God orders the affairs of man with consummate ease.

Grave errors

The most severe credit crunch in living memory struck terror into policymakers in the USA and the UK from September 2007. Europe had shown resilience until it was revealed that the projected fall in GDP for Germany during 2009 was expected to exceed even that of the UK at 6%, the biggest drop since the war.

But the social evil of inflated housing cost is something that the Germans have effectively organised themselves against for decades, enforcing the rule that only 60% mortgages will be offered against residential properties. Unlike the UK, Germany has not relied on property bubbles to stimulate an otherwise sluggish economy. However, the countries of the EU did not escape damage in the gathering debt storm. Smaller East European nations such as Latvia were found to have loaded up on consumer credit denominated in Euros, only to find themselves fighting to service this debt in their own weaker currencies. Such was the scale of the problem that eventual membership of the eurozone could be the only way to avoid mass default.

Following the run on the UK's Northern Rock, the American Federal Reserve's immediate response of a one half percent interest reduction had looked like panic from an institution terrified of recession, or worse. That cut was *"a grave error"* and *"playing with fire"* according to Jim Rogers, former partner to George Soros ('the man who broke the Bank of England' forcing it out of the ERM in 1992). Soros himself considers that the dollar is set to collapse as well as the US Bond market. According to the Merrill Lynch September 2007 Survey of Institutional Investors there was more pessimism than following the 9/11 attacks, with spillover into the wider economy considered inevitable.

Debt Held by Foreigners

After Northern Rock, threats to the US economy suddenly appeared overwhelming. Saudi Arabia refused to cut interest rates in September in tandem with 'Helicopter Ben', as the beleaguered Fed chairman Ben Bernanke had been nick named in 2002 after a speech in which he'd expressed faith in Quantitative Easing (money printing, in effect) as a cure for deflation. The Saudi event highlighted the ever present danger that Gulf kingdoms, tired of importing US inflation, may soon break their peg with the dollar.

If and when that happens, there could be a stampede out of a currency already at a fifteen year low. Vast capital inflows covering America's monumental trade gap would be likely to dry up. Long term borrowing costs would then rocket and the standard of living nosedive. Financial Times economist Andy Xie, looking back on the crisis in April 2009, made a prognostication for the dollar:

> *"If global stagflation takes hold, as I expect it to, it will force China to accelerate its reforms to float its currency and create a single, independent and market-based financial system. When that happens, the dollar will collapse."*

The winners would be the oil producing states and former US Defense Secretary Donald Rumsfeld's 'Old Europe.' But that old Europe had already become the mighty EU with its Euro currency, a new Deutschmark in all but name, administered from Frankfurt. The obvious losers would be the English-speaking nations. An ancient prophecy concerning the rebellious Anglo-Saxons is gradually coming true:

> *"I will break the pride of your power; and I shall make your heaven as iron, and your earth as brass." Leviticus 26:19*

By this point the prosperity of the English speaking peoples was in serious jeopardy.

A New More Dangerous Phase

Events soon took a dramatic turn. A few days after the collapse of the Wall Street giant Bear Sterns, described as America's Northern Rock by some, Gary Duncan writing in the March 17th 2008 edition of *The Times* under the title *"Time to put a break on the dollar's decline"* observed the following:

> *"The global credit crisis and the US economic upheavals that are driving it appear to have entered a new, more dangerous phase...(there is) a scary escalation of events...the bleak reality is that, for all the unprecedented aggression of the Fed's recent action, it is more mere palliative than panacea. The efficacy of monetary policy to quell the present turmoil, and to break the vicious downward spiral into which the US economy is locked, is more and more in question. Across the United States house prices continue to slump..."*

The deep and excessive cuts in American interest rates, a tactic repeatedly used and abused by the previous Fed incumbent Alan Greenspan, had only served to delay the big shake-out. By 2008 the dollar was perceived to be so weak that, in order to protect themselves from the much feared hike in inflation, traders were for the first time willing to lock into a negative return using US index-linked Treasury Bonds. The Fed's panic moves on interest rates will most likely be counterproductive, as *The Times* explained:

> *"The materialisation of a long-dreaded dollar collapse would be catastrophic. It would undoubtedly trigger a crash in the equity and bond markets. A huge sell-off of US Treasury bonds by foreign investors would undercut the Fed's policy, driving long-term market interest rates through the ceiling. A severe world recession would become unavoidable."*

At the end of June 2008 former US Federal Reserve Chairman Alan Greenspan stated that America was on the brink of recession (defined as two consecutive quarters of negative growth). Property in major cities had experienced declines of 20% - 30% in a year, led by Las Vegas, Miami and Los Angeles. US house prices were falling on a national basis for the first time since 1938. Los Angeles pawn shops were booming.

Meanwhile in the UK, auction houses witnessed thousands of buy-to-let investors dumping flats at little more than half of the prices paid two years previously. According to analysts, house prices were falling faster than at any time since World War Two. The IMF subsequently estimated that houses in the UK could be up to 40% overpriced. Most people's homes were losing value faster than their occupants could earn, with a loss on the average home of seven hundred pounds a week in value during the first four months of 2008. By August the rate of loss still ran at six hundred and fifty pounds per week.

By June 2008 the Halifax housing index showed an unprecedented recorded drop of 8.5% in the price of the average house in first six months of the year. The recorded fall in the three months April-June was a massive 6.1%. The Centre for Economics & Business Research in June was projecting an 18% plunge in the value of a house worth £216,000 in January, by the end of the year. No UK expert could see any bottom to the domestic property market. In one desperate move, an estate agent offered the gift of a £55,000 car as an incentive to buy. A Citygroup analyst concluded that *"We are in the worst housing slide for fifty years."* A Bank of England official and member of the interest rate-setting committee, Professor David Blanchflower, warned in July 2008 of a possible further drop of 30% in UK house prices. At the same time Citibank, the world's largest bank, warned that the fall in prices could be worse than that of the Great Depression. These unprecedented doom-laden pronouncements were heard a mere ten months after the Northern Rock debacle.

The Times of March 13th 2008 warned that *"One false step could precipitate an unthinkable chain reaction of banking failures."* The sober judgment of Sir William Rees-Mogg writing in *The Times* was that *"The 2007 to 2008 credit crunch has undermined the monetary base of the world economy."* The beginning of 2009 brought calamitous news from around the world providing a dramatic backdrop to the January discussions at the World Economic Forum, the annual event held in Davos, where discussions came *"close to common abuse"* according to *The Times*.

Bankers from Barclays and Merrill Lynch pulled out at the last minute. Goldman Sachs, famously described by online magazine *Rolling Stone* in July 2009 as a *"vampire squid sucking the face of humanity"* cancelled a drinks party. But champagne flowed freely at dozens of other parties around the town as delegates dreamt of *"Shaping the Post-Crisis World."* For bonus-snatching bankers it would soon be business as usual.

Four Years since Rome

It is worth reflecting once more on the timing of the unfolding crisis. Queues first appeared outside of the Northern Rock in September 2007, in the hours following the *Day of Trumpets*. This was a period of four annual cycles precisely since Rome was blacked out in 2003, according to the Holy Day system (the 1711171 pattern). It was both a symbol of and a catalyst for a unique crisis. From that exact point in 2007, the Anglo-Saxon world began a toboggan slide towards economic hell.

In the US, the shares of General Motors, the world's largest car manufacturer, hit a 54-year low. The closely followed Case-Schiller report talked of the *"biggest destruction of household wealth since the thirties."* By June 2008, UK savings had slumped to the lowest level in fifty years. In early July, *The Times* in an article entitled *High Street hit by full force of crunch* gave vent to fears of *"a severe slump."* The chairman of *Marks & Spencer*, a UK store regarded as a bell-weather for the economy, had never seen *"such a sharp and continuous downturn"* during his entire career. The seasoned UK economic commentator Roger Bootle predicted that UK unemployment figures could jump by a million. Stock markets on both sides of the Atlantic had caught up with reality and were in confirmed bear territory by July.

Yet it is a peculiar fact that dire economic developments have a way of unfolding as if in slow motion. All recessions and depressions proceed in much the same way: a daily dripping of bad news with periodic crises and sudden vertiginous plunges in values. This lagging effect was seen in the diversion of savings in favour of consumption during 2008. In the UK this ploy served to prop up faltering economic growth, delaying the technical onset of recession and postponing what had already been billed by some commentators as an approaching 'Armageddon.' But according to *The Times'* Economics Editor, City economists were sounding warnings that this virtual disappearance of personal savings *"raised the threat of an even more vicious economic downturn."*

Not only were savings drying up, but the unprecedented figure for UK personal debt of £1.4 trillion was ballooning exponentially by mid-2008. Personal indebtedness was at that point rising at the rate of a million pounds every five minutes, or £900 million each month. Unchecked, this astounding trend would lift personal debt in the UK from £1.4 to £2.5 trillion (£2.5 thousand billion) in only twelve months! Government borrowing was also out of control at this point, with tax receipts falling sharply. In July 2008, a *Times* headline trumpeted: *"Economic gloom deepens as public borrowing hits highest since Second World War."* A few days later the Ernst & Young Item Club suggested Britain faced *"an economic horror movie."*

Betting Against the USA

With catastrophe bearing down like a tidal wave on the UK, across the Atlantic a seismic tremor struck on July 11th. This event called into question the sovereign AAA credit rating of the USA itself: the share prices of the two largest 'building societies' (Savings & Loan institutions) in the USA, holding debts of $5.3 trillion, fell by 50% during intra-day trading. Panicked experts agreed that these institutions must be supported at any cost. Writing in *Emirates Business* 24/7 Frank Kane wrote:

> *"These are cataclysmic declines, reminiscent of the worst days of ...1929. The US financial system resembles a ship that has sprung a hundred leaks, each one of which is capable of sinking the vessel. The regulators are scrambling around desperately trying to plug the holes, but are in danger of being overwhelmed...it is difficult to avoid the conclusion that the troubles of Freddie and Fannie could be the two decisive holes below the waterline."*

285

At that point, the value of the loans on books of 'Fannie & Freddie' far exceeded the sale prices of properties mortgaged with them. Standing as guarantor for mortgage bonds sold on to institutions, these two 'S&Ls', the two largest financial institutions the world has ever known, could only expect to see a further deterioration in the quality of their loan portfolios as house prices continued to slide. By the end of March 2008 Freddie had, according to *The Economist*, a 'fair value' net worth of *minus* $5.2 billion. That position had arisen because the institution benefited from, and had for years been abusing, an implicit government guarantee. The July 19th *Economist* reported at the close of an article entitled *End of illusions* that traders in the credit-default market had begun placing bets that the American Government may default on its debt.

On July 11th there had been a further hammer blow that mirrored the calamitous Northern Rock bank run. In the prosperous Californian city of Pasadena police were called to control early morning queues forming outside of IndyMac Bancorp. The scary thing about this development was that the Federal Authorities had compiled a list of banks at risk, but IndyMac was not on it. The bank was instantly taken over by the Federal Reserve. It was the third largest Bank to fail in US history.

On September 7th the markets finally got what they wanted: the bailout of Fannie and Freddie. At an unknown cost to tax payers, but thought to be up to $100 billion (later estimated at twice that figure), Secretary Paulson became 'history's greatest communist' by effectively nationalising these two vast loan companies.

Mysterious Worldwide Forces

The 81 year old Sir William Rees-Mogg, anticipating the nationalisation of Fannie and Freddie, brought the benefit of his experience to a mid-July *Times* article entitled *Recession could easily tip into a depression*. Ruminating over the fact that unfolding events were always proving to be worse than anyone expected, he commented in respect of Fannie Mae and Freddy Mac that *"you cannot save six trillion dollar institutions without suffering on a large scale."* His conclusion was a sobering one:

"There is now a momentum of negative events sweeping away financial flood defenses; in the 1930s that force overturned democratic governments as easily as it overturned banks."

Within two days of Rees-Mogg's comments, the bourses of Karachi, Lahore and Islamabad were ransacked and burned by investors incensed by the bursting of Pakistan's stock market bubble. And in the political sphere, developments during 2008 in continental Europe continued to alarm all freedom loving people. Italy reappointed Silvio Berlusconi as Prime Minister to head the most right wing administration in Europe since the war. In June 2008 it was announced that Italy's estimated 152,000 gypsies were to be finger-printed in a racially defined and broadly popular policy. This move of the Berlusconi government provoked comparisons with that of Mussolini in the 1930's.

Meanwhile, by 2008, Germans seemed sufficiently comfortable in their own skin to install a wax work of Hitler in a newly opened Madame Tussauds by Berlin's Brandenburg Gate, only to see his head ripped off by a protestor on opening day. That almost comical protest not withstanding, the economic bell weather of rising anti-Semitic sentiment was fast rising in Germany.

In the USA the mood was less jocular. The Washington Post in July 2008 said of the advancing economic calamity:

"There is a general anxiety that we are in the grip of mysterious worldwide forces" (emphasis added).

Fraud

August brought a revelatory moment when America's New York's Attorney General announced plans to charge the nation's largest bank Citigroup with fraud and the destruction of subpoenaed documents. It is suggested that they falsely represented auction-rate debt as a safe cash equivalent.

How can the USA a nation shake off the perception that they have brought the world economy to the brink of ruin through fraudulent dealing? Deceit had been at the root of the growing crisis. Mortgage fraud was endemic from the street corner loan shark right up to the 'see no evil' executive on the boards of the biggest companies.

Writing in *The Times* July 15th Gerard Baker in the *American View* said:

> *"Fannie and Freddie morphed over the years from being a benign source of support for the American housing market into nothing short of a racket...the Government was indirectly subsidising the US housing market through interest rates that were lower than they would otherwise have been...the bill for that massive subsidy is finally coming due."*

Dan Atkinson of the *Mail on Sunday* described the crisis in this way:

> *"Like spendthrifts shifting debts from one credit card to another, the Anglo-Saxon economies are on the fast track to ruin...the Anglo-Saxon model of turbo-charged finance...has crashed into insolvency on both sides of the Atlantic..."*

Such dramatic events, amounting to *"An economic natural disaster"* to quote the Prime Minister of Iceland Geir Haarde, were in a sense pre-determined. The prophetic aspects of the Grid, as explained in the earlier chapters *The Prophetic Grid* and *Decline and Fall*, underpin already well understood and unerringly accurate prophecies concerning the Anglo-Saxon tribes, as explained in the chapters *Germani*, and *Samaria on Thames*. Britain must again become the sick man of Europe as America goes into a huge economic depression, turning once again to protectionism. By 2008, trade conflicts between the USA and the rest of the world were already growing (witness the July cancellation of a massive US defense contract with Airbus in favour of Boeing).

Once in a Century

Then in September 2008 tumultuous events that would have seemed impossible weeks earlier became inevitable. In the words of Alex Brummer, City Editor of the Daily Mail:

> *"It is as if an Ian Fleming-style baddie had grabbed the controls of the Anglo-Saxon economies and switched them off."*

Terrifyingly, the banking system had started locking up even before any recession had been formally announced. 'Meltdown Monday' was the stock markets' 15th September reaction to the bankruptcy of

Wall Street giant Lehman Brothers, was according to the 82 year old Alan Greenspan:

"A once in a century event ...outstripping anything I've seen."

The front page of *The Times* reported one senior banker in London as saying:

"The world is on the brink. The market is puking all over us. There's no capital left in the world."

The following day US Treasury Henry Paulson used $85 billion of treasury money to rescue the once mighty insurance group AIG, by then tainted over the February prosecution of former executives for fraud. AIG had been the major insurers of sub-prime mortgage debt. Immediately following that - the first American nationalisation of a financial company since the Great Depression - the FBI launched an enquiry into all four institutions: AIG, Lehman Brothers, Freddie Mac and Fannie Mae. The suspicion was that ratings agencies had been encouraged to inflate those institutions' credit ratings for higher fees, keeping the party going. Separately, senior Fannie Mae executives were named in an investor lawsuit alleging that the share price had been illegally boosted by false claims about its financial position.

Untrammeled Power

If all these developments seemed apocalyptic, what was to follow staggered the imagination. After what *The Times* called *"The biggest fire sale of financial companies since the Depression"* world stock markets had swung wildly. On September 19th, in a move described by Anatole Kaletsky of *The Times* as *"a swaggering demand for untrammeled power"* the US Treasury Secretary Henry Paulson submitted to Congress a three page demand for a $700 billion fund. This was to act as a financial landfill for the toxic 'sliced and diced' debt weighing down the balance sheets of banks worldwide. The previous day, members of the Senate Banking Committee had, according to *Emirates Business 24/7*, been shocked into uncharacteristic silence when briefed on the catastrophic potential of the meltdown by Federal Reserve Chairman Bernanke:

"There was dead silence...the oxygen went out of the room."

A few days later, at an emergency White House meeting of congressional leaders President Bush remarked:

"If money isn't loosened up, this sucker could go down."

If it was not clear which 'sucker' the President was referring to *The Times* of London clarified the matter:

"President Bush spelt out what was at stake. According to senior advisors briefed on the detail of the meeting, he told the assembled leaders that it was not just the US financial system that was in peril. He made it clear that the entire architecture of the global economy was at stake."

After one abortive vote in Congress and the enlargement of Paulson's plan by a further 450 pages, the $700 billion was agreed within days. But it did not stop the panic. *The Times* lamented that Henry Paulson was to the world of finance *"what Donald Rumsfeld was to military strategy."* The shareholder wipeout following Paulson's nationalisation of Fannie Mae had *"triggered unintended consequences around the world, resulting in the death-spiral of financial values."* There was no longer any incentive to buy shares in a bank. Banks were going down the plughole fast and many Governments had no choice but to buy them up, or guarantee their deposits. This they were reluctant to do. Of the EU nations, the Irish Republic jumped first, promising to honour all bank deposits by way of state funds. Other countries quickly followed suit in a move considered by many to be bordering on the immoral. Looking back on the crisis Martin Wolf of the Financial Times in March 2009 observed:

"The crisis has broken the American social contract: people were free to succeed and fail, unassisted. Now, in the name of systemic risk, bail-outs have poured staggering sums into the failed institutions that brought the economy down."

Total Capitulation

By October 8th 2008 The Times was describing the crisis as:

"unfathomably massive... (It) has metastasised into an infection that has the potential to wipe out wealth in every corner of the globe."

There then followed an audacious announcement from the UK that overshadowed the Paulson plan. *"£16,000 EACH"* headlined the *Daily Mail* as Prime Minister Gordon Brown announced a plan with contingency liabilities for the UK taxpayer totalling £500 billion, a greater sum than Paulson's scoop to rescue a country with five times the population! Leading banks RBS and HBOS were, in effect, nationalised immediately. Yet the banking system continued to malfunction. Within days of Brown's announced package UK parliamentarians protested that small businesses were beginning to see their overdrafts cancelled. At the same time, spiraling business loan costs were reaching credit card levels. The markets reacted badly to the package with leading indices plunging 8% or more in a day, only to bounce back the following day and collapse again the next. *"The banking sector has imploded ...everyone is running scared"* remarked one London based equity strategist. *"Total capitulation"* observed another: *"so much worse than the 1987 crash...fundamentals don't count anymore."*

An Economic Pearl Harbour

If the renowned Warren Buffett can pronounce, as he did, an Economic Pearl Harbour before recession has even had a chance to bite, then this must indeed be Alan Greenspan's 'once in a century event.' The count for new jobless claims in the US exceeded 600,000 for three months in a row, from December 2008 to March 2009. The Dow Jones index lost 25% of its value in the first two months of 2009. The Authorities revised their projections almost monthly. The IMF was predicting an annual fall in GDP for the UK of 2.8% in January. By March of 2009 that was revised to 3.8%. An editorial of The Times commented that

> *"Few foresaw how complete would be the rout of the Western financial system."*

Commenting on the need to stimulate the world economy by coordinated action between governments the *Financial Times* considered that this was:

> *"probably the global economy's last chance to avoid a Depression."*

But in March a new factor was introduced by policymakers: Quantitative Easing. This was much the same measure as the 'seigniorage' deployed by Germany's Weimar Republic in 1921-23. The Weimar government had avoided tax increases to pay war reparations by invoking seigniorage - money printing - for 50% of planned spending. This resulted in hyperinflation, leading to the social chaos that was a contributory factor in the rise of Hitler. The currency commissioner of that time, Hjalmar Schacht, wrote *The Lost Science of Money* in 1967, blaming speculators for the downfall of the currency; but John Maynard Keynes observed that:

> *"The various belligerent governments, unable, or too timid or short- sighted to secure from loans or taxes the resources they required, have printed notes for the balance."*

Of the new policy of Quantitative Easing, or money printing, Jim Rogers, former business partner of George Soros of bank of England fame, lamented:

> *"...they flooded the world with money."*

One un-named Chinese official, speaking to the Financial Times in February 2009 put it this way:

> *"We hate you guys. Once you start issuing $1 trillion, $2 trillion...we know the $ is going to depreciate."*

This he was saying of a currency that had already lost 33% of its value since 2002. In effect he meant that there is no such thing as a free lunch; the Chinese will have to pay! The QE policy remains highly controversial. The banking Editor of The Times observed that extension of the QE program adds to the longer-term risk of inflation. Economist Frank Shostak stated:

> *"It is the repayment of debt and then savings that are needed. The mal-investments of the boom must be liquidated."*

Doug French, another economist, said of the policy:

> *"You can't print prosperity...policy makers are trying to tell us we're not borrowing enough!"*

US Congressman Ron Paul said in July 2009:

"The Obama administration continues to do the things that created the problem in the first place. The more we do to interfere with the correction, the longer it lasts."

Martin Hutchinson, author of Great Conservatives, commenting recently on QE:

"Under Weimar, the profits from "seigniorage" financed 50% of public spending...in the US today the figure is 15%...in Britain the figures are more alarming...75 billion pounds of gilts (were) purchased over three months, an annual rate of 300 billion – 65% of government expenditure...the Bank of England may not repeat its gilt purchases...[but] it is disturbing that Britain is currently 'printing money' faster than Weimar Germany."

The German government was highly critical of US policy regarding QE. Clearly willing to sacrifice short-term economic recovery for long-term economic stability, Angela Merkel, the German Chancellor, was accused in June 2009 of breaking an unwritten ban on German leaders commenting on monetary policy:

"We must return to independent and sensible monetary policies, otherwise we will be back to where we are now in ten years' time."

Charles Dumas, director of world service at Lombard Street Research, responded by saying:

"Merkel and the German elite are divorced from the realities of the global economy...German policies, behaviour and outspokenness ...are a menace to the world economy."

These sharp disagreements bring into focus the difference in thinking between the present day Anglo-Saxons and their German antagonists.

Inflation or Deflation?

But how had US policymakers responded to the Wall Street crash of 1929 and ensuing Great Depression? The main obvious difference from the 1930's is that the authorities had refused to expand the economy by either cutting interest rates, granting social security payments to the poor or expanding the monetary base.

There was no state intervention to save banks, car manufacturers or anyone else in the 1930's. Moreover, the USA maintained a gold standard until 1934 and this ruled out QE.

For the Anglo-Saxons, QE is an experiment that has never been tried until now. The outcome of hyper-inflation is by no means certain. Japan used QE for five years until 2005 to counter deflation, a general fall in prices. They have had no inflation since and deflation has been reappearing. But many influential people still fear inflation. Professor Arthur Laffer, economic advisor under the Reagan administration, in a June 2009 article in the Wall St Journal entitled *"Get Ready for Inflation and Higher Interest Rates"* said:

> *"It's difficult to estimate the magnitude of the inflationary and interest-rate consequences of the Fed's actions because, frankly, we haven't ever seen anything like this in the US. To date what's happened is potentially far more inflationary than were the monetary policies of the 1970s."*

The Governor of the Bank of England, Mervyn King, speaking to bankers at London's Mansion House in March 2009, summed up what had just happened to the world as:

> *"An extraordinary, sudden, severe and simultaneous downturn of activity and trade in every corner of the world economy."*

An April 2009 editorial in *The Times* observed that:

> *"In raw numbers, the global recession of 2009 is as severe as the Depression. Industrial output is collapsing...[there has been] a sustained fall in prices...[that if it continues] would be catastrophic..."*

Then suddenly, after nineteen months had elapsed since the Northern Rock meltdown, the storm seemed to abate. The 'mysterious worldwide forces' spoken of by the Washington Post seemed to evaporate. Share prices were recovering rapidly, while blanket guarantees on retail deposit accounts had calmed the worst fears of the general public. By June UK property prices appeared to have stabilised. The rate of increase of unemployment on both sides of the Atlantic had slowed markedly.

So will prosperity soon return? The world has repeatedly depended on the USA to pull it out of recession. This it has been able to do until now. But the greatest geo-political power shift since the last war with Germany is well under way. After that war, a bankrupt Great Britain ceded world dominance to the USA, but now a dangerous power vacuum is developing. In 2008 the USA maintained 865 facilities in more than forty countries and overseas U.S. territories. For example, on the Japanese island of Okinawa 50,000 troops remain as a legacy of the Second World War. This global presence costs an estimated one quarter of a trillion dollars annually. To a crippled economic power this is unaffordable.

No empire can maintain its primacy on a sea of ever growing government debt, or private usury. Usury may be the second oldest profession. Returning from captivity to rebuild the temple of Solomon, the Israelites were upbraided by the prophet Nehemiah for trying to enslave one another with debt:

> *"And there was a great cry of the people...we have mortgaged our lands, vineyards and houses that we might buy corn...other men have our lands and vineyards...I said unto them, We after our ability have redeemed our brethren the Jews, which were sold unto the heathen; and will ye even sell your brethren?...Let us leave off this usury...Then said they, we will restore them..."* Nehemiah 5:1-12

Debt is a way of life in Britain and America. A June 2009 article in the Financial Times recounted:

> *"The Walton family, of Wal-Mart fame, is wealthier than the bottom third of the U.S. population put together – about 100 million people...the benefits of economic growth have gone into the pockets of plutocrats...why has there been no revolution? Because there was a solution: debt."*

Anglo-Saxon policymakers have only two conventional ways to pay for bailing out their banks: higher taxation or more foreign borrowing. But with money printing as an ongoing policy there must come a time when the Beijing's and Abu Dhabi's of the world, already gagging on US Treasury Bills, decide to invest elsewhere.

Only one European nation, still in 2009 the world's greatest exporter, has the resources and the flair to leap into the gap; in doing so they - not China - will eclipse the USA. In the words of former Japanese finance minister Eisuke Sakakibara *"The American age is over."* Now a global power vacuum is opening up as the USA is drained of money and credibility. Some nation must fill this gap. This Germany will most certainly do, unopposed, and clothed in the fine mantle of EU respectability.

Yet someone must be held accountable for the ruination of the world's first experiment with globalisation. Without doubt there is much pain to come. Mark Mobius, chairman of Templeton Asset Management in Asia and rated as one of the world's one-hundred most influential people thinks:

> *"Definitely we are going to have another crisis...a very bad crisis may emerge within five to seven years as stimulus money adds to financial volatility."*

Things are in no way back to normal. Frederik Reinfeldt, Sweden's Prime Minister, speaking in July 2009 said:

> *"We are warning that we are not through the financial crisis. There is a still a crisis affecting the financial sector..."*

Massive headaches for European policymakers lie ahead. The Financial Times in August 2009 reported that:

> *"...policymakers in Brussels...doubt the capacity of many European banks to absorb future heavy losses."*

Unemployment, already dangerously high, is still rising worldwide. In the longer term, the ongoing malaise is bound to affect the debt-addled Anglo-Saxon economies. The rich will get richer as the poor get poorer. Lagging behind in vital wealth creating industries, they will be devastated and their currencies debased as never before.

Currency stability is very much on the minds of policymakers worldwide. The Gulf Cooperation Council, a club of half a dozen wealthy nations in the region, has discussed for several years the idea of a new Middle Eastern currency backed by gold for Gulf and Arab countries.

Other regions such as the Union of South American Nations and West Africa Monetary Zone are examining similar arrangements. China, the fifth largest sovereign gold holder, is promoting the settlement of business deals outside of the dollar. Bloomberg News reported on July 6th 2009 that for the first time three Shanghai companies agreed to settle import/export contracts in yuan, not dollars. Of this move Nizam Idris of UBS commented:

> *"This is the first step towards making the yuan a global reserve currency."*

The trend towards ditching the dollar is unmistakable. The ambitious Chinese are buying up commodities and the means to produce them as fast as they can. They, like the EU, are racing to secure their future as an emergent superpower. But the second decade of the 21st century must surely see Europe's economy above all and the Euro, the new Deutschmark, rein supreme. In the meantime, will world growth resume? Nobel Prize winning economist Paul Krugman said in August 2009:

> *"We seem to have avoided the Great Depression 2.0...[but] unless we can find another planet to export to, we cannot have an export-led recovery from this global financial crisis, which means we have a serious difficulty."*

Back in January the utter uniqueness of what has happened to the world was rammed home by former US Federal Reserve chairman Paul Volker:

> *"Crises are old but this crisis is different...it's more global in scope than any previous crisis. [It has been] the greatest boom in asset prices in all of human history."*

The Germans economy was by July 2009 the first major Western nation to revert to growth. Yet Europe's banking system still carried more toxic debt than that of the USA, and where did most of that toxic debt originate?

German finance minister Peer Steinbrück voiced harsh criticism of the USA, laying the blame for the crisis at its door:

"The world will never be as it was before the crisis. The United States will lose its superpower status in the world financial system."

A resurgent Germany, dominating the institutions of the EU, is going to blame the Anglo-Saxon races for the woes of the world.

42

TWO AMERICANS
"...God hath numbered thy kingdom and finished it." Daniel 5:26

The stimulus to write this book came from the unsolved puzzle of Bullinger's 7's and 11's. He saw that these numbers were attached to the two leading personalities of the Bible. This led to the discovery of the Grid and its three defining numbers 7, 11 and 44.

Two signs with respect to the number 44 appeared as a result of the activities of two Americans in the UK. They each have the same initials: D.B. Both were operating in London in a manner that relates to the discoveries of this book. London is effectively the centre of the continents because the world still runs according to Greenwich meantime. The happenings concerning these two Americans in London were quickly followed by the most violent events on the surface of the Sun in 144 years, signaling the number 11; the Sun's identity number (where 144 is the 11th Fibonacci number, see Appendix VIII).

This question must again be posed: were David Blaine and his magic powers powerful enough to cause eight power cuts around the world? If so, did Blaine also cause the solar storm of 11 giant flares? Could he have induced a fellow American with the same initials to write a blockbuster book? Yet the activities of David Blaine and Dan Brown were highly significant. The 44 day fast was David Blaine's *fourth* stunt, and that The book *The Da Vinci Code* was translated in to 44 languages and was Dan Brown's *fourth* novel. Thus both men have unwittingly signalled the number 444, shown to be the number of *empire* in the chapters *A brief history of History* and *Decline and Fall*.

The (essentially anti-God) activities of these two Americans in 2003 signalled an impending crisis. That year both men were marked by the number 444. These numbers tally with the number pattern underlying the 'writing on the wall' at the fall of Babylon, shown earlier as follows:

Letters	**4**	**4**	**4**	8
	MENE	MENE	TEKEL	UPHARSIN
	mêne	mêne	t ªqal	(pªrac) [U-] pharsin

The summary for the occurrence of the 444 pattern can now be expanded to show five instances:

The fall of Babylon in 539 BC, mêne, mêne, t ªqal, (letters) 444
The Axis alliance between Hitler and Mussolini (in weeks) 444
The Tehran siege that released Iran from US control (days) 444
The *fourth* stunt of David Blaine, a fast lasting 44 days 444
The *fourth* book of Dan Brown translated into 44 languages 444

Obviously the 444 pattern of the Grid relates to the rise and fall of kingdoms and empires. A *double* occurrence of the number 444 in 2003 is quite simply a sign of the impending fall of the *two* 'empires' of the English-speaking nations either side of the Atlantic. Of course a *sign* of a fall is not the same thing as the fall itself.

What might have been signalled by the events of 2003? A clue is found in the initials of the two Americans, D and B. The *gematria* principle by which letters have a numerical value would, if applied to the English alphabet in the simplest way, suggest the numbers 4 and 2, or the number 42[1]. This is the number of the greatest calamity in history - the Great Tribulation - which is to last three and a half years, or 42 months. We know from several prophesies that what Daniel called *"the crisis at the close"* (Daniel 12:1 Moffat translation) is a 42 month period.

> *"...and the holy city shall they tread under foot forty and two months. v3 And I will give power unto my two witnesses, and they shall prophesy a thousand two hundred and threescore days [where a prophetic year equals 360 days], clothed in sackcloth."* Revelation: 11:2

> *"...a time of trouble, such as never was since there was a nation...at that time thy people shall be delivered...v4 But*

thou O Daniel seal up the book, even to the time of the end: many shall run to and fro, and knowledge shall be increased...v7 ...it shall be for a time, times and an half [three and a half years] ... Daniel 12

"And the woman fled into the wilderness, where she had a place prepared of God, that they should feed her there a thousand two hundred and threescore days." Revelation 12:6

"And they worshipped the dragon (Satan) which gave power unto the beast...who is able to make war with him?...v5...and power was given unto him to continue forty and two months." Revelation 13

The four Bible passages above show that this crisis period lasts forty-two months. As mentioned in earlier chapters, there is vital statement in the Olivet Prophecy concerning this time of crisis:

"For as in the days that were before the flood... Noah entered into the Ark...so shall also the coming of the son of man be." Matthew 24:27-28

But there is one difference between today and events at the time of Noah: in this crisis more people will survive. The fourth passage above states that *"at that time thy people will be delivered."* Isaiah refers to the same idea:

"...I will bring again the captivity of my people...v7...it is even the time of Jacob's trouble; but he shall be saved out of it." Jeremiah 30

The descendents of Jacob[2] are to be saved out of this calamity. Other tribes and nations will also be preserved:

"...every one that is left of all the nations which came against Jerusalem shall even go up from year to year to worship the King..." Zechariah 14:16

Thus it is seen in scripture that a representative remnant of humanity will survive the coming crisis to live under a new world government, controlled by Jesus Christ. It seems remarkable that God would use (or permit to be used) a popular magician and a pulp fiction writer to signal the imminence of such a crisis. The author's own involvement will certainly surprise some, though it was shown in the chapter

Unearthing the number 28 that God once spoke through the mouth of a donkey.

The logic and consistency of the manner in which the signs of 2003 were given can be appreciated from the following chart:

Grid no.	Depicting	The Phenomena
44	Pictures the reproductive *power* of the Father	Marked by 8 *power* cuts in 44 days and an American with '*powers*' and the initials D.B. (David Blaine fasts for 44 days)
7	Pictures the 'Word' or Jesus Christ	A book on the 1st of July about *Christ* full of *words* also by an American with the initials D.B. (Dan Brown's 4th book translated into 44 languages)
11	Pictures the Father	11 giant flares on the face of the Sun, 11 is the number of the Father

The *three defining numbers* of the Grid were signalled by *three worldwide events* in the year 2003. The eight power cuts drew a giant figure of 8 across the earth, centered on Rome. This signals a new beginning for a world freed from the tentacles of a false Christianity. Noah's number 64 is 8 concentrated, the sign of the end of an age. The present age will be brought to a close not by a flood, but by *fire*. The prophet Isaiah refers to this time:

> "...therefore the inhabitants of the earth are burned and few men left" Isaiah 24:6

This harmonises with the prophecy in Revelation regarding a 'fourth angel' pouring out a vial on the Sun:

> "And the fourth angel poured out his vial upon the sun; and power was given unto him to scorch men with fire. And men were scorched with great heat, and blasphemed the name of God, which hath power over these plagues: and they repented not..." Revelation 16:8-9

This prediction concerning the 'end time' is amongst the litany of disasters described in Revelation. It appears to have credence in some scientific circles (See Appendix XIV for further ideas about Global Warming).

The full findings with respect to the 'Signal' of 2003 contain not only the defining numbers of the Grid, but the Grid itself. The Signal contains by implication the 1711171 pattern, the related number 111 and the 6+7 pattern in the *Sabbaths* showing action of the *Lord of the Sabbath*. Dominant over the whole scheme is the number 127, the 31st prime number, which is the number of deity. The pattern begins with the London launch of *The Da Vinci Code* and closes at the end of the sixteen days of solar activity. The observations can be summarised as follows:

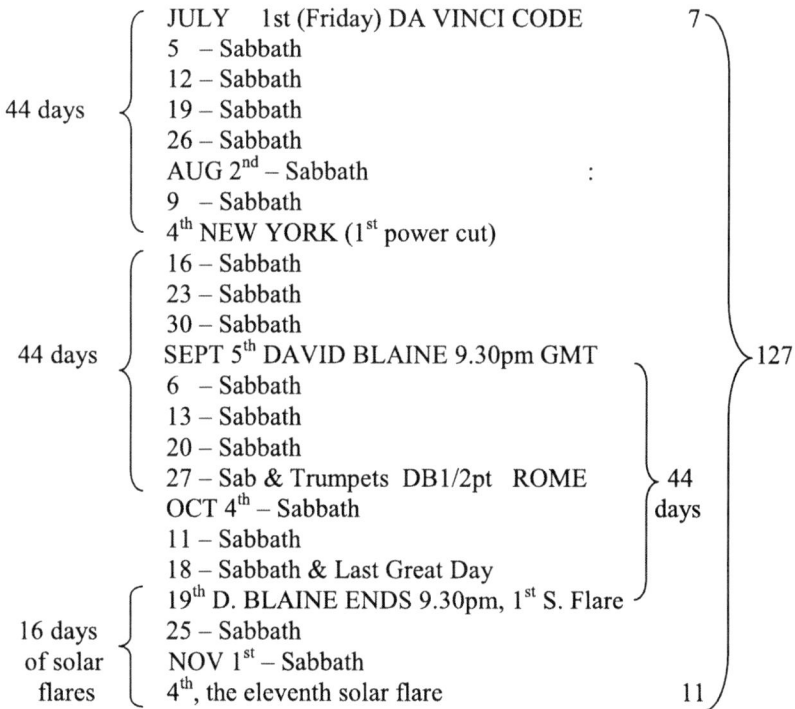

```
                   JULY   1st (Friday) DA VINCI CODE          7
                   5  – Sabbath
                   12 – Sabbath
     44 days       19 – Sabbath
                   26 – Sabbath
                   AUG 2nd – Sabbath                  :
                   9  – Sabbath
                   4th NEW YORK (1st power cut)
                   16 – Sabbath
                   23 – Sabbath
                   30 – Sabbath
     44 days       SEPT 5th DAVID BLAINE 9.30pm GMT                    127
                   6  – Sabbath
                   13 – Sabbath
                   20 – Sabbath
                   27 – Sab & Trumpets  DB1/2pt  ROME       44
                   OCT 4th – Sabbath                        days
                   11 – Sabbath
                   18 – Sabbath & Last Great Day
                   19th D. BLAINE ENDS 9.30pm, 1st S. Flare
     16 days       25 – Sabbath
     of solar      NOV 1st – Sabbath
     flares        4th, the eleventh solar flare              11
```

The whole pattern from 1st July to 4th November comprises 127 days where127 is the 31st prime number

The signal begins with a blasphemous pseudo-historical challenge to Jesus Christ as the son of God, suggesting that he was Mary Magdalene's lover. It then incorporates the activities of a magician who has made a lucrative career masquerading as Jesus Christ.

Those childish two-fingered salutes are convincingly answered by a supernatural display of worldwide power cuts. The leadership of the Great Counterfeit Church is humiliated, and the finale is a display of the Father's wrath signalled by record solar eruptions.

Two signs answered by two other signs amounts to four. These followed a pattern in days of 44, 44, 44 and 16, suggestive of 3 to the power of 4. All these features are contained within a time frame marked with the number 31, a general signification of deity, but starting and finishing with 7 and 11.

43

THE GOD BRAND

"...with feigned words [they] make merchandise of you..." 2 Peter 2:3

Every religion and all notable political ideologies have annual days of commemoration. It has been shown that the authentic Bible system of worship is based upon a cycle of such days. These are encapsulated in the 1711171 pattern.

The method of teaching a group or congregation via a set of special days spread throughout the year is a sound one, although it can be utilised to teach false knowledge just as readily as the truth. Historical Christianity has often done just that, substituting Christmas and other anti-Christian festivals for the genuine article. The apostasies mentioned by Paul, Peter and Jude have spawned a kaleidoscopic variety of denominations, sects and movements. All claim to represent Jesus but, strangely, not many will actually do as he says. Today's 'Christian' landscape of religious confusion, a wilderness of the blind leading - but more often than not competing with and splitting from - the equally blind does not anymore sanction the torture and slaughter of 'heretics.'

Some religious movements are of a stupefying odiousness and folly, demonstrating their 'faith' by handling fatally poisonous snakes during a 'service'; others indulge in weird and unnatural behaviour, smacking the keys of a piano in a state of hysteria. The so called 'God channel' is one TV window on this world of man-made madness. An example from this source serves to illustrate a growing phenomenon. In a not untypical public meeting involving a leader within in a charismatic sect, the following spectacle was recently broadcast:

A portly middle-aged man in a dark suit faces a large audience in the

Netherlands. Standing at the front of a gymnasium with his arms at his sides he eyes the audience intently. Shrieks of laughter can be heard from various parts of the hall but he does not speak. He continues to scrutinise the rows in front of him. Laughing people can be heard leaving the hall, stumbling over chairs as they go, but the camera remains trained on the man in the suit. Gradually the laughter subsides until only sporadic chuckles are heard.

After five minutes the necessary level of tension is achieved and the man in the suit begins to speak. What he says does not matter. He invokes the 'holy spirit' and mentions a 'revival' but gives no explanation. He then moves rapidly from one audience member to the next. It becomes apparent that the earlier hilarity did not involve normal laughter, but was a form of hysteria. *"Give him a double dose Lord"* he says as he touches a man on the forehead. Assistants catch the man as he falls backwards to the floor. *"Give her a double dose"* he exclaims as a woman collapses at his feet like a sack of potatoes. Then a cameraman is touched and both he and his equipment clatter to the floor. The speaker holds sway over this audience but gives no biblical teaching: he does not instruct the audience from the Bible at all. He only tells his audience what they want to hear.

The proceedings are chaotic. Some in the meeting claim they felt a wind in their faces as they fell backwards: is this God at work? This teacher and his audience fulfill the description given in Isaiah:

> *"This is a rebellious people, lying children, children that will not hear the law of the Lord: v10 Which say to the seers, see not; and to the prophets, prophesy not unto us right things, speak unto us smooth things…" Isaiah 30:9*

The entertainment spectacle in a Dutch gymnasium described above may seem far removed from the dignified altars of Catholicism. However, it has been demonstrated that the Catholic Church fulfills the prophecies concerning *The Lady* in the earlier chapter of that name. Revelation chapter seventeen describes this great false church as *"The Mother of Harlots"* meaning that there must be in existence *daughters*: these too are harlots. As shown in the later chapter *The Mark*, these daughter churches carry much the same doctrinal baggage as their distant mother church.

Before the Reformation, only one church dominated the West, but today thousands of denominations, sects and cults provide religion to suit almost every taste. Many of these sects originate in the USA. In the past thirty years the so called 'born again' coalition has brought about a rightward lurch in American politics that has had a global impact. The odious tone adopted by this movement could be summed up as *"I'm saved, you're not."* This highly vocal and politically motivated 'moral majority' is led to a great extent by charlatans and political hotheads. These constantly trumpet the idea that they are 'saved', but Christ said

> *"... he that shall endure to the end, the same shall be saved."*
> *Matthew 24:13*

The born-again coalition is a money making machine headed by 1600 televangelists[1] who beg for handouts from a gullible public. In taking cash to preach a demonstrably false gospel, backed only by the doctrines of Constantine, many of these are little more than swindlers and snake oil salesmen. For such people the commandments of God are of no effect:

> *"For there are certain men crept in unawares, who were before of old ordained to this condemnation, ungodly men, turning the grace of our Lord into lasciviousness..." Jude 4*

These preachers major on the subject of *grace*, but do not understand it. The subject is simple enough: to be 'let off' from the death penalty, the Bible's penalty for sin, in respect of past transgressions is one thing. The penalty for 'old sin' was paid by the death of Christ, wiping the slate clean. Provision for falling short in the *future* was also then made: all this comes under the heading of grace. But it continues to apply only to the person who remains committed to obeying God, which is in this world a *sacrifice*; this is the catch that sticks in the craw of the typical evangelical preacher meddling in matters he does not comprehend.

Grace does not give leave to continue in sin. But to many pastors, no sin has any consequence so long as you believe, pay up and obey the minister. Such preachers reason that as it is impossible to cease from sin (though Christ as a human managed it) that it is then OK to 'come as you are.' But Christ said:

"Be ye therefore perfect, even as your Father which is in heaven is perfect." Matthew 5:48

In today's America Jesus Christ would be branded a Judaising legalist.

The Catholic innovation of the 'indulgence' is another money spinner. Priests would sell a license permitting a sinful act. This is just one more way men have turned *"grace into lasciviousness."* The idea seems so obviously bogus as to appear more of a joke than a deception, yet behind it is the serious delusion that the Pope has been put on Earth *in place of Christ.* So it is the Pope and his representatives who forgive you! The traitorous tolerance of the 'come as you are' commandment-damning false preachers is just a more insidious version of turning grace into lasciviousness. Both of these philosophies fulfill the scenario Peter gave regarding false teachers:

> *"...many shall follow their pernicious ways; by reason of whom the way of truth is evil spoken of." 2 Peter 2:2*

What is the way of truth? The commandments of God are the way of truth. King David attested that *"all thy commands are righteousness"* in the Psalms:

> *"v142...thy law is the truth. v151 ...all thy commands are truth. v152 Concerning thy testimonies, I have known of old that thou hast founded them forever. v172 ...all thy commandments are righteousness." Psalm 119*

Much of what one hears from some preachers is an unremitting vilification of God's law. None of today's 'charismatic' preachers praise it as King David did. Neither will you hear them say with Christ:

> *"If you would enter into life, keep the commandments." Matthew 19:17*

No one above the age of twelve who first heard that statement by Jesus Christ could have been in any doubt *which* commandments were being referred to. Various complex and arcane arguments have been put forward by theologians ever since to try and overcome Christ's plainness of speech.

Many such leaders, seizing the moral high ground in American politics, invariably oppose the commandments of God in order to keep their own traditions. In a cynical crowd-pleasing *marketing exercise* they say 'come as you are' rather than teach the people to turn from sin. This should be contrasted with the position taken by John the Baptist who told the undoubtedly similar religious teachers of his day:

> *"...O generation of vipers, who hath warned you to flee from the wrath to come? v8 Bring forth therefore fruits meet for repentance." Matthew 3:7*

Paul had always taken precisely this line:

> *"...to the Gentiles, that they should repent and turn to God, and do works meet for repentance." Acts 26:20*

The 'born again' evangelical movements preach 'smooth things' to please the maximum size of audience. They never promote 'works' or 'fruit,' yet Christ is to reward us *"according to your works"* (Matthew 16:27, Revelation 2:23) and every tree that bears not fruit is to be 'cut down and thrown into the fire' - as John 15 says in essence. But like all good businesses these churches are desirous of the largest possible market for their wares. They dare not put people off by talking tough or by telling the truth! In the USA charitable contributions have long been tax deductible. Preaching popular crowd-pleasing messages is an efficient route to a well-salaried life on the golf course. Twenty five percent of the US population consider themselves to be evangelical[1]. Due to the meddling of the 'moral majority' of so called 'born again' Christians of America and their place-men in the political system, the world is at its most dangerous pass in eighty years.

The born again movement is ambivalent about Catholicism: it cannot be anything else as any denunciation of that corpus of error could expose the reality of its own faulty DNA. The analogy of biological inheritance is appropriate in this case. In paternity disputes the DNA of parent and offspring are compared to find a match. If a sufficient number of similarities can be demonstrated a match is admitted. Consider the following list of thirteen similarities between Catholicism and the apparently dissimilar evangelical-born again movements:

1. The Cross, as a symbol
2. The Trinity
3. Christmas
4. Easter and its attendant false Sunday resurrection
5. The denial of Christ's only sign as the Messiah (three days/nights in the grave)
6. Sunday observance, supplanting the Bible's Sabbath
7. Belief in an immortal soul and going to heaven immediately upon death (an error)
8. Infant 'baptism'
9. Belief in a continuous everlasting punishment, not the 'second death' of the Bible
10. The Ten Commandments - tampered with, done away, superceded or set aside
11. The assumption that Christianity should either be sponsored by or be in control of the state
12. Architecture with gargoyles, Sun symbols and/or phallic symbols
13. The cannibalisation of Christ's annual sacrament in a weekly 'holy communion'

These thirteen beliefs and practices - the DNA of Christendom - are at odds with the teaching of the Bible. They sum up the essence of all *so called* Christian churches, from the Catholic Church on the one extreme through to evangelicals on the other. Most churches have turned 'grace to lasciviousness' in one way or another, either through indulgences, or by claiming that the church can overrule scripture, to changing or doing away with the laws of God.

Repentance as in the commonly accepted meaning of 'turning from sin' to obey the law of God, is constantly undermined. It has been downgraded and diluted with glib, seductive arguments. One evangelical minister revealed himself to be a 'man of sin' when he published the following on the subject of *Repentance* in the February 2002 *Plain Truth* magazine. It amounted to a full frontal attack on

Bible truth. In a three page article, this influential evangelical failed to mention a single one of the dozens of New Testament scriptures directly dealing with the subject of repentance:

It is a common mistake to think of repentance as ceasing from sin...everything that needed to be done for human forgiveness and salvation has already been done...Repentance is...seeing God as the centre...when...you come to see that and believe it you have repented. Repentance is not about morals... [not about] your chastity, your honesty, your obedience, your devotion, your spiritual disciplines...It is not about teeth-clenching straining to put 'sin out of your life' ...Your eternal future is assured... Repentance is not just another worn-out, hollow, moth-eaten commitment to be a good boy or girl...God forgives our sins – all of them – past, present and future...when you put your trust in God you have also repented...repentance, you see, is not bringing forth some good and noble work...It is believing...The only thing we contribute to the process of our resurrection is being dead...Repentance is accepting the fact that you are dead...repentance is saying 'yes' to the gift of forgiveness... repentance is a change from being on the side of yourself...It means facing the fact that we are no better than anybody else...nothing can come between him and you – no, not even your wretched sins." J.M. Feazell

So presumably Mary's chastity was of no value to God. According to this man's reasoning, God would have been perfectly content for the Christ child to have been born to a prostitute. Admittedly, Christ mixed with publicans and sinners, preferring their company to the religious leaders of that day (as he undoubtedly would of religious leaders of the present day), but that is hardly the point: a distinction is to be drawn between the holy and the profane. The above position on repentance is a gross contradiction of the Bible. It admits to no advantage in virtuous conduct or thought. Human life becomes pointless, unless one finds the mindless praising and endless re-praising of a fictitious trinity a good reason to live.

This perverted approach in formulating doctrine affects nearly everything many ministers say and do. Of them Christ gave the following prophecy:

> *"Many shall come in my name, saying, I am Christ [that Jesus is the Christ] and deceive many." Matthew 24:5*

Preachers will not *en masse* be claiming to be Jesus Christ themselves: that never did advance a preaching career. Frequently possessed of a root of bitterness, they claim to *represent* Christ yet promote the doctrines of the Devil. Of them Christ says:

> *"...in vain do they worship me, teaching as doctrines the commandments of men." Matthew 7:7*

Of them Christ asks:

> *"Why do ye also transgress the commandment of God by your tradition." Matthew 15:3*

The contrast between the above deceitful diatribe against repentance and many clear statements of the Bible is stark:

> *"...vipers, who hath warned you to flee from the wrath to come...v8 Bring forth fruits meet [worthy] of repentance... v10...every tree that brings not forth good fruit is hewn down, and cast in to the fire." Matthew 3*

> *"Be ye perfect, even as you Father which is in heaven is perfect." Matthew 5:48*

> *"Not everyone that saith unto me, Lord, Lord, shall enter into the kingdom of heaven; but he that doeth the will of my Father which is in heaven." Matthew 7:21*

> *"...all manner of sin and blasphemy shall be forgiven unto men: but the blasphemy against the holy spirit shall not be forgiven unto men." Matthew 12:31*

> *"But he that shall endure unto the end, the same shall be saved." Matthew 24:13*

> *"... they should repent and turn to God, and do works meet for repentance." Acts 26:20*

> *"...not the hearers of the law... but the doers of the law shall be justified." Romans 2:13*

> *"But be ye doers of the word, and not hearers only, deceiving your own selves...v25...the perfect law of liberty...2:17...faith, if it*

hath not works, is dead...v19 Thou believest...the devils also believe, and tremble...v20...faith without works is dead...v21 Was not our father Abraham justified by works?...v26...faith without works is dead..." James 1:22

"...priests have violated my law...they have put no difference between the holy and the profane." Ezekiel 22:26

"...choosing to suffer affliction with the people of God, than to enjoy the pleasures of sin for a season." Hebrews 11:25

"If you will enter into life, keep the commandments." Mat. 19:17

"For the Son of Man shall come in his glory... then shall he reward every man according to his works."   Matthew 16:27

"Take heed, brethren, lest there be in any of you an evil heart of unbelief, in departing from the living God... v14...made partakers...if...steadfast to the end."   Hebrews 3:12

"Looking diligently lest any man fail of the grace of God...v16 Lest there be any fornicator, or profane person as Esau who for one morsel of meat sold his birthright...v17...[he] found no place of repentance though he sought it carefully with tears." Hebrews 12:15

"If any man see his brother sin a sin which is not unto death, he shall ask, and he shall give him life for them that sin not unto death. There is a sin unto death: I do not say that he shall pray for it." I John 5:16

"But your iniquities have separated between you and your God, and your sins have hid his face from you, that he will not hear." Isaiah 59:2

In splitting away from Rome in 1533 Britain was able to liberate its people from many of the excesses of Catholicism, yet in certain respects Rome's Mother Church has conformed to the scriptures more closely than Protestants and Evangelicals.

One important example of this is seen in the matter of the role of women:

"...God is not the author of confusion...v34 Let your women keep silence in the churches [Gk. ekklesia, calling out, a

*religious congregation]: for it is not permitted unto them to speak; but they are commanded to be under obedience, as also says the law. v35 And if they will learn anything, let them ask their husbands at home: for it is a shame for women to speak in the church. v36 What? Came the word of God out of you? Or came it unto you only? v37 If any man thinks himself to be a prophet or spiritual, let him acknowledge that **the things I write unto you are the commandments of the Lord.**" I Corinthians 14:33-37*

Like it or not, this is the Christianity of the Bible. It is obvious that there are more than Ten Commandments. No other scripture countermands the above commandment of God concerning women addressing a church gathering. No scriptural authority can be found for any type of woman – old, young, married, educated or not - in charge of other women or not, running a business or not, appointed a deaconess or not - to preach in a congregation or church. In this passage even those who attained the status of married women are to *"ask their husbands at home"* regarding the word of God. How strange, how 'cringe-worthy' to most Western ears these words must sound; they are *alien* to our culture, except perhaps for Muslims. This is because *God has been an alien to this world.*

One statement of Paul's is used to justify the use of women speakers, but the prophetesses of I Corinthians chapter eleven are not prophesying in a church meeting or scripture would contradict itself:

"But every woman that prayeth or prophesieth with her head uncovered dishonoureth her head..." I Corinthians 11:5

The context proves that hair length is the matter under discussion (see verse 15). Whereas the Catholic Church will from time to time admit to usurping the Bible with their own ideas, it is more generally the Protestant sects who claim to derive their beliefs from the Bible; in this they are frequently hypocritical. The ordination of women, placing them in authority *even over their own husbands*, is a decadent Anglo-Saxon innovation. How can it be reconciled with the command given by God through the Apostle Paul:

"Wives, submit yourselves unto your husbands, as unto the Lord. v23 For the husband is the head of the wife, even as Christ is the head of the church...v24...as the church is subject unto Christ, so let the wives be to their own husbands in everything."
Ephesians 5:22 - 24

One Christian minister remarked that his church did not *"completely understand"* the command for women to *"keep silence in the churches"* in I Corinthians chapter fourteen. What he meant was they did not wish to obey it. One cannot normally negotiate with God. Moses did not haggle over the Ten Commandments; Mary did not debate over whether she would carry the infant Jesus. Where there is an option, the choice is whether to obey God or not. Mary did not debate over whether she would carry the infant Jesus. Where there is an option, the choice is whether to obey God or not. Catholicism can at least claim a more biblical stance for one area than its liberal cousins in Protestant churches, that of roles for women. But all denominations are in gross error regarding the pivotal subject of the New Testament: the Gospel of the Kingdom of God.

The major denominations of Christianity teach an incorrect gospel, a false version emphasizing the *person* of Jesus but not his actual message, the Gospel of the Kingdom:

"And this gospel of the kingdom shall be preached in all the world as a witness unto all nations; and then the end shall come." Matthew 24:14

The major denominations of Christianity teach an incorrect gospel, a false version emphasizing the *person* of Jesus but not his actual message, the Gospel of the Kingdom:

"And this gospel of the kingdom shall be preached in all the world as a witness unto all nations; and then the end shall come." Matthew 24:14

Analysing the above statement, it is apparent that Christ anticipated that the preaching of his true message across this world would be an aberration – a departure from the norm! For over nineteen hundred years that norm has been a bastardisation, substituting the messenger for the message. The 20th-century philosopher Marshall McLuhan promotes the error as follows:

> "Christ came to demonstrate God's love for man and to call all men to Him through himself as Mediator, as Medium. And in so doing he became the proclamation of his Church, the message of God to man. God's medium became God's message."

In contrast to that nonsense Christ considered the gospel to concern primarily 'the Kingdom' because he called it *"this gospel of the Kingdom."* It is true that Christ himself will be ruling in that kingdom. To that extent he is included in the message. Paul described the gospel in various ways but at the conclusion of *Acts* he could be seen preaching the same gospel *"of the Kingdom."*

Historical Christianity has totally subverted the message of Christ by substituting the messenger for his message. There can be only one true religion. Jude exhorts: *"...[that] ye should earnestly contend for the faith **once** delivered" Jude v3.* The faith was delivered once by Jesus Christ in and around Jerusalem; no subsequent delivery was made to Rome.

Can there be any doubt that today's major denominations are apostate? They do not represent the 'faith once delivered' or anything remotely resembling it: they keep it in *name only*. They have usurped God's 'brand' in Jesus Christ and are guilty of 'passing off.' In their diversity in perversity they offer multiple 'channels of distribution' through which various flavours of the same old error can be funneled into the minds of the unsuspecting.

The position really is this bad: humanity has been victimised by a satanically orchestrated marketing campaign promoting religions as ludicrous as a 'flat earth' society, where black is white and white is black. *Isaiah* referred to this phenomenon:

> *"Surely your turning of things upside down shall be esteemed as the potter's clay: for shall the work say of him that made it, He made me not?" Isaiah 29:16*

The underlying cause is given in the same chapter:

> *"...they are drunken, but not with wine...v10 For the Lord hath poured out upon you the spirit of deep sleep...v11 And the vision*

of all is become unto you as the words of a book that is sealed...v13...this people draw near me with their mouth, and with their lips do honour me, but have removed their heart far from me, and their fear toward me is taught by the precept of men." Isaiah 29:9

Isaiah describes a religion for the spiritual sleepwalker, the mentally indolent sloth that places his infantile trust in people round about. Their religion is a feel-good, self-serving, salary inflating sham. Within such movements there are generally leaders capable of dismissing the entirety of the Old Testament as 'Hebrew poetry' containing prophecies written merely to 'comfort the people of the day'; prophecies that will never come true. Prophesies that cannot come true, that must not come true because that would overturn their politically motivated apple carts and show up this world's religions for what they are: *counterfeit.*

The curtain came up on the squalid reality of contemporary Christianity in an August 2007 article in the *Times.* A consultant to the ailing Anglican Church suggested that 'exit polls' be conducted at the premises of competing churches to discover what they were doing 'right.' It would never cross their minds to boldly preach the unalloyed truth and let the chips fall where they may. They seem unperturbed at the absurdity of rival factions of 'Christianity' competing like supermarkets for the same footfall and revenue.

To spot a counterfeit, one must be an expert in the original. Only one true system of worship is confirmed in the hidden structures of the Bible here revealed. That system has always been there, on display, and backed up by the example of Christ himself. But the churches of the world *will not stand corrected by scripture.* Of them Paul in effect says:

> *"For whom the Lord loves he chastens...v8 But if you be without chastisement...then ye are bastards, and not sons." Hebrews 12:6*

The protestant churches are Rome's illegitimate offspring. Only one religion is acceptable to God: the 1 7 11 pattern of the Grid confirms the single authentic system of worship in the Bible; it has never been rescinded. This is the system through which the true doctrines and the commandments of God are explained and understood. It was a formula observed by the New Testament church of Paul, Peter and the apostles. It is encapsulated in the 1711171 pattern, 19 days in all of which 7 are high days, or days of rest (see the later chapter *The 1711171 system*). Nothing else is admitted in the Scriptures. The use of trees in worship, derived from paganism, is not accepted by God. Any man made days appointed by man according to his reasoning act as counterfeits, in opposition to the God given Holy days:

> *"Learn not the way of the heathen...for one cutteth a tree out of the forest...with the axe. v4 They deck it with silver and with gold..." Jeremiah 10:2-4*

However you understand the above passage, it should give pause for thought for those who substitute man's days for God's. The Grid discovery with its pattern 1 7 11 is a six inch nail in the coffin of an already moribund, motley collection of derivative sects and religions. They are 'drunken, but not with wine.' None of them have the truth: none of them can 'save' anyone.

God's warning to this world to turn from false religion, and his alert to the impending return of Christ and the calamitous events that will rock the globe at that time, appears to be encoded in the giant figure of eight and its attendant holy days pattern. This pattern returns every year because it is a system of annual days. However, in 2003 it was linked to the global figure of eight by way of the fast of a magician, an event that received worldwide publicity. The Northern Rock calamity, ushering in worldwide economic panic, struck exactly *four* years later. This was not according to the Roman calendar, but according to the 1711171 pattern.

Eight power cuts in a figure of eight were attached to the first global event, the Pope's conclave. That was struck by the eighth and final power cut. Is something else very strange developing from that event in 2003? Consider again the precise way in which Rome was blacked out:

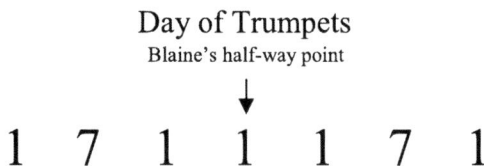

Day of Trumpets

Blaine's half-way point

↓

1   7   1   1   1   7   1

It was on the Day of Trumpets, which is the central event in the 1711171 pattern (depicting the return of Jesus Christ), that both events began, mere hours afterwards in the night. In the darkness, immediately following the close of the day of trumpets 2003 Pope John Paul II and thirty-one cardinals were blacked out in history's biggest power cut. In the darkness after the day of trumpets 2007, the English speaking peoples were plunged into a great financial crisis.

Does the giant figure of eight across planet earth in 2003 parallel another pattern of eight: an eight year time table starting and ending on the Day of Trumpets? If so, what might occur in the hours following the Day of Trumpets 2011? It is already apparent that there is a powerful prophetic theme around the number 4. Seeing that relationship, the Day of Trumpets' four year pattern deserves scrutiny. What might it suggest?

Scripture shows that it is not wrong in principle to ask God for a *sign*. The evil Pharisees were rebuked for doing this but they still got what they asked for. Christ gave them the *three days and three nights in the grave* as the sign that he was the Messiah. In another example, Gideon asked for a sign on two consecutive days and each time this was granted.

If it is the case that God Himself is behind the events of 2003 and 2007 (and the timing of the Northern Rock queues were therefore not a coincidence) then, logically, He may inspire or permit an event at the eight year point in 2011. What might that event be?

Assuming an event so timed were to happen, clues to its nature might be seen the previous two events. The first two events, the blackout of the Vatican's key ceremony and the calamitous blow dealt to Anglo-Saxon economies, share deception as a common denominator. On the one hand, there was the deceptive miss-selling and miss-labeling of financial assets, and on the other hand the pope's high profile conclave, which was all part of the continuing papal masquerade. Deception and *counterfeiting* are both themes of 2003. The whole world is deceived and has been enslaved by false religion, but it will be delivered from that condition at the return of Jesus Christ.

An eight year pattern, harbinger of the age in which the Creator will appear, may complete itself as follows:

8

Day after D.O.T. 2003: Pope John Paul II, his thirty-one Cardinals and all of Italy were blacked out in history's biggest power cut.

Day after D.O.T. 2007: The greatest ever worldwide financial crisis; marked deception and fraud, a UK bank run beginning in the night hours.

Day after D.O.T. 2011: a world changing event involving a great deception, or fraud?

One could expect that its final fulfillment would involve, like the first two events, a *great deception*. It would need to be worldwide in newsworthiness in order to match the events of 2003 and 2007.

It is worth remembering that magicians have been a part of the narrative.

At a time when people lack faith in leaders and institutions, a powerful new leader could emerge. In a crisis, a new leader is the one thing able to rekindle confidence. This would not be a Chinaman. Only Europe has the critical mass to become the next superpower, and only the Vatican has sufficient secular authority to bang political heads together and unify the dissimilar *"iron and miry clay"* (Daniel 2:33) of the nations of Europe.

Perhaps it will take a magician to unify Europe and galvanise it towards a final quest for world domination.

---

1. 25% in the USA consider themselves evangelical; page 181 *God won't save America* by George Walden.

2. There are 1600 televangelists operating in the USA, ibid.

44

THE MARK

"It is clear that thought is not free if the profession of certain opinions makes it impossible to earn a living." Bertrand Russell

What will happen to the citizens of Europe in the coming decades? The book of Revelation speaks of a 'mark' that 'the beast' causes everyone under his power to receive in their hand or forehead, the means by which one earns a living:

> *"And he had power to give life unto the image [Vatican] of the beast [EU], that the image of the beast should both speak, and cause that as many as would not **worship the image of the beast** should be killed. v16 And he caused all, both small and great, rich and poor, free and bond, to receive a mark in their right hand, or in their foreheads: v17 And that no man might buy or sell, save he that had the mark…" Revelation 13:15*

Here the *"image of the beast"* is the Vatican government, itself an *image* of the original Roman senate. The 'mark of the beast' is connected with trading and earning a living. Some have speculated that this mark will be an electronic tag or chip inserted into the body. Such technology is already utulised in a Californian club for the payment of bar tabs. Such a means may later be widely adopted for various financial transactions; the satellite monitoring of traffic tolls and for general surveillance purposes. It could ultimately be the *means* by which the 'mark' is enforced, though it will not itself constitute the mark of the beast.

This mysterious mark must also have a **religious component.** It will be the criteria by which the *"image of the beast"* (the Vatican) decides to act against those who resist a specific aspect of religious observance because this *"image of the beast"* will*"…cause that as many as would not worship the image of the beast should be killed."*

The death penalty is to apply to all who refuse, in effect, to worship the Vatican! If this death penalty were to involve a Vatican-enforced observance of a day of rest that would not be without precedent, as death was the penalty for those found 'Judaising' (keeping the Sabbath) at the time of Constantine and also at the council of Laodicea, AD 363-4.

Not only will the 'mark' be linked to money but this mark must also have a *religious aspect* if it is to fulfill the description in Revelation chapter thirteen. Enforcement of Sunday observance is the one scenario that fits both history and prophecy. It must therefore involve the disputed *day of rest*, the Sabbath of the Bible vs. Sunday, the day of the Sun, or sun-worship. There has already been a recent bid to legally impose Sunday worship in modern Europe. In 1991 there was an attempt to amend the then EU constitution to include a ban on Sunday working:

> *"The EU is pretending to be God. It is about to decree that EU citizens must take Sundays off.* **The culprit is Germany***…" The Economist 19th October 1991.*

Germany was only doing the bidding of its secret partner, the Vatican. In the new Europe, as laws to control the working week are eventually introduced and the noose is tightened it will become a question of obedience: either to obey God or the Catholic Church. Those who keep the day prescribed by the 'image of the beast' will be deemed by God to be *worshipping the Vatican*. Rome's DNA is already seen in Protestantism in the form of Sunday observance. Catholic scholars themselves attest to this clear violation of God's law:

> *"The Catholic Church for over one thousand years before the existence of a protestant, by virtue of her divine mission, changed the day from Saturday to Sunday…the Christian Sabbath is, therefore, to this day, the acknowledged offspring of the Catholic Church as the spouse of the Holy Ghost, without a word of remonstrance from the Protestant world." Cardinal Gibbons, Archbishop of Baltimore, 23rd September 1893 writing in The Catholic Mirror.*

Evidently the Catholic Church took matters into their own hands:

"We observe Sunday instead of Saturday because the Catholic Church transferred the solemnity from Saturday to Sunday." Martin J. Scott in Things Catholics are asked about (1927), page 136.

Protestant scholars still meekly acquiesce to Catholic sovereignty over their manner of worship:

"To me it seems unaccountable that Jesus...never alluded to any transference of the day...but what a pity that it (Sunday) comes branded with the mark of paganism and christened with the name of the sun god, when adopted and sanctioned by the papal apostasy, and bequeathed as a sacred legacy to Protestantism." Dr. Edward T. Hiscox, author of The Baptist Manual before a 13th November 1893 New York ministerial conference.

It is a pity that Protestants submit to Rome, even knowing God's will to the contrary:

"But, the moral law contained in the Ten Commandments, and enforced by the prophets, he [Christ] did not take away. It was not the design of his coming to revoke any part of this. This is a law which never can be broken...Every part of this law must remain in force upon all mankind, and in all ages..." John Wesley in The Works of the Rev. John Wesley, sermon 25, volume 1, page 221.

Theologians are aware that the Sabbath was even a part of creation:

"The Sabbath was binding in [the garden of] Eden and it has been in force ever since. This fourth commandment begins with the word 'remember', showing that the Sabbath already existed when God wrote the law on the tables of stone at Sinai. How can men claim that this one commandment has been done away with when they will admit that the other nine are still binding?" D.L. Moody, Weighed and Wanting (Fleming H. Revell Co.: New York), page 47.

Even though they refuse to teach it, theologians know that the matter of spiritual law is a question of all or nothing:

"Until, therefore, it can be shown that the whole moral law has been repealed, the Sabbath will stand...The teaching of Christ

confirms the perpetuity of the Sabbath." T.C. Blake D.D., Theology Condensed, page 475.

It is clear that Christ knew his true followers would be *keeping the Sabbath* at the time of the Great Tribulation at the end of this age:

"...and then the end shall come...v20 But pray that your flight be not in the winter, neither on the Sabbath day. v21 For then shall be great tribulation, such as was not since the beginning of the world to this time, no, nor ever shall be." Matthew 24:14

One major deception used to justify Sunday observance is the supposed resurrection of Christ on Sunday morning. He was in fact resurrected on Saturday afternoon as will be shown. The *only sign* Christ gave to show that he was the Messiah was his predicted time period in the tomb:

"...an evil and adulterous generation seeketh after a sign; and there shall no sign be given to it, but the sign of the prophet Jonas: v40 For as Jonas was three days ands three nights in the whale's belly; so shall the Son of man be three days and three nights in the heart of the earth." Matthew 12:39

There can be no misunderstanding the time period, that it was 72 hours. This is confirmed in the Old Testament:

"...And Jonah was in the belly of the fish three days and three nights." Jonah 1:17

It is not possible to compress 72 hours into a "Good Friday" crucifixion and an Easter Sunday resurrection. These vain traditions are a *denial* of the only sign Jesus gave that he was the Messiah. Bullinger gives the truth of the matter in his *Companion Bible*:

"The first day of the Feast [of unleavened bread] 'the high day' 15ᵗʰ Nisan (Our Wednesday sunset to Thursday sunset)...the first day in the tomb...It follows, therefore, that the Lord was crucified on our Wednesday; was buried on that day before sunset, and remained three days and three nights in the tomb as foretold by him in Matthew 12:40...Thus the resurrection of our Lord took place at our Saturday sunset, or thereabouts, on the third day." Appendix 179, page 197

This is a significant admission from a protestant theologian. The 72 hour period identifies Christ as the Messiah. As a numerical factor it fits his symbol of the Moon as the Sun and the Moon each occupy $1/720^{th}$ of the sky (see the chapter *Heavenly Bodies* and *The Prophetic Grid*). Further support for the continuing relevance of the weekly and annual Sabbath system is the birth date of Christ. This is given by Bullinger, on page 197, in *The Companion Bible,* as:

> *"15th Tishri, the 29th September 4B.C." Appendix 179*

The first day of the Feast of Tabernacles (the second 7 in the 1711171 pattern), picturing the Millennial rule of Christ on Earth, begins on that date:

> *"Speak unto the children of Israel, saying, The fifteenth day of this seventh month [Tishri] shall be the feast of tabernacles, for seven days unto the Lord..." Leviticus 23: 34*

Would God send the Messiah into this world on an arbitrary day? To fail to engineer such an event with the greatest precision would be entirely out of character for God. In fact the arrival of Jesus as a newborn heralded his kingdom on Earth as pictured by the Feast of Tabernacles (feast of temporary dwellings, i.e. the human body). His circumcision *eight days* later pointed to an almost equally great event – the general fleshly resurrection (see *Ezekiel* chapter 37) of all of humanity. That event is pictured by the Last Great Day, a high or holy day, a last *eighth day* of the Feast of Tabernacles and a holy day in its own right. It is represented by the last 1 in the 1711171 pattern. Jesus spoke on that day:

> *"In the last day, that great day of the feast, Jesus stood and cried, saying, if any man thirst, let him come unto me, and drink. v38 He that believeth on me, as the scripture hath said, out of his belly shall flow rivers of living water." John 7:37*

The last great day of the feast was highly significant to Jesus Christ. He had been circumcised on that day, an event that looked forward to the eventual spiritual circumcision of the human family in the general resurrection – the taking away of their stony hearts! The time of Christ's death is, as one would expect, very meaningful. The Julian calendar system moves directly from 1 B.C. to 1 A.D., there being no year zero. This gives an age of 30 for Christ at the beginning of his ministry. He was born 4 B.C. in the autumn; therefore that ministry began in 27 A.D. at about the same time of year. His death occurred in the spring of 31 A.D. three and one half years later. Thus Christ's death is marked with the number of deity, the number 31, although he lived to the age of 33.

Why have the churches failed to make this information available? Why do the Protestant churches follow Catholicism in perpetuating myth and fraud, denying the only sign Jesus gave that he was the Messiah? John gives the answer:

"...Satan, which deceiveth the whole world." Revelation 12:9

Many will scoff at the idea of such a personality. They will consider themselves far too sophisticated to be deceived by or to believe in a devil, a state of mind Satan himself has craftily cultivated in the schools of higher learning. But this malevolent power has cunningly maneuvered to replace the commandments of God by pagan tradition. Historical Christianity is the Devil's Trojan horse. For humanists who doubt the existence of 'evil,' some of the best evidence is to be found in ecclesiastical history. In a soon coming inquisition, Europe will revert to the barbarism of the dark ages but God will avenge his servants:

"...the ten horns [leaders]...they shall hate the whore...and burn her with fire. v17 For God hath put it in their hearts to fulfill his will..." Revelation 17:16

The schemes and wiles of the Devil will ultimately fail. It may seem strange that he persists when the conclusion of everything is already written in the Bible. The explanation is that he is under the same delusion he sends to his followers. Like many in politics he believes his own propaganda:

"...the working of Satan with all power and signs and lying wonders. v10 And with all deceivableness of unrighteousness in them that perish: because they have received not a love of the truth." 2Thessalonians 2:9

Of course, one cannot be indicted for not loving something never received. The ordinary people have not known the truth, but to the professionals who know what they are doing - promoting their own

Satan-inspired agenda at the expense of plain Bible facts - Christ warns them at the end of the Olivet prophecy:

"But and if that evil servant shall say in his heart, My Lord delays his coming; v49 And shall begin to smite his fellow servants, and to eat and drink with the [spiritually] drunken; v50 The Lord of that servant shall come in a day when he looketh not for him, and in an hour that he is not aware of, v51 And shall cut him asunder, and appoint him his portion with the hypocrites: there shall be weeping and gnashing of teeth [the second death]." Matthew 24:48

Despite many frightening global trends and abominable and immoral present day practices, the material luxury enjoyed by many on planet earth is unprecedented. There is a purpose for this material advancement and technology, because *"two witnesses"* in Revelation chapter 11 will be preaching to the world during the military rampage of the *"Beast and the False Prophet"* - two opposing two. These human messengers will expose the deceptions of a miracle working leader and his *mark*, Sunday observance. Modern communications via satellites, the internet and mobile phones will facilitate the fulfillment of the following prophecy:

"And when they shall have finished their testimony the beast...shall kill them...kindreds and tongues and nations shall see their dead bodies three days and a half...shall rejoice over them, and make merry, and shall send gifts to one another because these two prophets tormented them that dwelt on the earth. And after three days and a half the spirit of life from God entered into them and they stood upon their feet; and great fear fell upon them that saw them." Revelation 11:7

Such events could not occur until the present day when live worldwide news coverage via mobile phones into the remotest parts of the planet is, for the first time, possible. Today there exists the technology for *"kindreds and tongues and nations"* to see those dead bodies exactly as described. That event will not be intended to 'convert' people. There will be no millions switching over to Sabbath observance. In the face of a circumstance where *"no man may buy and sell"* the masses will submit. Eventually, in this fourth

Reich of totalitarian rule, the point will be reached where *"as many as will not worship the image of the beast should be killed."*

People assume that there is safety in numbers. If they avoid being in an awkward minority then they will survive. Many derive comfort from rituals and 'sacraments' and rousing music with pious arm-waving displays of outward zeal. They usually declare to God in songs that 'I am thine' and 'Here I am Lord' and that he is their 'Triune God.' They claim to 'highly respect' scripture - damning it by faint praise - then follow the doctrines and traditions of men. In most cases such worshippers already carry the *mark* of the beast, Sunday observance. Their obedience to Rome will only win for them a short reprieve. But the coming 'cashless society' first mooted by Greece's finance minister in February 2010, will be a devastatingly efficient means of outlawing Sabbath keepers at the push of a button.

Those who prostitute themselves to Rome and its mark will be able to partake of the riches of this coming Babylonish kingdom, by submitting to the Devil's system - the 'pleasures of sin for a season' - until it is destroyed in an instant:

> *"And the kings of the earth, who have committed fornication and lived deliciously (luxuriously) with her, shall bewail her, and lament for her, when they shall see the smoke of her burning...v15 The merchants...made rich by her shall stand afar off for the fear of her torment, weeping and wailing...v17 For in one hour so great riches is come to nought...v23...thy merchants were the great men of the earth; for by thy sorceries were all nations deceived. v24 And in her was found the blood of prophets, and of saints, and of all that were slain upon the earth." Revelation 18:9*

The crisis that began at the Northern Rock in 2007 has a purpose. A blow has been struck that will result in Churchill's 'Hun' taking the world by storm, one more time. A German-led United States of Europe will shortly reassert its time honoured place amongst the nations, a place it enjoyed for many centuries during the *dark ages* of European history:

> *"The German Reich long endured as the oldest political institution in Europe - older than the government of France or*

328

England by centuries. The German people called their Reich the Holy Roman Empire. It bore rule over Europe for a thousand years. This 'Holy Roman Empire of the German People' was officially designated by the Church in the Middle Ages as 'The Kingdom of God on earth.' Its citizens, the Germans, felt themselves true Romans and bearers of the Christian Reich or Kingdom. THEY were therefore the CHOSEN PEOPLE of the Christian era, entrusted with a world mission to be the protectors of Christianity. The German leaders and philosophers have never forgotten this notion of the Middle Ages that the German, in place of the Jew, has a special mission from God. German politicians know that their dream of a world Empire can be created only if they maintain this claim of a world-mission. This strange concept, which lies behind political thinking in Germany today, is plainly stated in a German work which I have before me as I write this article. The book is entitled 'Die Tragödie des Heiligen Reiches' or, translated, 'The Tragedy of the Holy Roman Empire.' It is by Friedrich Heer. It is a remarkable volume. It lays bare the reason for the secret moves of the German government towards a United States of Europe." Dr. Herman L. Hoeh, The Plain Truth Magazine, December 1962

A United States of Europe is now at the doors. As shown in the chapter *A brief history of History,* this new power-bloc is the seventh and final resurrection of the Holy Roman Empire. It will be 'ridden' by a persecuting religious entity. It will have a *mark* given to it by that church. Catholic theologians have long understood that Sunday is a **mark** of their temporal power:

> *"Of course the Catholic Church claims that the change [from Saturday to Sunday] was her act. And the act is a **mark** of her ecclesiastical power and authority in religious matters." C.F. Thomas, Chancellor of Cardinal Gibbons. [emphasis added]*

Catholics claim that *they* are the authority, not God!

> *"Sunday is our **mark** of authority...**The Church is above the Bible**, and this transference of Sabbath observance is proof of that fact." Catholic Record September 1st 1923 Ontario. [emphasis added]*

The *mark* of the beast is Sunday observance. It is to be enforced by a Catholic influenced, German dominated, United States of Europe - a nightmarish modern Babylon. This political power will ensure that those who do not submit to the dictates of a Satan-inspired church leader will pay the ultimate price:

> *Revelation 13:15 "And he [the beast] hath power to give life unto the image [Vatican] of the beast, that the image of the beast should both speak [make pronouncements], and cause that as many as would not worship the image of the beast should be killed."*

Those identified as Sabbath keepers will come under the same death penalty decreed in the fourth century at the council of Laodicea:

> *"The seventh day Sabbath was…solemnized by Christ, the Apostles and primitive Christians, till the Laodicean Council did in a manner quite abolish it…The Council of Laodicea (about 364A.D.) …first settled the observance of the Lord's day, and prohibited…the keeping of the Jewish Sabbath under an anathema [effectively a **death penalty**]." William Prynne in his Dissertation of the Lord's Day, 1633, pages 33,34 and 44.*

History is going to repeat itself in the matter of Sabbath observance. This most peculiar of all the commandments was designed as a test of faith and obedience as no other. The Apostle John exhorts all true Christians regarding this soon coming but short lived earthly kingdom:

> *Revelation 18:4 "Come out of her, my people,*
>
> *that ye be not partakers of her sins,*
>
> *and that ye receive not of her plagues."*

PART FOUR: INSIGHTS

45

SUBTERRANEAN NUMBERS

*"...to Cyrus...I will give thee the treasures of darkness,
and hidden riches..." Isaiah 45:1-3*

How could it be that a book as universally circulated, reviewed, critiqued, and minutely analysed as the Bible could contain a yet undetected system of numbers?

How could this Book of books for which countless martyrs, theologians, politicians and preachers have contended for millennia possess a never before glimpsed secret architecture? That is the power of the present discovery. Here is now revealed an amazing secret system. But why has it lain dormant for dozens of centuries? The answer lies in part in a statement made to the prophet Daniel over 2500 years ago:

> *"But thou O Daniel, shut up the words and seal the book, even to the time of the end: many shall run to and fro and knowledge shall be increased." Daniel 12:4*

Knowledge has been increased because this is the time of the end! This is the era in which mankind will be visited by his creator. Now it is time for the proof of the inspiration of the Bible to be released and made freely available. It is time for a world drunken on materialism to know the purpose of human life.

It has already been noted that the significant numbers in this discovery have a basic form, but also a higher form or version. This can be seen in the case of 44 and 88, and 28 and 280. The larger number is the greater expression of the lesser. For example, 44

331

denotes reproduction, as evidenced by both biology, the Bible's structure and number values in both languages of the Bible. But its greater expression 88 points to the end result of God's spiritual reproduction, eternal life. Here is a summary of numbers and their greater expression:

BASIC FORM	GREATER EXPRESSION
4 Power of the Father	16 Power concentrated
11 The Father	111 The Father's plan depicted
28 The teaching function	280 The teaching programme
40 The works of Christ	400 The Work of God
44 Reproduction	88 Eternal life
1711 spirit, Son & Father	1711171 their entire plan

It is now obvious that a system of submerged or 'subterranean' numbers has been concealed in the Bible. Not all have a higher expression: 7 and 19 may in different ways indicate completeness and as a result may not have any higher expression. But here and there some numbers are seen like islands jutting above sea level. Apparently nobody has suspected that a whole new continent lay beneath. The Grid is the kernel of this newly discovered landscape.

Certain numbers of this system are commonplace Bible numbers, such as the number 7. Others like 11 or 28 have not been noticed by theologians (only Bullinger noticed 11). This amazing subterranean system has never before been expounded by any author until now.

Within this general system it can be shown that there is a *main system* containing twelve whole numbers, including the six numbers of the Grid. The other six numbers relate closely to the Grid. The main system is comprised of the following twelve numbers in ascending order:

1, 4, 7, 11, 19, 28, 40, 44, 88, 107, 111, 280

This set of twelve totals 740 (20 × 37, the 12th prime). Why are there twelve numbers in this secret system? In the realm of commonly known Bible numbers, twelve is the number of organisational beginnings. This can be deduced from the fact that the Church was founded by drafting in the twelve apostles.

Symmetry is evident above, in that half of the numbers belong in the Grid and half do not. Furthermore, the distribution of the six Grid and six non-Grid numbers is again symmetrical. The Grid numbers are here underlined:

1, 4, 7, 11, 19, 28, 40, 44, 88, 107, 111, 280

This symmetry can be represented in the following simple pattern where Grid numbers are denoted by 'G' and non-Grid numbers by 'N':

G G G G   N N   G G   N N N N

Another feature is that the first 7 numbers total 110, a result suggesting 7 and 11. The positional numbers of the Grid are also fascinating, as their total is 5², the importance being that five is the number of the Temple. This is because the top line numbers of the Grid 1 7 11 appears in places1, 3 and 4 = 8; the bottom line in places 2, 7 and 8 = 17. So the grand total is 25, or 5².

There is another temple related factor, because the only primes in the series are 7, 11, 19 and 107. These total 144! This is the number of the first fruits (first resurrection) who marry Christ in a heavenly temple. There are sufficient properties in this set of twelve numbers to demonstrate the existence of a previously unknown system of Bible numerology.

Will the churches of Christendom wake up from their drunken slumbers in the light of these revolutionary discoveries? Will they begin to look afresh at the book they claim to base their beliefs on? Will they begin to treat it as *the authority* by which they live their lives and run their congregations?

46

A SECRET ARCHITECTURE
"...the mystery that was kept secret since the world began."
Paul, Romans 16:25

Not only is the Grid pattern found in eleven passages of the Bible, but it makes an appearance in the familiar order of its sixty-six books.

The total of sixty-six for the books of the Bible has been challenged. Orthodox Jews regroup the books of the Old Testament to create 22 instead of 39. But the maximum number of books in the cannon available as separate entities remains 66 for good reason.

The book of Isaiah illuminates the subject. It is considered to be a mini-Bible in view of the 66 chapters of which it is composed. Its opening verses address the issue of rebellion just as do those of Genesis one, once properly appreciated; the closing verses of its sixty-sixth chapter look forward to the new heavens and the new earth much in the same manner as Revelation 22. There is also a division in Isaiah matching the break between the Old Testament and the New.

The pattern is seen where a lengthy section of four chapters dealing with the affairs of the ancient king Hezekiah comes to a close at the end of chapter 39. Abruptly the narrative switches to matters concerning, once understood, the New Testament Elijah-type work for which John the Baptist was also a forerunner and metaphor. This was to be a work crying out in a wilderness (verse 3) of religious confusion. From this point, the book of Isaiah becomes orientated towards a more New Testament understanding. In this way the 39/27 structure of the 66 book of the Bible is mirrored within Isaiah, the twenty-third book.

But why is this secret hidden within that particular 23rd book? The reason for this becomes apparent when the books of the Bible are arranged in columns of 22 as follows:

Genesis	**Isaiah**	Romans
Exodus	Jeremiah	1 Corinthians
Leviticus	Lamentations	2 Corinthians
Numbers	Ezekiel	Galatians
Deuteronomy	Daniel	Ephesians
Joshua	Hosea	Philippians
Judges	Joel	Colossians
Ruth	Amos	1 Thessalonians
1Samuel	Obadiah	2 Thessalonians
2Samuel	Jonah	1 Timothy
1Kings	Micah	2 Timothy
2Kings	Nahum	Titus
1Chronicals	Habakkuk	Philemon
2Chronicals	Zephaniah	Hebrews
Ezra	Haggai	James
Nehemiah	Zechariah	1 Peter
Esther	Malachi	2 Peter
Job	Matthew	1 John
Psalms	Mark	2 John
Proverbs	Luke	3 John
Ecclesiastes	John	Jude
Song of Solomon	Acts	Revelation

The book that provides the key to the 39/27 structure of the Bible is at the centre top in this arrangement. It is a cornerstone for the structure of the Bible. This symmetrical positioning of Isaiah begs the question as to what other features there might be. Immediately, the theme of reproduction on the bottom line is obvious in the light of previous discoveries. Acts, the forty-fourth book, contains the first reference to the fiery tongues of the spirit (44). The Song of Solomon is a romantic and at times explicit metaphor of Christ and the Church producing offspring. Revelation presses home the purpose of the two Deities described therein more powerfully than any other book:

"He that overcomes shall inherit all things; and I will be his God, and he shall be my son." Revelation 21:7

So just like the Grid, these columns offer a bottom row stamped with the theme of reproduction. The three books from left to right depict, progressively, three stages of reproduction: romance, begettal and birth (see more detail in Appendix IX).

If the bottom line of the Grid is evident, what of its top line? This also makes a more than cryptic appearance. The spirit (1) makes its compelling debut in the second verse of Genesis, *vibrating* (Interlinear Bible) upon the face of the waters. In Isaiah the sufferings of Christ (7) are vividly and uniquely illustrated in a lengthy prophecy of fifteen verses. No other Old Testament book contains a comparable passage. In Romans the number of the Father (11) is highlighted in an obvious way, because all sixteen chapters of Romans contain references to Jesus Christ by name, except one. This lengthy eleventh chapter speaks of the Father from beginning to end. Not only that, but mention is made of the 7,000 men of 1 Kings 19:18 *"...who have not bowed the knee unto Baal."* These were disciples of Elijah, who was sent to reveal the true God. That true God is a family headed by the one whose identity number is 11. Thus, once again, the Grid makes its appearance:

1	7	11
Genesis	**Isaiah**	**Romans**
Exodus	Jeremiah	1 Corinthians
Leviticus	Lamentations	2 Corinthians
Numbers	Ezekiel	Galatians
Deuteronomy	Daniel	Ephesians
Joshua	Hosea	Philippians
Judges	Joel	Colossians
Ruth	Amos	1 Thessalonians
1Samuel	Obadiah	2 Thessalonians
2Samuel	Jonah	1 Timothy
1Kings	Micah	2 Timothy
2Kings	Nahum	Titus
1Chronicals	Habakkuk	Philemon
2Chronicals	Zephaniah	Hebrews

Ezra	Haggai	James
Nehemiah	Zechariah	1 Peter
Esther	Malachi	2 Peter
Job	Matthew	1 John
Psalms	Mark	2 John
Proverbs	Luke	3 John
Ecclesiastes	John	Jude
Song of Solomon	**Acts**	**Revelation**

Reproduction

There need not be any more debate concerning the number of books of the Bible, in view of this multifaceted proof. The 39/27 structure of the Bible stands validated. It has appeared all along in the fourth segment of the Deity string of pi:

The Father reproducing   The human   The divine teacher   The Bible

3.14159265358979323846264333383279

11digits totalling 44   totals 33, lifespan   totl. 28-teaching. 27books NT

(Ιησους)   **(χριστος)**   39 books OT

6 Greek letters   7 Greek letters

47

THE RAINBOW

*"...a rainbow round about the throne,
in sight like unto an emerald." Revelation 4:3*

It will surprise some that the first agreement, or in biblical parlance *covenant*, made between God and man in scripture concerned the sign of the Rainbow.

It is self-evident that the Sun and the Moon enjoy primacy of place in the Bible as *signs* or symbols. They are the first to be mentioned. But the Rainbow also has primacy of place as the *subject of the first ever covenant*. That fact suggests an importance for the Rainbow that has thus far been entirely overlooked.

Consider again certain facts regarding the Sun and the Moon. They are of course circular in appearance and when in alignment with the Earth produce the spectacle of a solar eclipse, revealing the Sun's Corona, which in the Grid corresponds to the 'spirit.' This is the same spirit that appeared as tongues of flame on the heads of the disciples (a plainer symbol to represent the emanating spirit of the Father could hardly be imagined). Upon the founding of the New Testament Church, these flames were the *first* symbol used to represent the spirit.

The Sun and Moon appear as identically sized circles. Rainbows are circular; they are the *image of the Corona,* seen only when water is present as an atmospheric vapour. In all, this gives four circular symbols, the Sun, the Moon, the Corona and its image the Rainbow. Whereas the two solid objects, the Sun and the Moon, represent personal beings, the two more ephemeral circles, the Corona and the Rainbow, represent the spirit. Seven facets of the spirit (Revelation 1:4) are pictured by the seven colours of the Rainbow. This can all be summarised as follows:

SUN	**CORONA**	**MOON**	**RAINBOW**
solid	ephe-meral	solid	ephe-meral
Father	**spirit**	**Christ**	**Coronal image**

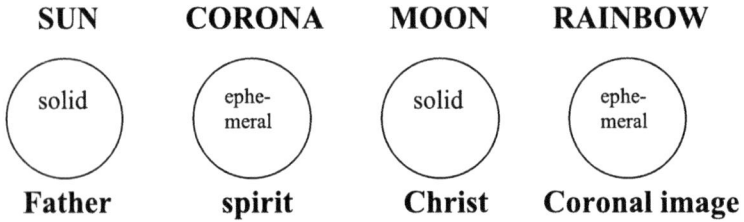

A further example of the 3^4 law (see the chapter *Numbers: Figment or Fact*) emerges. Three of the above are heavenly; the fourth is of the Earth. Three are fundamental; the fourth is an *effect*. Moreover, the fourth has 7 facets, and that in turn implies the number 28.

The circular Rainbow links the first and last books of the Bible in which it predominantly appears. Apart from a single mention in Ezekiel, only Genesis and Revelation feature the Rainbow. Ezekiel is the one book where a vision of God's moveable throne is recorded: it has four wheels, or circles composed of eyes - a surveillance system par excellence. The Rainbow fits not only the symbolism of the two great lights, but their doctrine. This is apparent when one considers that Christ is the:

"...express image of his [God's] person." Hebrews 1:3

And also the statement:

"If you have seen me you have seen the Father." John14:9

The human Jesus was a faithful representative of the Father, his express image, just as a rainbow is an image of the Corona.

This theme of representation is consistent with the ways the four symbols relate symmetrically:

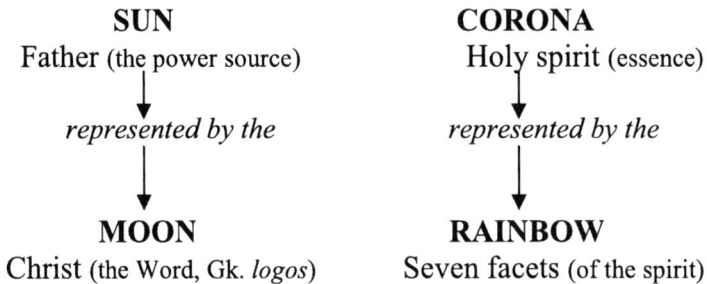

SUN	**CORONA**
Father (the power source)	Holy spirit (essence)
↓	↓
represented by the	*represented by the*
↓	↓
MOON	**RAINBOW**
Christ (the Word, Gk. *logos*)	Seven facets (of the spirit)

As stated above, the Rainbow was the subject, or token, of the first ever covenant. This first covenant and its token are mentioned 8 times in total:

1. The pre-flood introduction: Genesis 6:18 "...a flood of waters upon the earth, to destroy all flesh, wherein is the breath of life, from under heaven; and everything that is in the earth shall die. 18 But with thee [Noah] shall I establish my covenant and thou shalt come into the ark..."

2. The post-flood establishment: Genesis 9:9 "Behold I establish my covenant with you and with your seed after you v10 And with every living creature that is with you...to every beast of the earth. v11 And I will establish my covenant with you; neither shall all flesh be cut off any more by the waters of a flood; neither shall there any more be a flood to destroy the earth."

3. The token: Genesis 9:12 "And God said this is the token of the covenant which I made between me and every living creature that is with you, for perpetual generations. v13 I do set my bow in the cloud[2] and it shall be for a token of a covenant between me and the earth.

4. The remembrance: Genesis 9:14 "And it shall come to pass, when I bring a cloud over the earth, that the bow shall be seen in the cloud v15 And I shall remember my covenant which is between me and you and every living creature of all flesh; and the waters shall no more become a flood to destroy all flesh. v16 And the bow shall be in the cloud and I will look upon it that I may remember the everlasting covenant between God and every living creature of all flesh that is upon the earth."

5. The conclusion: Genesis 9:17 "And God said unto Noah, this is the token of the covenant which I have established between me and all flesh that is upon the earth."

6. The throne of God: Ezekiel 1:28 "...as the appearance of the bow that is in the cloud in the day of rain, so was the appearance of the brightness round about."

7. *The throne of God*: Revelation 4:3 *"And he that sat was to look upon like a jasper and a sardine stone: and there was a rainbow round about the throne, in sight like unto an emerald."*

8. *The new beginning*: Revelation 10:1 *"And I saw another mighty angel come down from heaven, clothed with a cloud: and a rainbow was upon his head, and his face was as it were the sun, and his feet as pillars of fire: 2 And he had in his hand a little book open: and he set his right foot upon the sea, and his left foot upon the earth...v6 And sware by him that liveth for ever and ever...that there should be time no longer."*

The number 8 not only signifies redemption, but a new beginning - as the 8th day is the *beginning* of a new weekly cycle. (This meaning for the number 8 is discussed in the earlier chapters *The Key Pattern 888* and *Rome in the Crosshairs*.) The time setting for the 8th and final mention of the Rainbow (point 8 above) is evidently just before the return of Christ. That point in history is referred to in Revelation as follows:

"And the seventh angel sounded: and there were great voices in heaven, saying, The kingdoms of this world are become the kingdoms of our Lord, and of his Christ, and he shall reign for ever and ever...v18 And the nations were angry, and thy wrath is come...(thou) should destroy them that destroy the earth." Revelation 11:15

In view of the fact that a giant figure of 8 has been drawn across the world, starting and ending at the political and religious power centres of the world, New York and Rome, can the time described above be all that far into the future? There is a reason that Ezekiel is the only book, other than Genesis and Revelation, to contain a mention of the Rainbow. The last mention of the Rainbow in Revelation contains a reference to a little book, and so does Ezekiel – it is the *same* book. This is evidenced by the description in each case:

"...and lo the roll of a book was therein...and there was written therein lamentations, and mourning, and woe. 3:1 Moreover he said unto me, son of man, eat that thou findest; eat this roll, and

go speak unto the house of Israel...v3...it was in my mouth as honey for sweetness." Ezekiel 2:9

"And I went unto the angel [with the Rainbow upon his head] and said unto him, give me the little book. And he said unto me, take it, and eat it up; and it shall make thy belly bitter, but it shall be in thy mouth sweet as honey." Revelation 10:9

Evidently, the *same book* is described in both Ezekiel and Revelation. It contains a message from the Rainbow festooned throne of the Father. It may be becoming clearer why the covenant of the Rainbow is of first importance in the Bible. A second worldwide destruction was foretold by Christ:

"...except those days should be shorted there should no flesh be saved, but for the elect's sake those days shall be shortened..." Matthew 24:22

Yet the Rainbow carries with it a token for good. It is actually a promise from the throne of God that human life will be preserved. The pattern in pi contains a reference to the flood. The 31 digits of the Deity String total 150, the number of days of the Great Flood. These are the last days of Satan's rule. The world is going to be started all over again. This will not take the form of an almost total annihilation as in the Flood, but will amount to a take-over of existing structures:

"...the kingdoms of this world have become the kingdoms of our Lord..." Revelation 11:15

There will however be a great deal of death and destruction:

"Behold the Lord maketh the earth empty, and maketh it waste, and turneth it upside down, and scatters abroad the inhabitants thereof. 6...the inhabitants of the earth are burned and few men left..." 34:2 "For the indignation of the Lord is upon all nations, and his fury upon all its armies: he hath utterly destroyed them..." Isaiah 24: 1

After the 42 month period of the Great Tribulation (followed by the Day of the Lord, see Revelation chapter sixteen) there will be a rebuilding program:

> *"And they shall build the old wastes, they shall raise up the former desolations, and they shall repair the waste cities..."* Isaiah 61:4

> *"And it shall come to pass that everyone that is left of all the nations which came against Jerusalem shall go up from year to year to worship the King, the Lord of hosts, and to <u>keep the Feast of Tabernacles</u>... v17 ...whosoever will not come up...there will be no rain. v18 And if the family of Egypt go not up, and come not, that have no rain; there will be the plague...v19 This shall be...the punishment of all nations that come not up to keep the Feast of Tabernacles."* Zechariah 14:16

Thus the administration of true worship will be enforced. The human family will by that time have had millennia to experiment with their own religions and reap the bitter fruits. An idyllic life style will be now developed, exemplified at the Jerusalem world HQ:

> *"The wolf shall also dwell with the Lamb...the young Lion and the fatling together, and a little child shall lead them."* Isaiah 11:6

This will be a millennium, a one thousand year period of peace, under Christ who will rule from Jerusalem with his own 'cardinals':

> *"And I saw thrones... (for those) who had not worshipped the beast, neither his image (Vatican)...and they lived and reigned with Christ a thousand years."* Revelation 20:4

The Rainbow is the promise that God will not flood the world again. However, a cataclysm is coming: it will be brought about by a German-dominated United States of Europe.

48

THAT THEY MAY BE TAKEN

"Men are so simple and so much inclined to obey immediate needs that a deceiver will never lack victims…" Machiavelli

Why has the Bible been written in such a way that almost nobody can understand it? For what reason did Christ speak in parables?

Misunderstandings of the Bible abound. The statements of Paul regarding *the law* often appear contradictory. In one passage the 'law' appears to be 'obsolete' but then in another it is *"holy, just and good."* There are many apparent conundrums spread across the New Testament. *James* warns that breaking the 'law' in one point is tantamount to have broken all points, to be in fact *"guilty of all."* Yet the apostle Paul observes that *"a man is not justified by the works of the law"* in which case there would not seem to be much point in worrying about keeping every point of it. But which law is referred to? Is it a spiritual law, a sacrificial law, a civil law, an agricultural principle or what?

The Bible is frequently opaque and not at all straightforward. It is laden with symbols, similes and metaphors many of which serve to throw the reader off the track of truth completely. A further problem is the existence of competing translations. Yet another is copyist error, even fraud! Yet scripture is so cleverly constructed, with the truth on any topic so cunningly woven into dozens of passages spread across centuries, that to overturn its true meaning is nigh on impossible.

One consequence of the way the Bible has been written is the scope it gives for dishonest interpretation. Because so many statements can be used selectively to promote almost any given bias, the theologians have plenty of rope with which to hang themselves. If a reader of the Bible, be he theologian or layman, wishes to follow a certain popular line based on the selective use of scripture - to which a certain 'spin' may have been applied - he can feel free to do so, proclaiming it loudly from the pulpit and quoting scripture. Jeremiah wrote that the human heart is *"deceitful above all things..."* (Jeremiah 17:9)

One lady haughtily informed the author that hers two sons - both ministers of religion - could make the Bible say anything they wanted. That conversation swiftly closed when he enquired whether they could make it command the observance of Christmas.

A Colossal Deception

Anyone can quote scripture including the Devil. Anyone can manipulate commas that don't appear in the original, or verses known to have been tampered with (of which there are very few). A key example of the deceitful use of scripture is the common misinterpretation of Colossians 2:16-17. Almost all mainstream Churches subscribe to something like this popular misunderstanding. The correct rendering powerfully vindicates the keeping of the 19 days of the 1711171 pattern. This is in keeping with the recorded example of Christ, Paul and the early Church. A correct rendition of the disputed passage in Colossians would read as follows:

> *"Let no man therefore judge you in meat, or in drink, or in respect of a Holyday, or of the new Moon [the appearance of which regulates the timing of holydays] or of the Sabbath - which are a shadow of things to come - but only the Body (Greek: soma) of Christ [the Church].*

Paul makes the point that it is the Church's place, through its ordained ministry, to pass any necessary judgment about *how these days are to be observed* by its members. No one else can fairly do that as they are not a party to this 'church' and presumably not called to become a part of it in the present age. Such an interpretation is entirely in keeping with the rest of the Bible. But the Churches do not want to submit to the Bible system of religious observance pictured by the number pattern 1711171. They prefer the competing and counterfeit system of Sunday observance, Christmas, Easter, All Saints Day and other traditions of pagan origin. These

346

days are the hallmark of Rome, in particular the change from the Saturday Sabbath of Christ (who was "Lord of the Sabbath") to Sunday. The weekly cycle has never been interrupted. Indeed, Cardinal Gibbons proclaimed that this change to Sunday, and the later conformity of Anglicans and their offshoots to that day, demonstrated the power of the Catholic Church:

> *"You may search the scriptures from Genesis to Revelation but find no basis for Sunday observance. The Bible enjoins Saturday, a day we have never reverenced." Faith of our Fathers, page 76, 1980 edition. C. Gibbons*

The approach of such preachers and their various churches is similar to that of the Pharisees, to whom Christ prophesied the following:

> *"...in vain do they worship me, teaching as doctrines the commandments of men. v9...you reject the commandment of God so that you may keep your own traditions." Mark 7: 7*

That is the cause of all of the pain and wretchedness of this world! It is the choice that has led to all of the wars, assignations, pogroms and holocausts of a human history defined by war. It is in this stiff-necked mindset that historical Christianity has produced a fraudulent rendition of Colossians 2:16-17, in various translations, roughly as follows:

> *"Let no man judge you in meat, or in drink, or in respect of a holyday, or of the Sabbath days - which are a shadow of things to come - but the substance [Greek: soma] is Christ."*

The word *is* has been added in the English by the translators. Theologians are trying to convey here that "Christ is the important thing." They imply the holy days and the Sabbath are unimportant - a mere shadow! They are portrayed as shadows in comparison to Christ, but no such comparison is being made. Certainly the holy days are shadows in that they *foreshadow* future events, but an utterly false translation has been cunningly crafted by 'spinning' the Greek word for body, *soma*. In this context it should, to be consistent with other renderings, be translated *the Church* (e.g. the 'body' of Christ). Instead it has been twisted to say Jesus himself, rather than his Church.

347

On that point, let the Bible interpret itself. Chapter one of Colossians shows the correct translation of soma in verse eighteen: *"He [Christ] is the head of the body [soma], the <u>church</u>…"* In this nearby passage *soma* is correctly rendered and this, in turn, was dictated by the context. Again, consistency is vital. If words are translated differently within the same epistle of Paul, without a contextual reason, this is suspicious. Furthermore, the resulting negation of the Sabbath and annual Sabbaths (which stand or fall together) is completely at odds with the example of Christ and the early Church. The Sabbath is one of the Ten Commandments! Notice that in the false version of Colossians 2: 16-17 not only has the word *"days"* been added by most translators but, critically, the word **"is"** as in "the substance **is** Christ" is *not in the original*. That added word 'is' totally alters the meaning of the sentence.

In the false version we no longer see the word SOMA rendered as *the Church* - **the Body of Christ** - but just 'Christ.' There has been a switch: what was irrefutably correct as 'the church' a few verses earlier is now translated, in effect, as 'Christ.' In chapter one verse eighteen it says: *"And he is the Head of the body [soma], the Church…"* In that passage "body" is correctly understood to refer to *the body of Christ* - the Church - as it could not possibly be anything else!

It is only *context* that can safely determine the meaning in such cases, not political agendas. But the weight of tradition was too strong for the scholars to resist. By adding the word "is" and rendering *soma* in a manner inconsistent with the rest of *Colossians* they placated only their earthly masters. See again a truthful version of this passage:

> *"Let no <u>man</u> therefore judge you in meat, or in drink, or in respect of a Holyday, or of the new Moon [that regulates holy days] or of the Sabbath - which are a shadow of things to come - but [only] the Body of Christ [the Church].*

Thus it is seen that Paul *expected* the members of the Church to be observing these days; days made holy by God that had never been rescinded! It was only the Church that could voice an opinion on what was an appropriate activity on those days. Outsiders were not qualified to judge whether or not sunbathing, chess playing, eating in restaurants or playing musical instruments would be suitable activities for observing holy days, but the Church alone. It is no wonder that historical Christianity has gone to such elaborate and corrupt lengths to wrest the true meaning of Colossians chapter two.

Preordained Enemies

God has determined that men have a choice. The overwhelming majority are not yet called, but those who choose to entangle themselves in God's affairs and deliberately sabotage his truth cling to a terrible fate. It was preordained that there will be enemies of the truth:

> "...Christ...a rock of offence to them that stumble at the word, being disobedient, to which they were appointed." I Peter 2:8

A common ploy of evangelical preachers is to claim that the Old Testament (77% of the Bible) is nothing more than poetry, thereby freeing them to preach what they like. For example, the Old Testament record of Jonah's entombment inside a giant fish for three days and three nights can be ignored, as this was 'poetic.' As a result, the confirmation of the only sign that Christ gave as Messiah - three days and three nights in the grave - by way of a comparison to Jonah, is made of no effect. Thus the path is cleared for an Easter Friday/Sunday scenario in which it is, of course, impossible to fit three days and three nights.

Because the Bible has been constructed as a coded book it is not obvious that its two most overlooked symbols, the Sun and Moon, denote deities; apparently no one has ever noticed this before. The Bible has not been intended to be widely understood until the present age. The book of Isaiah gives the most fundamental principle of Bible study in chapter twenty-eight (the teaching number):

> "But the word of the Lord was unto them precept upon precept, precept upon precept, line upon line, line upon line, here a little and there a little that they might go, and fall backward, and be broken, and snared, and taken." Isaiah 28:13

The principle of 'line upon line' is not followed by most of today's preachers. Their method is to highlight one passage to which they may apply the maximum distortion, ignoring other verses to the contrary. Pastors often read the first half of a verse, but omit the second half if it contradicts their point. God holds them responsible, seeing that they set themselves up as Authorities, for the sins of the people they are teaching.

Dressed in fancy religious garb they pose as Christ's representatives but they teach against him. Attired in the billowing robes of the Pharisees they proclaim themselves 'holier than thou' but their righteousness is faked.

The Sun symbol

It is important to address one particular error of Catholicism. It is a Catholic idea that the Sun is symbolic of Christ. Like so many religions they seek to put Christ above his Father. Then they put the supposed 'person' of the holy spirit above them both (as Anglicans and Evangelicals often do) by singing praises to and praying to 'him.' This is hardly in accord with the principles prescribed in the *Lord's Prayer*.

The idea that the Sun represents Christ is derived, by false assumption and careless reasoning, from one statement in the book of Malachi:

> *"But unto you that fear my name shall the Sun of righteousness arise with healing in his wings; and you shall go forth and grow up as calves of the stall." Malachi 4:2*

In this pictorial language, God's followers are told that they will grow vigorously in the manner seen typically of calves thriving on their mother's milk. This is an obvious allusion to spiritual growth. Nowhere else are disciples of Christ compared to growing calves and at no other time is Christ likened to the resplendent Sun, though the patriarch *Jacob* is likened to the Sun in the dream of Joseph (Genesis 37). Presumably that did not put Jacob into the Godhead with Jesus!

To take the argument further, because calves in the book of Malachi depict followers of God, can we therefore reason that Aaron's idolatrous golden calf represented true Christians! That would be a ludicrous suggestion. Clearly, Bible symbols are interchangeable;

both Satan and Christ are likened to the lion. Another animal used pictorially is the dove. The holy spirit was likened to a dove alighting on Christ at his baptism. This is not an exclusive symbol either, as the Church is likened to a dove in Song of Solomon. In that passage (Chapter 4:1 *"…thou hast doves' eyes…"*) a parallel is drawn between that author's romantic ardour and Christ's yearning for his espoused Church.

Thus it is seen that many symbols and metaphors are not fixed and unalterable in the Bible. The single use of a metaphor for growth, heat, warmth or any other concept associated with the Sun cannot overcome the mountain of evidence concerning the Two Great Lights of Genesis.

There is another difficulty with the claim that the Sun depicts Christ: if he is pictured by the Sun then how is the Father pictured? There is no answer to this that squares with scripture. And if Christ is pictured by the Sun who is pictured by the Moon? Some may argue that that is the Church, but the Church is only a temporary institution that ceases to exist at Christ's return.

The two Great Lights, as viewed from the Earth, are *monumental* and self-evidently of an **equivalence to one another**. The Church is in no way equivalent to Christ or the Father, so the Moon cannot represent it. They are permanent and *pre-eminent over all*. The Church, the 'little flock', is not pre-eminent. The Sun and the Moon appear in the first few verses of scripture and are its first symbols. The idea that one of them pictures an almost anonymous *"little flock"* who are *"strangers and pilgrims"* to this world (Luke 12:32, Matthew 11:25, Hebrews 11:13) is risible. That small and persecuted group, as described in the Bible, has certainly not been a great or continuous light to the world, but rather driven underground for centuries at a time (for 1260 years, Revelation 12:6). God's true believers have not been pre-eminent, but mostly scattered, usually persecuted, and *certainly not monumental*. Obedience to the commandments of God does not endear one to society. The lot of such people is described in Hebrews:

> *"…others were tortured, not accepting deliverance: that they might obtain a better resurrection…v37 They were stoned, they*

were sawn asunder…they wandered about in sheepskins and goatskins; being destitute, afflicted, tormented v38 (Of whom the world was not worthy) they wandered in deserts, and in mountains, and in dens and caves of the earth." Hebrews 11: 35

The True Church is a temporary spiritual institution which has never, for very long, been influential and has not been monumental in this world. The Moon, emphatically, does not picture the Church of God. The Sun must therefore picture the Father and the Moon his Christ. The equivalence of the Sun and Moon as viewed from Earth pictures the equivalence of The Father and Christ, as both are deities. The light of the Sun's flames is reflected by the lesser being, the messenger of the New Covenant, who made it clear that *"My Father is greater than I."* and that *"I have kept my Father's commandments."* and *"Not my will but thine [His] be done."* A messenger cannot be as great as the one for whom he carries a message. The symbolism fits perfectly and in only one way. The Two Great Lights are here now revealed for all fair minded men and women to consider.

The value of pi

A final point regarding the use of scripture will be helpful to the reader. In a 2007 public lecture on the subject of the number pi, sponsored by Gresham College, London it was stated that the Bible gives a value for pi of three. This is completely untrue. The Bible never refers by name to pi. Neither does it give instructions for the fabrication of any object that requires a value of pi as 3.0. The lecture was supported by a handout which states:

> *"…the Egyptians found the area of a circle…a value of pi of about 3.16…these values are better than the Biblical value given a thousand years later. In I Kings VII, 23 and II Chronicles IV, 2, we read: 'Also, he made a molten sea of ten cubits from brim to brim, round in compass, and five cubits the height thereof, and a line of thirty cubits did compass it round about.' This gives a value of $\pi = 30/10 = 3$."*

The standard of academic rigour in the above statement approximates to the lowest levels of tabloid journalism. This particular slur is often cast upon the scriptures in an attempt to discredit them.

Firstly, there is no *"biblical value given"* for pi. Instructions had been given to artisans for the construction of a large ornamental bowl about fifteen feet wide and taller than a man. Decimals are nowhere used in scripture; they would not have been understood circa 1400 B.C. when the passage was written. The universal Bible convention for all instructions given for the building of tabernacles, temples and associated artifacts is the use of round (whole) numbers. A specified diameter of the round top (from *"sea to sea"*) of 9.5493 cubits would have mystified everyone.

This giant bowl was to be an object of beauty in a temple. Such an enormous artifact would need to be styled in some way so as to avoid the appearance of a miniature gas-works. Such styling would have been highly likely to incorporate an overhanging, or outwardly curving, lip. The obvious feature of a curving lip could augment the otherwise 9.5493 cubits of the wall diameter, creating the required 10 cubit span liquid surface. So from *"sea to sea"* the specified width with *liquid brimming over* creates exactly the necessary ten cubits. An outwardly curved edge of a mere four inches from the vertical (about 10cm) would suffice. In that way, ten cubits would also comprise the true maximum width vis-à-vis the removal of the object, which is a useful piece of information. By the use of an overhanging lip the conundrum is resolved and the instructions can be followed precisely.

One cannot discredit a source by quoting a value never given by it, nor prove a source wrong if the information required to do so isn't explicitly or unmistakably stated in it. It cannot be said in truth of the Bible: "This gives a value of π..." yet such a distortion, using the very such wording *"biblical value given"* is made by those who claim to be authorities in their subject. But no 'biblical value' is given at all! It is those who use these methods that stand discredited.

David Blatner in his book *The Joy of Pi* states *"The Bible is very clear on where it stands regarding pi."* But how can that be right when no properly constituted court could possibly agree with him. In court, written evidence must always be read accurately. It would be convenient for those who treat the world's greatest historical document with contempt if it explicitly stated '$\pi = 3.0$' but it does not. Were it to say "the ratio between a circle and its diameter is three, exactly" it would wrong, but it doesn't say this either.

The Bible does not explicitly state any value for pi, yet pi is encoded in its first 31 verses. That was humanly impossible in 1500 B.C.

49

THE 1711171 SYSTEM

*"I am a Jew, but I am enthralled by the luminous figure
of the Nazarene." Albert Einstein*

More than a year after solving the Grid puzzle the author finally
realised something about 1 7 11: the top line of the Grid is an
abbreviation for of the remarkable 1711171 Bible holy day pattern.
That relationship is obvious when the Grid top line pivots around a
central digit 1 as shown below, again invoking circles:

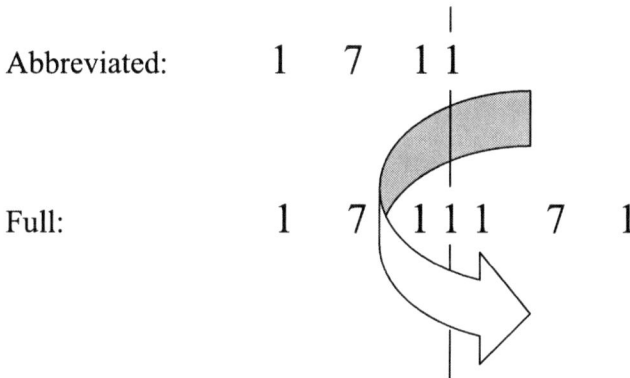

Abbreviated: 1   7   1 1

Full: 1   7   1 1 1   7   1

Again it is seen that a circular principle unlocks a mystery. The
1711171 pattern encapsulates a set of 19 days first given in *Leviticus*
chapter twenty-three. These 19 include 7 days designated by the
Bible as 'holy' days, or days of rest. Historical Christianity claims
that these annual holy days, or annual Sabbaths regulated by the
Hebrew calendar, are done away. This is untrue, as these days were
observed by the early Church in conjunction with the seventh day
Sabbath:

	Ordinary days	rest days
Passover	1	
Days of unleavened bread (2 holy)	7	2
Day of Pentecost (holy)	1	1
Day of Trumpets (holy)	1	1
Day of Atonement (holy)	1	1
Feast of Tabernacles (first day holy)	7	1
Last Great Day (holy)	1	1
Total number of days	*19*	*7*

The following excerpts prove that these days were observed in New Testament times and upheld in New Testament writings:

1 Passover – <u>death of Jesus Christ</u>. Luke 22:15 "With desire have I desired to keep this Passover with you before I suffer...v19...this do in [annual] remembrance of me." The day, first instituted at the time of the exodus, was *already part of an <u>annual</u> system* of days. Christ said in effect 'keep on keeping it.'

7 Unleavened bread – <u>putting out sin</u>. Acts 12:2 "...he [Herod] killed James the brother of John...v3...he proceeded to take Peter also. Then were the days of unleavened bread." These days are taken for granted by that author. Acts 20:6 "And we [Paul's party] sailed away from Philippi after the days of unleavened bread..."

1 Pentecost – the "first fruits" the founding of the Church, (the small harvest). Acts 2:1 *"And when the day of Pentecost was fully come, they were all with one accord in one place. v2 And suddenly there came a sound from heaven as of a rushing mighty wind...v3 And there appeared unto them cloven tongues like as of fire..."* The Church was founded on this day, hardly an indication that the day was part of an obsolete package. *Acts 20:16 "...he [Paul] hasted, if it were possible for him, to be at Jerusalem [for, or on] the day of Pentecost."*

1 Trumpets – <u>Return of Christ</u>. *I Thessalonians 4:16 "For the Lord himself shall descend from heaven...with the trump of God: and the dead in Christ shall be raised first." I Corinthians 15:52 "...the last* *trump: for trumpet shall sound, and the dead shall be raised incorruptible." Revelation 11:15 "And the seventh angel sounded [the seventh trumpet]: and there were great voices in heaven saying the kingdoms of this world are become the kingdoms of out Lord, and of his Christ; and he shall reign forever and ever."*

1 **Atonement** – <u>removal of the Devil and his works.</u> (a day of fasting) *Revelation 20:1 "And I saw an angel come down from heaven ...v2 And he laid hold on the dragon... Satan, and bound him a thousand years." Acts 27:9 "Now when much time was spent, and when sailing was now dangerous, because* **the fast** *was now already past, Paul admonished them."* It was at or close to autumn, the season for the Day of Atonement, and the weather was no longer conducive to safe sailing.

7 **Tabernacles** – <u>the millennial rule of Christ on Earth</u>. *Acts 18:21 "I [Paul] must by all means keep this feast that cometh in Jerusalem..."* This was not done to appease angry Jews. More than 18 months earlier Paul had told the Jews, *"Your blood be upon your own heads: I am clean: from henceforth I will go unto the Gentiles."* Evidently, Paul had not been in appeasement mode with the Jews when attending 'their' feast.

1 **Last Great Day** – <u>White Throne Judgment & general resurrection</u> *John 7:37 "In the last day, that great day of the feast, Jesus stood and cried, saying, if any man thirst let him come unto me and drink." Revelation 20:11 "And I saw a great white throne...v12 And I saw the dead, small and great, stand before the throne, and the books were opened..."*

These are as shown above *"...the feasts of the Lord...in their seasons"* Leviticus 23:4. The last five of the 7 digits in the 1711171 pattern are prophetic. The Day of Pentecost, the third digit in the pattern (17**1**1171), when fully understood, is a prophecy concerning a small first spiritual harvest reaped during the age of man, during which Satan is the 'god of this world' (2Corinthians 4:4). This prophecy is finally fulfilled in the 144,000 (Revelation chapter 14) at the return of Christ. At the end of the pattern (171117**1**), the much greater harvest of humanity is produced during the Great White Throne judgment period (Revelation 20). Unless you are the Devil, there is no logic in abolishing a system of days depicting God's plan, to substitute days of pagan origin that obscure the truth.

It has been shown that the symbols, or signs, of the Sun and the Moon unlock the entire scheme by which a pattern is revealed in the number pi. The *Intercalated Hebrew Calendar* runs according to the motion of the Moon. It is the same calendar system that regulates the seven annual Sabbaths. Furthermore, the seven annual Sabbaths and the weekly Sabbath, taken together as a package, constitute the only authentic Bible-based system of worship. Jesus Christ observed weekly and annual Sabbaths, both of which are grounded in the Torah (the first five books of the Bible). The weekly and annual Sabbaths stand or fall together. The days observed by Christendom are mostly pagan in origin, in particular Christmas. This feast is described in the Bible whereas Christmas trees are denounced:

> *"...learn not the way of the heathen...v3 For the customs of the people are vain: for one cutteth a tree out of the forest, the work of the hands of the workman, with the axe [Strongs 4621 maastad, to hew, an axe, tongs]. v4 They deck it with silver and with gold; they fasten it with nails and with hammers, that it move not." Jeremiah 10:2*

The main problem with these pagan feasts is that they displace the knowledge of God's feasts, by which his plan is understood. Christmas is a counterfeit holy day. By the imposition of false religious festivals the true festivals have been suppressed. This turns true religion on its head. In the genuine God given plan it is the death of Christ (Passover) that is to be commemorated, not his birth.

The book of *Acts* records the approach of the Apostle Paul to all of the Sabbaths, both weekly and annual. Clearly he kept them, but it has been claimed that he only observed these days to placate the Jews so as to be 'all things to all men.' Paul was certainly no appeaser of the Jews. That notion soon collapses in the light of the Bible evidence:

> *"...Paul testified to the Jews that Jesus was Christ. 6 And when they opposed themselves, and blasphemed, he shook his raiment*

358

*and said unto them, Your blood be upon your own heads: I am
clean: from henceforth I will go unto the Gentiles." Acts 18:5*

Yet approximately eighteen months later Paul makes this statement:

*"...I must by all means keep this Feast [of Tabernacles] that
cometh in Jerusalem." Acts 18:21*

Paul kept the Feast of Tabernacles because it remained an integral
part of the only true religion, regardless of Jewish practices. Paul
was not a man given to appeasement. He publicly rebuked the
leading apostle for favouritism:

*"...I [Paul] withstood him to the face... v12...(Peter had)
separated himself, fearing them which were of the
circumcision." Galatians 2:11*

This incident shows how little Paul was prepared to be 'all things to
all men.' His tolerance did not extend to accepting wrong doctrine or
favouritism. The Sabbath command was established from the
beginning of the age of man:

*"And on the seventh day God ended his work...and he
rested...v3 And God blessed the seventh day, and sanctified it..."
Genesis 2*

To sanctify something means to set it aside for holy use. Both Christ
and his apostle Paul observed the Sabbath as a day of rest. Today's
Orthodox Jews exhibit much the same approach as Paul's opponents
in their approach to the Sabbath. The Pharisees had devised their
own regulations and frequently clashed with Jesus Christ over his
conduct. Christ was not earning a living on the Sabbath, but he
refused to conform to their rules and regulations.

Today, an Orthodox Jew typically believes that driving a car is
wrong on the Sabbath. Some may stone your car if you drive on the
Sabbath in Jerusalem. They have invented something called an *eruv*.
This ingenious device nullifies the purpose of God within any
geographical area designated by the rabbi. A 'sin free' conclave, at
least in respect of the fourth commandment, is created by
positioning metal poles and wire to create a perimeter within which
the Sabbath can be safely 'broken.' According to their own
regulations cars can then be driven, toilet paper torn, lift buttons

pushed and shopping carried. Apparently, they feel safe from God's 'wrath' in their *eruv*, like children playing hide and seek in a dolls house. Fixated by trivia and distracted by legalism, they are inoculated from any understanding of the real purpose of the Sabbath day. What could be more ludicrous?

Another example of this stultifying mindset was reported in the The Times of June 17th 2009 which reported that a Jewish couple were suing their neighbours. Motion sensors turning on lights in a communal stairwell made it 'impossible for them to leave their flat' during the Sabbath. Lights were coming on as soon as the couple set foot outside their front door, making them responsible for switching them on in the minds of their religious teachers. Chanie Alperowitz, director of Bournmouth Chabad, an Orthodox Jewish group was quoted as saying:

"On the Sabbath there are thirty-nine forms of creative activity which are forbidden. Among them is the prohibition of lighting a fire. When using electricity, one causes a fire as there are sparks created by the electricity. If the light is switched on by someone stepping outside of their door, their actions have caused it."

This is the same turgid mentality Jesus Christ encountered when seen picking food during a Saturday stroll. The Pharisees accused him of sin, but his reply was:

"The Sabbath was made for man, not man for the Sabbath."

If the Orthodox Jew succeeds in bringing the Sabbath, and himself, into disrepute it was the Catholic Church that utterly usurped the day, changing it to Sunday - the ancient day of Sun worship. This was enforced in 321 AD with the edict of Constantine:

"Let all judges, inhabitants of the cities, and artificers, rest on the venerable day of the Sun..." Constantine, Chambers Encyclopedia, 1882 edition, vol.8, page 401

The edict of Constantine, enforcing *disobedience* to the commandment of God, has endured through to the modern age. The Catholic priest T. Enright, President of the Redemptorist Father's College, posed the following in a lecture:

"Which church does the whole civilized world obey? The Bible says: 'Remember to keep holy the Sabbath day,' but the Catholic Church says, 'No, keep the first day of the week,'... the world bows down in reverent obedience to the mandates of the Catholic Church." 1893 lecture in Des Moines, Iowa, by T. Enright.

The apostle Paul normally preached on the seventh day regardless of whether or not he was in a synagogue:

"... (Paul) went into the synagogue on the Sabbath day... v16... Paul stood up...v44...And the next Sabbath day came almost the whole city together to hear the word of God." Acts 13:14

This would have to have been an open air event as it is not possible to fit *"almost the whole city"* inside a synagogue; however, Paul did not switch to Sunday. Later in Acts 17:2 Paul preached *"three Sabbath days"* in Thessalonica. Had Sunday been the correct day of worship and the system of 19 holy days obsolete, Paul would have had to make this known. In 2 Corinthians chapter twelve Paul recounted at length from verse 1 his *"...visions and revelations of the Lord"* showing that he did not lack guidance from God.

The huge controversy in *Acts* regarding circumcision would have been dwarfed by any commotion as a result of any doing away, superseding or setting aside any of the Ten Commandments of which the Sabbath is one. The abolition of the seven annual holy days would, as with the supposed introduction of a 'trinity,' have been just as controversial. And a vast tumult would have erupted had Paul began to teach the notion that 'days don't matter.' Preachers today will glibly intone that 'Christianity is not about days.' But to God, Christianity is about anything and everything he wants to include. No such changes are recorded in scripture: there are no valid grounds for claiming that the seven annual holy days or the weekly Sabbath are biblically obsolete. Paul would have known of and fearlessly taught any such 'new truth.'

The cycle of 19 days encapsulating the seven annual Sabbaths given in Leviticus chapter twenty three, and kept by Christ, Paul and the early Church is the teaching programme *par excellence* by which the truth of God's plan can be understood. These days encapsulate the most inclusive of all religions. Following the exodus from Egypt, God made the following agreement in which this pattern is seen:

> *"...I will bring you out from under the burdens of the Egyptians, and I will rid you of their bondage, and I will redeem you with a stretched out arm, and with great judgments: v7 And I will take you to me for a people, and I will be to you a God: and you shall know that I am the Lord your God, which bringeth you out from under the burdens of the Egyptians. v8 And I will bring you in unto the land, concerning the which I did swear to give it to Abraham, to Isaac and to Jacob; and I will give it to you for an heritage: I am the Lord." Exodus 6:6-8*

The points in this passage fit the 1711171 pattern as follows:

1	Passover	*"I will bring you out"*
7	Days of u/bread	*"I will rid you of their bondage"*
1	Day of Pentecost	*"I will redeem you"*
1	Day of Trumpets	*"I will take you to me for a people"*
1	Day of Atonement	*"I will be a God to you"*
7	Feast of Tabernacles	*"I will bring you into the land"*
1	Last Great Day	*"I will give it to you for a heritage"*

The provision made for all people who have ever lived, a general resurrection, is pictured by the Last Great Day. The teaching behind this particular day is sometimes called 'divine perseverance.' This is not the same as the teaching of 'universal salvation.' In the first case, all people receive a fair chance which in almost every case occurs during a general physical *fleshly* resurrection on Earth (evidence for this is given in Appendix XV). The other idea of universality is at odds with the scriptures, as it is clear that the incorrigibly wicked are to be consumed. They will be ashes under the feet of the righteous (Malachi 4:3).

According to this God-inspired programme pictured by the numbers 1711171 no one depends on the patronage of missionaries, by whom they might have been 'saved' on their deathbed, but in whose absence they are arbitrarily consigned to 'eternal torment.' The truth is that nobody needs to be absolved by a priest and the dead do not need to be prayed for. Mary can't hear prayers because she is unconscious in death together with Noah, Moses and King David: all of these await a future resurrection.

An understanding of the annual holy days will sweep from one's mind any idle concepts concerning going to heaven or burning for ever in hell, or rosaries and indulgences. A fair opportunity awaits all. The holy days are the *skeleton on which to hang the truth* of the Bible.

The Deities of the Grid are *identified* by the system of worship instituted by them. The pattern 1711171 represents God's system of true worship, keeping the plan of God in remembrance. That is anathema to the apostate, politicised churches of this present world. They will be anything but comfortable at the return of Christ to this Earth. Isaiah describes that pivotal event, depicted by the centre digit of the 1711171 plan. He will not come as gentle lamb the second time around, but as a roaring lion:

> "...they shall go into the caves...for fear of the Lord...
> when he ariseth to shake terribly the earth." Isaiah 2:19

50

THE DEITY PRIME SEQUENCE

"[Regarding] order in the sequence of primes...we have reason to believe that it is a mystery into which the mind will never penetrate." Leonard Euler

Firstly, in this next discovery, the defining numbers of the Grid, 7, 11 and 44 can be seen within a *Deity Prime Sequence* of prime numbers. In this observation the Deity Prime Sequence constitutes the first 32 primes up to and including the prime number 131. These 32 primes are as follows:

2, 3, 5, 7, 11, 13, 17, 19, 23, 29, 31, 37, 41, 43, 47, 53, 59, 61, 67, 71, 73, 79, 83, 89, 97, 101, 103, 107, 109, 113, 127, 131

Many of the above form pairs separated by a gap of four, the first instance being 7 and 11 themselves. There are 11 such pairs in this sequence as highlighted:

2, 3, 5, <u>7 11</u>, <u>13 17</u>, <u>19 23</u>, 29, 31, <u>37 41</u> , <u>43 47</u>, 53, 59, 61, <u>67 71</u>, 73, <u>79 83</u>, 89, <u>97 101</u>, <u>103 107</u>, <u>109 113</u>, <u>127 131</u>

The defining numbers of the Grid, 7, 11 and 44, are evident within this set of eleven in following ways:

1. The first pair separated by four is 7 and 11

2. The 11 pairs possess a '4-gap' encoding the number 44

3. The 11th pair begins with the 31st prime, the number 127. And 31 is itself the 11th prime.

Intriguingly, 7 and 11 of themselves cryptically suggest 44, as 7 is the fourth prime and 11 is found four integers further on (7, 8, 9, 10,

11). But all across the range of primes from 2-131 key numbers pertaining to the Grid discovery are compellingly in evidence. For example, the last eight primes in the series (comprising the last four pairs) total 888.

However, what is really striking is that the first pair possessed of a 4-gap, 7 and 11, complete a set of the first five primes 2, 3, 5, 7, 11. These total 28, the teaching number. Here there is a definite pattern because the last five primes in the range (107, 109, 113, 127, 131) begin with the 28th prime, 107:

First five total 28

2, 3, 5, 7, 11, 13, 17, 19, 23, 29, 31, 37, 41, 43, 47, 53, 59, 61,

Last five begin with 28th prime

67, 71, 73, 79, 83, 89, 97, 101, 103, 107, 109, 113, 127, 131

The last 8 primes total **888**

Intriguingly there are twenty-two primes remaining in the centre section between the two sets of five (ranging 13 to 103, a total of 1236 which is 12 ×103). Pairing these twenty-two numbers symmetrically, as in the manner 13+103, 17+101, etc. produces the following eleven numbers:

116, 118, 116, 112, 112, 110, 110, 112, 110, 108, 112

The number 4 is found by adding and subtracting the value of the spaces in this series, which from the left would be plus 2, then minus 2, then minus 4 and so on. This produces a net figure of 4, or minus 4 when counting is started at the right hand end. As 11 pairs of primes encapsulate the number 4, then 44 is encoded.

But there is a simpler way in which 44 is represented within the first thirty-two primes, as the series falls neatly into two squares suggesting pictorially - as squares of 4 - the number 44:

2	3	5	7	59	61	67	71
11	13	17	19	73	79	83	89
23	29	31	37	97	101	103	107
41	43	47	53	109	113	127	131

The Grid pattern is also seen in the following vertical arrangement of the eleven 4-pairs. This shows again the symmetry that exists within this Deity Prime Sequence, confirming the fundamental nature of the Genesis Grid and its relationship to a pattern in prime numbers not glimpsed until now:

1	2	3	4	5	6	7	8	9	10	11	
7	13	19	37	43	57	79	97	103	109	**127**	(31st)
11	17	23	41	47	71	83	101	107	113	131	

Here, seven and eleven head up a set of eleven pairs, the last of which codes the deity number. In this way the Grid and *The Signal* of 127 days (previous chapter) are linked. Evidently the Periodic Table of elements and music (containing as they do the defining Grid numbers 7, 11 and 44) are therefore also linked, via the Grid, to number theory.

The *Deity Prime Sequence* exhibits symmetry nowhere else seen in the primes. The defining numbers of the Grid 7, 11 and 44 are clear and present in this set. As if any more evidence was required, the last eight primes in the set total 888.

In summary, the defining numbers of the Grid, 7, 11 and 44, and the 'surrounding' number 28 are seen in the Deity Prime Sequence in the following ways:

7	This appears in the first of the 4-pairs.
11	This also appears in the first of the 4-pairs.
44	There are eleven 4-pairs, highlighting the number 44. Also, 44 is seen in the arrangement of two four-squares. Then the 11th 4-square is highlighted by the fact that it codes the number of deity, suggesting 44.
28	This marks the first five and last five of the set. In doing so it 'surrounds' all of the other numbers, just as 28 surrounds the Grid. Also, five plus five is ten, suggesting 280.

Finally, linking the discovery to the hidden structure of Genesis chapter one there is one more number:

888	This appears at the end of the series. The sum of the last 8 primes is 888.

51

THE PROOF OF ELEVEN

"It would be very difficult to explain why the universe should have begun just this way except as the act of a God who intended to create beings like us." Professor Stephen Hawking

The Supreme Being, long hidden from the religions of this present evil world, has now been identified by the number 11. This number is found in his creation and in his book the Bible.

The Grid via its associated numbers and patterns appears in *eleven* Bible passages. This it does in different ways, for example in John it appears in its entirety. But in Genesis the last Grid number pi appears first. Genesis chapter one is built on pi.

The second way the Grid appears is again in Genesis chapter one. This is as the pattern 11/44. Its double appearance is seen within two commands, both comprised of 11 words and 44 characters. These are the first instructions to mankind in the Bible and appear in the 28th verse, the teaching number of the Bible! In Exodus, the Grid's derivative numbers 107 and 88 construct the Ten Commandments. Later in that book there is seen the 7 and 11 of the Tabernacle, surrounded by 28's and 280 – just like the Grid itself. Then the book of Deuteronomy links together 107 and 44, as does The Psalms.

Isaiah contains an allusion to the theme of reproduction. This has not been previously mentioned here. But like the Acts of the Apostles it encodes the number 44, because the 44th chapter contains - uniquely for Isaiah - a reference both to seed and the womb. Furthermore, Isaiah takes the form of a mini-Bible, its chapters reflecting not only the Bible's 66 books but its division into the Old (39) and New (27) Testaments. The ways in which the Grid appears in the Bible have already been shown detail in earlier chapters, but can be summarised here:

1. Genesis: structure of chapter one based on pi, the 7th Grid number
2. Exodus, the Ten Commandments. Related numbers 88 and 107
3. Exodus, the Tabernacle. Grid numbers 7, 11, 28 and 280
4. Deuteronomy, primes; relationship to Psalms highlights 107/44
5. The Psalms; their relationship to Deuteronomy highlights 44
6. Isaiah; its 44th chapter and position in the 3/22 Bible structure
7. Matthew's Sermon on the Mount; built on Grid related numbers
8. The Gospel of John; it encodes the Grid in its entirety
9. The Acts of the Apostles, holy spirit (flames) and the number 44
10. Romans,11th chapter; its position in the 3/22 Bible structure
11. Revelation, the seven spirits of God give Grid top line 1 7 11

The Grid and related numbers appear in the 66 books as follows:

Genesis, pi, 11, 44, 28	Isaiah, 44	Romans 7, 11
Ex. 107, 88, 7, 11, 28	Jeremiah	1 Corinthians
Leviticus	Lamentations	2 Corinthians
Numbers	Ezekiel	Galatians
Deut. 44, 107	Daniel	Ephesians
Joshua	Hosea	Philippians
Judges	Joel	Colossians
Ruth	Amos	1 Thess.
1 Samuel	Obadiah	2 Thess.
2 Samuel	Jonah	1 Timothy
1 Kings	Micah	2 Timothy
2 Kings	Nahum	Titus
1 Chronicals	Habakkuk	Philemon
2 Chronicals	Zephaniah	Hebrews
Ezra	Haggai	James
Nehemiah	Zechariah	1 Peter
Esther	Malachi	2 Peter
Job	Matt. 7, 11, 44, 111, 400	1 John
Psalms 107, 44	Mark	2 John
Proverbs	Luke	3 John
Ecclesiastes	John ENTIRE GRID & 107	Jude
Song of Solomon	Acts, 44	Revelation 1 7 11

As there is a double occurrence of the Grid in Exodus (the Ten Commandments and the Tabernacle) these eleven occurrences are found in ten books. These ten fall into two sets: five in the Old Testament and five in the New. The ten books' positional numbers total 288, or 2 × 144.

These eleven occurrences of the Grid are bound together in a twelfth: the overarching three-column Grid structure of the sixty-six books of the Bible shown in the earlier chapter *A Secret Architecture*. Twelve is the number of organisational beginnings, hence the twelve apostles.

Once more, the Grid is shown to be real and with a proof that substantiates the order and number of the sixty-six books of the Bible.

52

THE KEY PATTERN 888

"No one can read the Gospels without feeling the actual presence of Jesus. His personality pulsates in every word. No myth is filled with such life." Albert Einstein

A casual visit to a charity shop resulted in finding the first ever pattern in pi.

The clue was in a small secondhand book *The Joy of pi*, by David Blatner; price, fifty pence. Some unevenly printed digits drew the author's attention to the pattern 888, the key that unlocked the first 31 digits of pi, as will be explained in this chapter.

Why should 8 suddenly be so important when its numerical neighbour 7 has always seemed to hog the limelight? Could 8 knock 7 from its perch of importance, and if so might the 7 of the Grid as Christ's identity number be wrong?

In the book *Number in Scripture* Ethelbert W. Bullinger asserted that 8 is the 'dominical' number; or, that which pertains to Jesus Christ as his principal number. Bullinger considered the number 8 to be more characteristic of Christ than the number 7, but he was mistakenly following in the mainstream evangelical thinking. Objectively, the number 7 is more fundamental than 8, because it is only the completion of a 7-day-cycle that determines any meaning for 8 as a 'new beginning.' But is it more fundamental to Christ?

The suspicion must be that it is the presumption of a Sunday resurrection, according to the Easter Sunday tradition, that really underlies the evangelical claim of 8 as the dominical number (where Sunday would be the 8th day). Yet a Sunday resurrection cannot be reconciled with the tradition of a crucifixion on Friday. Christ gave as his only sign the literal fact (cross referenced to Jonah) that he would be buried for *three days and three nights*, a time frame that does not fit an Easter Sunday resurrection. The only scenario that fits all of scripture and history is a Wednesday crucifixion and a Saturday resurrection, shortly before sunset in each case, a period which would encompass the necessary 72 hours. The dates given for this by Bullinger in the *Companion Bible* are Wednesday the 25th and Saturday the 28th of April 31 A.D. (page 197, Appendix 179, *Parallel datings of the times of our Lord*). These dates are correct.

The Grid shows that 7 is the principal 'identity number' of Christ, as do many plain Bible facts. Christ was 'Lord of the Sabbath,' the *seventh* day. His title 'Christ' has 7 letters in the Greek. Yet while it is necessary to take issue with Bullinger over the claimed pre-eminence of the number 8 for Christ, that number does have meaning with regard to the name *Jesus*. However, the Messiah aspect is greater and more important than his brief human incarnation. Therefore the number 7 must have primacy of place.

A bias toward the number 8 is not surprising in a theologian who, as a part of the nineteenth century academic establishment, would not have kept the Saturday Sabbath observed by Christ, Paul and the early Church. But the number 8 has, very logically, been linked with the idea of redemption, or the 'saving' of humans from their old selves through a religious conversion. The number 8 provides a ready meaning of 'redemption' because Sunday - as we call it today - starts the *new* week. Noah and his family, who together totalled 8 in number, started the *new* human family. The gematria of his name is 64. Therefore 8 can denote a new start.

The Sunday keeping tradition, imposed by the Catholic Church many centuries ago, imposes the 8th day in place of the 7th day. But the Grid contains the number 7 as Christ's identifying number and he is *Lord of the Sabbath*, the 7th day. Thus the Grid is in harmony with the scriptures and tradition is at variance with them.

Yet there is the well known numerical label of 888 for Jesus. But that, it bears repeating, is his name as a *human being*. That is the part corresponding to 888, whereas the *Messiah aspect* corresponds to the number 7 (where the Greek for Christ has seven letters: χριστος). A key point is that it is *humans* who can be marked with the number 8. It is human beings who are in need of redemption and a new beginning, *not divine personages*. That is what Bullinger overlooked. He also omits to mention that the eighth day of The Feast of Tabernacles is in fact an entirely separate festival - *The Last Great Day*. It is the eighth day of that festival season and pictures the redemption of humans in a general (physical) resurrection following the millennial rule of Christ on the Earth. The number 8 pertains therefore to the 'saving' of humans; it is to do with those in the flesh - having been *resurrected in the flesh* - not deities. That resurrection is pictured by the last day of the cycle of holy days. Denoted by the number 1, it completes the symmetrical 1711171 pattern -19 days in all. Again, that pattern is as follows:

Passover	1
Days of unleavened bread	7
Day of Pentecost	1
Day of Trumpets	1
Day of Atonement	1
Feast of Tabernacles	7
Last Great Day	1
Total number of days	19

Seven is pre-eminent within this scheme in several ways. Firstly there are seven digits of which two happen to be 7. Then 7 annual holy days are incorporated within this cycle of 19 days regulated by the Hebrew calendar, not eight. These holy days are referred to as Sabbaths in the Bible, annual Sabbaths, and like the weekly Sabbaths no servile work was performed on them. Christ as *Lord of the Sabbath* observed weekly and annual Sabbaths. Thus it is obvious that the number 7, as in the 7th day and 7 annual Sabbaths, is fundamental to and characteristic of Christ the divine teacher and his teaching program. The number 8 is only characteristic of his *human* name Jesus.

However, once the 888 key to pi was tested, it was the 6+7 aspect to the name Jesus Christ that was crucial to eventually deciphering the 31 digit Deity String. The number that applies to the complete name *Jesus Christ* is therefore 13, as there are 6 Greek characters in *Jesus* and 7 Greek characters in *Christ*. He was the antidote to rebellion and apostasy, denoted by 13. Bullinger observed of Jesus' home town Nazareth (*"He shall be called a Nazarene"* Matthew 2:23) that the name occurs 13 times in the New Testament:

> *Number in Scripture, page 29 "Ναζωραῖος (Nazarethan) occurs 13 times."*

Bullinger comments on the number 13 as it pertains to Jesus Christ by way of '6+7' as follows:

> *"When we remember that six is the human number, and seven the Divine, can we doubt that we are thus pointed to the fact that Jesus was both Son of God and Son of Man? His two names have the same significant stamp and seal: for Ἰησοῦς, Jesus, the birth name of His humiliation as Man, is composed of six letters; while Χριστός, Christ, His Divine title as the anointed of God, is composed of seven letters." Number in Scripture, page 161*

Jesus Christ, possessing 13 Greek letters in his name, 'became sin' to pay the death penalty for the sins of humanity. This is point is made in the second book of Corinthians:

> *"For he [God] hath made him [Christ] [to be] sin for us, who knew no sin." 2 Corinthians 5:21*

He was killed to pay for the sins of humanity and this is *proof of Christ's status as a deity.* It seems logical that the death of a God would be sufficient to pay for the sins of billions of humans.

Further confirmation of the importance of the numbers 6 and 7 as linked to Jesus Christ is found in his two genealogies. The royal genealogy traces the *descent of kings* in the book of Matthew; the human genealogy traces the *ascent of a man* in the book of Luke. One gives the royal line, that is, the line of legal succession through Solomon; the other gives the line of natural descent through Solomon's elder brother Nathan[2].

The genealogy of Luke begins with God at one end and Jesus at the other: it contains exactly 77 names (11×7). It is significant that the *royal* lineage in Matthew through Solomon contains 66 names (11×6). That genealogy begins not with God but Abraham. Jesus Christ the 'Son of man' is therefore linked to both 6 (royal but human, from Abraham) and 7 (divine, from God) through the two genealogies, which again marks on him the number 13 (6+7).

The adversaries of God are also stamped with the number 13 as Bullinger goes on to show:

Page 220

Genesis chapter three where the Devil is first mentioned and revealed. The opening words: "The serpent was more subtle than all the beasts of the field"

= 1521 (13² × 9)

Page 224-6

*John 12:4 "Judas Iscariot, he that should betray him" = 4511 (13 * 347)*

*OTHER ADVERSARIES Simon Magus = 1170 (13 * 90)*

Yet Christ is marked with the number 13. Bullinger explains:

Page 228

THE CONNECTION OF THE NUMBER THIRTEEN WITH SUBSTITUTION AND ATONEMENT. The Saviour, though without sin, was "made sin", or a sin offering, for his people. He was "wounded for our transgressions", and bruised for their iniquities. He was, in fact, "NUMBERED WITH THE TRANSGRESSORS" (Isaiah 53:12) Therefore this number is not only the all-pervading factor of SIN, but of sin's atonement. It is not only the number that brands the sinner as a rebel against God, but it is the number borne by the sinner's Substitute. His very names in the Old Testament, before the work of atonement was entered on or accomplished, are all multiples of 13…

*Jehovah = 26 (13 * 2), Adonai 65 (13 * 5),*

*Ha-Elohim = 91 (13 * 7)*

It was the 6 + 7 and 888 patterns that provide the key to unlocking the first 31 digits of pi. Those 31 digits have been figuratively staring out into an uncomprehending world for four hundred years from the gravestone of the Dutch mathematician who first calculated them, Ludolf van Ceulen.

A providential printing error resulted in the 888 pattern being noticed by the author. In the book *The Joy of pi,* by David Blatner, the first 32 digits of pi appear on page 112. Although that hardbound book is immaculately produced, the three 8's within the string are strangely misaligned to stand proud of their neighboring digits. If it were not for that printing imperfection, the discovery of the Genesis Grid may have taken a further four hundred years.

Further information about the 888 pattern is given in Appendix XV.

1. One can argue that the 1711171 pattern forms a loop, as the first "1" digit corresponds to Passover (both new and old) at which time Christ's blood (previously pictured by lamb's blood) covers the sins of the people. These sins are judged at the time of the great physical resurrection of Revelation chapter twenty. Because this time is pictured by *The Last Great Day,* corresponding to the last '1' in 1711171, the beginning joins the end and a circle is implied.

2. The two genealogies of Christ are explained more fully in *Number in Scripture* page 158.

53

THE ULTIMATE PLAN

*"He that overcometh shall inherit all things; and I will be his God,
and he shall be my son."* Revelation 21:7

The present investigation has uncovered powerful new proofs for the inspiration of the Bible. This has been done to a large degree through pattern recognition. There are no reams of movable data from which to fabricate such patterns. The patterns leap out from the page of themselves, as vivid and inevitable as the Periodic Table of elements.

Apart from the need to access remote values of pi, this discovery has been made without computers and using simple facts set, metaphorically speaking, in concrete. It is worth quantifying those facts and the simple manner in which they can be assembled:

1. The Sun and the Moon are the same diameter as seen from Earth, their relative diameters and distances are set in a ratio of 400:1 and they are the first symbols (or signs) mentioned in the Bible.

2. The number 11 suggests the Sun as a symbol due to its visible cycle of sunspots. The Sun's internal motion of four layers supports 4 as a secondary number. It was found that a 'law of 11' operates within number patterns (e.g. Pascal's Triangle) and that a *'three to the power of four law'* exemplifies the number 4 as the determinant factor in matter, both living and inert.

3. The Moon is marked with the number 7, as it has an albedo (light reflecting power) of 7, another factor relating to visibility; its secondary number 40 is supported by the lunar eclipse cycle of 173 days, this being the 40th prime number. Thus each heavenly body has a primary *identity* number

derived from *visible factors - specifically the emission of light* - while each secondary number is associated with power and motion.

4. A system of 'submerged numbers' on which the Bible is in part based includes a never before expounded number 28, the 'teaching number.'

5. A theme of duality is prominent throughout.

6. A correspondence between the Torah and the Psalms reveals the use of prime numbers. The Grid is validated by the total 107 which is the 28th prime number, a number play seen in the well known match between the two books.

7. The number 44 is shown to be the number of reproduction. The Greek word *sperma* (seed) occurs 44 times. This is reinforced by the Hebrew gematria of mother, father & child, these being 41, 3 and 44 respectively. The number is also marked on human reproduction in three ways; the dimensions of the ovum/sperm are in the ratio 1:44, DNA contains 44 autosomes and 2 sex chromosomes and the conjugal act encodes the number 88.

8. A clue for pi is seen in the observation of a circular motion of 44's in the Grid. It is derived from the outside columns - in effect a 44×2 pattern. Because the bottom line of the Grid totals 88, a circle and a line is suggested.

9. The Bible count for the 'spirit' of 308 of the 'spirit' tallies with the seven coloured Rainbow where $7 \times 44 = 308$. The Father's teaching program is $11 \times 28 = 308$. It tallies with 28 and 280 to unify a set of Grid derived numbers as in $308 = 280 + 28$.

10. The Grid is seen in pi in the form 1711 and 44404 in self-characteristic positions.

11. The full Grid string 171144404 is found in pi positions that are self-characteristic.

12. No further patterns can be found in subsequent pi positions of 1711 and 44404, or presumably 171144404 if subsequent

380

positions could be searched. In both cases positions 1 and 2, and their gap, show a pattern.

13. The 888, 6/7, 11/44 and 27/39 patterns encode the two Deities and the 66 books of the Bible in the first 31 digits of pi.

14. The structure of the first 31 verses of the Bible corresponds to the four segments of the first 31 digits of pi. These encode the two deities and their word. Four segments follow a 3^4 law also seen in the 81 stable elements, the structure of the atom and DNA.

15. The 28th verse of the Bible, the first command of God to the human family, contains a dual confirmation of the 11/44 pattern.

16. The decimal point in pi encodes the Great Breach of Genesis 1:1-2.

17. The three defining numbers of the Grid 7, 11 and 44 appear within a musical system encompassing 88 notes.

18. The scheme of world history in prophecy is structured according to the Grid numbers 1, 7, 11 and 4. The Grid pattern 444 occurs at major historical turning points from the fall of ancient Babylon to the United States of Europe. The number 444 is found to be the 'kingdom number' that appears at a time of change or the imminent fall of a world power. Two Americans signalled this number 444 in London during 2003, the year of the Iraq invasion.

19. Events in 2003 signalled the numbers 7, 11 and 44. A magician's fast of 44 days overlapped symmetrically with a period of 44 days. The exact 50:50 overlap defined a period of 66 days from the first manifestation of a power cut that started in Manhattan, and finished at the end of the Blaine fast. Blaine lives in New York. From the launch of *The Da Vinci Code* to end of the 16 days of solar flares the days elapsed were 127, the 31st prime number.

20. The fast of David Blaine corresponded with the 1711171 pattern. The centre of the pattern, the *Day of Trumpets* picturing Christ's return, coincided with Blaine's mid-point and the blacking out of Rome, interrupting the ordination of 31 new cardinals.

21. A period of 44 days and 8 power cuts marked out a figure of 8 across the globe, centered on Rome. The gematria of the name Noah is shown to be 8^2.

22. The Grid discovery is affirmed by the eleven Bible passages listed in the earlier chapter *The Proof of Eleven*.

23. The top line of the Grid is found to be an abbreviated form of the holy day pattern 1711171, a system of 19 days incorporating the 7 annual holy days of Leviticus regulated by the Intercalated Hebrew Calendar, the only authentic Bible system of religious observance.

24. The defining numbers of the Grid are seen in musical frequencies and intervals. The Grid pattern 1 7 11 is seen in the Periodic Table of elements twice. A symmetrical pattern in the first 32 prime numbers accords with the Grid.

These twenty-four findings demonstrate that the Grid is not only real, but logically must predate the physical universe. The pattern 1711171 is evidently the 'mind of God' in numerical shorthand. It affirms his plan which is also the theme of the Grid: *God is reproducing Himself.*

54

THE ALIEN

*"Who hath ascended up into heaven, or descended? Who hath gathered
the wind in his fists? Who hath bound the waters in a garment? Who hath
established all the ends of the earth? What is his name, and what is his
son's name, if thou canst tell?" Proverbs 30:4*

This investigation has uncovered fresh confirmation of the fact that
the world's churches, denominations and sects since the time of the
apostolic Church have veered off the track of truth. The writers of
the Bible anticipated that this would happen. It was predicted in
many prophecies. One might expect the nations, among which
billions claim allegiance to Christ, to rejoice at the skyward
appearance of a Deity coming down to Earth to enforce peace. This
great Personage would have announced his arrival beforehand. Why
then does Bible prophecy show that humankind will express *sorrow*
at his imminent arrival?

> *"And then shall appear the sign of the Son of man in heaven:
> and then shall all the tribes (nations) of the earth **mourn** (Gk.
> kopto: to beat the breast in grief), and they shall see the Son of
> man coming in the clouds of heaven with power and great
> glory." Matthew 24: 30*

There can only be one explanation for this rejection: the nations will
have been deceived into believing that the returning Christ is an
enemy, hostile to humanity - an *Alien* to this world! *The S.E.T.I.
mind-set is not about to disappear.* Such a response will be possible
because the nations do not know *who* or *what* God really is. The
churches have defaulted and will continue to utterly fail to teach the
masses, because this world is alienated from God. It does not know
who or what he is. It does not know his purpose. God is not calling
the world now. No mass conversions will yet occur. Christ declared:

"None can come to me except the Father draw him" John 6:44

Christ spoke in parables to **prevent** the masses in this present age understanding his message. He said:

"I thank thee O Father...thou haste hid these things from the wise and the prudent, and revealed them unto babes." Matthew 11:25

The world's great minds have never understood spiritual truth. Michael Drosnin[1] in the book *The Bible Code* recorded the following concerning Sir Isaac Newton's attempts to decipher the Scriptures:

"Isaac Newton was the first modern scientist; he worked out the mechanics of the solar system and discovered the force of gravity. He was certain, too, that there was a hidden code in the Bible that would reveal the future. He learned Hebrew, and spent half his life trying to find it. In fact, according to his biographer John Maynard Keynes, it became an obsession. When Keynes became provost at Cambridge University, he discovered there the papers that Newton had packed up in 1696 when he had retired as provost. Keynes was shocked. Most of the million words in Newton's own handwriting were not about mathematics or astronomy, but esoteric theology. They revealed that the great physicist believed there was hidden in The Bible a prophecy of human history. Newton, said Keynes, was certain the Bible, indeed the whole Universe, was a 'cryptogram set by The Almighty', and wanted to 'read the riddle of The Godhead', the riddle of past and future events Divinely Foreordained. Newton was still searching for the Bible code when he died. But his lifetime quest failed no matter what mathematical model he applied."

This inability of the great to understand God's purpose is referred to by the apostle Paul, writing to the self-important Corinthians:

"For you see your calling brethren, how that not many wise men after the flesh, not many mighty, not many noble are called; but God hath chosen the foolish things of the world to confound the wise; and God hath chosen the weak things to confound the mighty; and base things of the world, and things which are despised hath God chosen, and things which are not, to bring to

nought things that are: that no flesh should glory in his presence." I Corinthians 1:26

God is not calling the world now and he is not working through the Christian denominations and their theologians, archbishops, popes or other mighty or learned men. God does not need their human strength: it only gets in the way. The Apostle Paul commented on his many infirmities and the healing that was withheld from him in his second letter to the Corinthians:

"And he said unto me, my grace is sufficient for thee: for my strength is made perfect in weakness." 2Cor 12:9

In the present era God is dealing principally with the weak of this world. Even the twelve apostles were unlearned men, lacking in higher education. He has not set about 'saving' the world in this present age, using the mighty, or he would have accomplished it. That process will begin after the great conflagration in which man finally comes to see his utter helplessness in governing his affairs, and the futility of his own religions. Under the sway of the Great False Church and its daughter churches, the demonic forces of the Earth will cause the nations to join in fighting their Creator. This will be the *only* time in history, other than at Babel, that the nations will have been united:

"And I saw three unclean spirits like frogs come out of the mouth of the dragon (Satan), and out of the mouth of the beast and out of the mouth of the false prophet. v14 For they are the spirits of devils, working miracles, which go forth unto the kings of the earth and of the whole world, to gather them to the battle of that great day of God Almighty…v16 And he gathered them together into a place called in the Hebrew tongue Armageddon." Revelation 16:13

The battle by these newly *united nations* against the returning Christ will be incited by and through the one claiming to be the vicar (as, in place of) of Christ. This uniting of the nations to fight Christ at his return is described in the following prophecy:

"I will gather all nations against Jerusalem…v3 Then shall the Lord go forth and fight against those nations…v4 And his feet shall stand in that day upon the mount of Olives…v9 And the

Lord shall be king over all the earth…v12 And this shall be the plague wherewith the Lord will smite all the people that have fought against Jerusalem; their flesh shall consume away while they stand upon their feet, and their eyes shall consume away in their holes, and their tongue shall consume away in their mouth." Zechariah 14:2

Immediately after this massive slaughter of rebellious mankind the true religion of Jesus Christ will be implemented and all other religions abolished by decree:

"And it shall come to pass that everyone that is left of all the nations which came against Jerusalem shall even go up from year to year to worship the King, the Lord of hosts, and to keep the Feast of Tabernacles." Zechariah 14:16

Thus the 1711171 pattern of true worship picturing the plan of God will be reinstituted. This pattern represents the one true religion. In *Acts* chapter seventeen Paul preached by Mars' hill in Athens, a city *"wholly given to idolatry."* Like today's world, Athens was a Babylon of competing ideas, philosophies, theories, faiths and gods. To the people of Mars Hill, as today, the true God was an *Alien*. One altar caught Paul's eye:

"For as I passed by, and beheld your devotions, I found an altar with this inscription TO THE UNKNOWN GOD. Whom therefore ye ignorantly worship, him I declare unto you." Acts 17:23

The two who made this world have been *aliens* to every human society. They are the unknown family of God.

1. Michael Drosnin himself has been ambivalent about the claimed codes in his book *The Bible Code*. The book says, "I found the Bible Codes' prediction of [Rabin's] assassination myself. . . . When he was killed, as predicted, where predicted, my first thought was 'It's real' But in a CNN interview he said, "I don't think the code makes predictions. I think it might tell us about possible futures." (see Leaderu.com)

55

BULLINGER'S QUESTION

"Now this is not the end. It is not even the beginning of the end.
But it is, perhaps, the end of the beginning." Winston Churchill

The present book sets out to answer Bullinger's question regarding the meaning of the prime numbers 7 and 11 as they relate to Jesus Christ and the Father, and show why they are encoded in the Bible in this way. In finding the answer six things have been established:

1. The inspiration of the Bible is demonstrated through a new explanation of its structure. The description of the Bible in the introduction to the King James Version as *"that inestimable treasure, which excelleth all the riches of the earth"* is here vindicated.

2. It has been demonstrated that the Grid derived from '7 & 11' is fundamental to the structure of the universe and its laws. The Grid is glimpsed in the Periodic Table of elements, musical frequencies and intervals, the prime numbers and pi - not to mention the human reproductive system and the Bible. These and other patterns in nature (e.g. the four forces) align together under the general umbrella of a 3^4 law.

3. The present book confirms the true nature of God (as reflected by the structure of the Grid) and that God has always been a family of two Deities. The term 'God' in the Bible describes both individuals and a group. Moreover, the holy spirit is shown to be a force relating to both Deities, not a personage. It remains true that there is only one true God, the family of God.

4. This book gives timely warning of a miracle-working false teacher operating from within the auspices of the Great False Church. He must appear within the next several decades. This leader will, through the offices of a United States of Europe, trigger global conflict beginning with an invasion of the Holy Land. Anglo-Saxon nations will also at that time be attacked.

5. Prophecies of the Bible have been brought into focus to reveal the immediate future. The sign of Noah, the figure of 8, has appeared with its cross over Rome. Just as the corrupted world in Noah's day was destroyed, the coming nightmare of a supranational government fulfilling Hitler's dreams for the European continent will be cut short. A 42 month period of hellish warfare will lead into the *Day of the Lord*, a several months long period of intense punishment on the world, the forerunner for which was the plagues on Egypt.

6. This book confirms the stated purpose of God. It amounts to the greatest prophecy of the Bible. As a human being Jesus Christ pointed the way for others to enter the Family of God. He fulfilled the role of a prototype at the beginning of a process in which God is creating literal sons and daughters.

If these six points have now been demonstrated what is a final answer to Bullinger's question, paraphrased as follows:

What is the meaning of the prime numbers 7 and 11 in respect of Jesus Christ and the Father, and why are they encoded in the Bible in this way?

It can now be seen that the two Deities have been unknown to a world that has embraced trinities, monotheism and nihilism - in fact anything but the truth. These two 'Strangers' are now clearly identified. They are shown to be individuals in a hierarchy that forms the beginning of a family. Their religious system and its related prophecies are confirmed by the Grid. The strangers and their celestial symbols are mentioned, all four together, towards the very end of the Bible:

"And the city had no need of the sun, neither of the moon, to shine in it; for the glory of God did lighten it, and the Lamb is the light thereof." Revelation 21:23

In view of the discoveries shown here it is incumbent upon the Bible student to read with fresh understanding such words as: *'Grace unto you and peace and mercy from God the Father and his Son Jesus Christ.'* It is a greeting from *two* Deities that appears with only slight variations fifteen times in the New Testament. The 'triune' God is nothing but a myth displacing knowledge of the real thing.

Therefore to the six points above one can add a seventh finding: *God is an Alien* to this world. When the author lived in the Netherlands he was categorised as an *alien*. That was the word used to describe him by the authorities, despite his EU citizenship. An alien is a stranger, one we do not know. The world does not know who or what God is. It has dismissed his educational program and replaced it with a counterfeit. It has sidestepped, altered, ignored or done away with his commandments. It has taken to itself the prerogative to decide what is right and what is wrong; what is truth or error.

In the 88th verse of the Old Testament we read that *"Cain slew Abel"* but in the 88th verse of the New Testament it is first mentioned that *"...Jesus went about all Galilee...preaching the gospel of the kingdom."* What is that gospel, or 'announcement of good news' as it might be characterised? It is simply that humans are to be the inheritors of, and real children in, God's Family Kingdom. That is why we are made in his image and marked with his reproductive number, 44.

The true gospel *of* Christ has been rejected. It was usurped in the primitive Church after nineteen years (Galatians 1). A false gospel *about* Christ was implanted and has reigned ever since in Satan's present world, to which the true gospel is a poisonous 'blasphemy.' So exceptional is the preaching of this true gospel that Christ stated in the Olivet Prophecy of Matthew 24 that this activity would mark a specific time in what was, for him, the distant future:

> *"And this gospel of the kingdom shall be preached in all the world for a witness unto all nations; and then shall the end come."*

In a work largely ignored by the media and strongly resented by Christendom this prophecy was fulfilled in the 20th century by one God-inspired organisation, the modern day manifestation of a spiritual organism more ancient than the Catholic Church. It is the oldest organisation on earth today. The Jews are its physical counterpart. Like Noah it converted virtually no one. Like God it has been alienated from, and hated by, the society around it. In the twentieth century it brought the message of Isaiah chapter eleven concerning a future world society, controlled by the returning Christ:

> *"The wolf also shall dwell with the lamb, and the leopard shall lie down with the kid; and the calf and the young lion and the fatling together; and a little child shall lead them…and the lion shall eat straw like the ox…they shall not hurt nor destroy in all my holy mountain: for the earth shall be full of the knowledge of the Lord, as the waters cover the sea." Isaiah 11:6-9*

That idyllic scene will be supernaturally accomplished and governmentally enforced. Mankind has already had its chance to show what it can achieve alone, unaided by its estranged Parent. This post-Great Tribulation restitution of all things will commence with the regrouping of the tribes of Israel, or at least their remnants:

> *"And it shall come to pass in that day, that the Lord shall set his hand again the second time to recover the remnant of his people, which shall be left, from Assyria, and from Egypt…and shall assemble the outcasts of Israel, and gather together the dispersed of Judah from the four corners of the earth." v11-12*

Under this Christ-ruled system all property holdings will be suitably regulated and a God-devised system of finance will provide the 'social security' this world craves, but cannot achieve unaided:

> *"But they shall sit every man under his own vine and under his fig tree; and none shall make them afraid; for the mouth of the Lord of hosts hath spoken it." Micah 4:4*

Today the world is at the 'Matthew 24 point' of *"and then the end shall come."* It is at the last gasp of an age of hatred, greed, lust, blindness and their natural outgrowths, myths, false religion and war. But within this devilish age a new era began in 2003, the year of the Iraq invasion. It will be the last era of man's misrule. In 2003 the true Father sent a signal, the sign of Noah as a giant figure of 8, against the human representative of Satan and his own invisible false father. That spirit-composed being is soon to be put away. He is to be made *the Alien* instead of Jesus Christ.

Therefore a fearsome time draws near, mentioned in Revelation 12:12:

> *"Woe to the inhabiters of the earth and of the sea. For the devil is come down unto you, having great wrath, because he knoweth that he hath a short time."*

APPENDICES

APPENDIX I – concerning the claimed discovery of a code in the Torah, the first five books of the Bible. Such claims have been debunked.

(Relating to the chapter *Heavenly Bodies*)

Codes in the Bible have been 'discovered' before. There is a huge difference between those attempts to find patterns, at best inconclusive, and the present findings.

Recent books claiming to have found patterns in the Bible have utilised vast strings of Hebrew letters taken from the Torah (the first five books of the Bible). These have been arranged in rows to create rectangles of letters. The length of the rows is changed repeatedly until a word is formed vertically, or diagonally, as in this example:

144 Hebrew characters in lines of 36
KKKKKKKKKKKKKKKKKKKKKKKKKKKKKKKKKKKK
KKKKKKKKKKKKKKKKKKKKKKKKKKKKKKKKKKKK
KKKKKKKKKKKKKKKKKKKKKKKKKKKKKKKKKKKK
KKKKKKKKKKKKKKKKKKKKKKKKKKKKKKKKKKKK

But if this gives no result the data is shuffled:

The same 144 Hebrew characters in lines of 29
KKKKKKKKKKKKKKKKKKKKKKKKKKKKK
KKKKKKKKKKKKKKKKKKKKK*B*KKKKKKKK
KKKKKKKKKKKKKKKKKKKK*J*KKKKKKKKK
KKKKKKKKKKKKKKKKKKK**R**KKKKKKKKKK
KKKKKKKKKKKKKKKKKK*D*KKKKKKKKKKK
↑

Now the word BIRD is found! With luck the word that has been 'discovered' may be adjacent to another word that appears to be related. For example, in one place there might be a Hebrew word with the meaning of *tower*, or *two towers*, or arguably *twin towers*. This may occur near to, or even be touching, another Hebrew word for *bird*, or *big bird*. By repeatedly altering the lengths of the lines to create new permutations, one might eventually find words that touched, or pointed at one another. It is not difficult to see how a 'result' could be obtained in any text if there was sufficient data to reshuffle. *When one's remit is open ended, almost any result could be useful in building up a prophetic scenario.*

Fifty-five mathematicians from around the world, PhD qualified in Mathematics or Statistics, or faculty members in a Department of Mathematics or Statistics at a college or university, signed a petition (coordinated by bsimon@bigfoot.com) supporting the following view of the word-clusters 'discovered' in *The Bible Code* books:

> "...*word clusters such as mentioned in Witztum's and Drosnin's books and the so called messianic codes are an uncontrolled phenomenon and similar clusters will be found in any text of similar length. All claims of incredible probabilities for such clusters are bogus, since they are computed contrary to standard rules of probability and statistics. Among the signatories below are some who believe that the Torah [the first five books of the Bible] was divinely written. We see no conflict between that belief and the opinion we have expressed...*"

In the present Genesis Grid discovery there is no pack to be shuffled and no huge strings of Hebrew to be repeatedly rearranged. The evidence is specific and the themes narrowly defined and tightly interwoven.

APPENDIX II - the list of twenty-eight attacks on religious leaders contained in the four Gospels.

(Relating to the chapter *Unearthing the number 28*)

Certain explanations should be given regarding the list of twenty-eight. Firstly, the confrontation in John chapter ten appears to have been only with 'the people' rather than the leadership, and is omitted accordingly. The exchange with a stoning Jewish rabble at the Feast of Dedication also in John chapter ten is also excluded, as there is no evidence that religious leaders were witness to it. Finally, Christ's altercation with the high priest of John chapter twenty-eight is not included in the count, as no attack on any teaching, attitude or practice was made. The twenty-eight attacks are:

1. Leaders lacking in the required fruits:
Matt 3:7 But when he (John the Baptist) saw many of the Pharisees and Sadducees come to his baptism, he said unto them, O generation of vipers, who hath warned you to flee from the wrath to come? v8 Bring forth therefore fruits meet for repentance...v10...every tree that bringeth not forth good fruit is hewn down, and cast into the fire.

2. Leaders described as evil trees:
Matt 12:24...the Pharisees...v33...the tree is known by his fruit. v34 O generation of vipers, how can ye, being evil, speak good things?...

3. The leaders of an adulterous generation:
Matt 12:38...certain of the scribes and of the Pharisees answered... we would see a sign from thee. v39 But he answered and said unto them, An evil and adulterous generation seeks after a sign...

4. Leaders teaching tradition in place of God's commands:
Matt 15:1 Then came to Jesus scribes and Pharisees...v3...He answered...why do ye also transgress the commandment of God by your own tradition? ...v9 But in vain do they worship me, teaching as doctrines the commandments of men.

5. Blindness in the leaders:
Matt 15: 12…knowest thee that the Pharisees were offended…
v13…Every plant which my heavenly Father has not planted shall be
rooted up. v14 Let them alone, they be blind leaders of the blind…

6. Day of visitation not discerned by the leaders:
Matt 16:1 The Pharisees also with Sadducees came…v3…O ye
hypocrites, ye can discern the face of the sky; but can ye not discern
the signs of the times?

7. Religious teachers of the day to be last into the kingdom:
Matt 21:23…the chief priests and the elders came unto
him…v31…verily I say unto you, That the publicans and the harlots
go into the Kingdom of God before you…

8. Pharisees to have their 'birthright' torn from them
Matt 21:42 Therefore I say unto you, The Kingdom of God shall be
taken from you, and given to a nation bringing forth the fruits
thereof…v43…the chief priests and the Pharisees…perceived that
he spoke of them.

9. Example of the leadership not to be followed:
Matt 23:1 Then spake Jesus to the multitude, and to his disciples, v2
Saying, The scribes and the Pharisees sit in Moses' seat: v3 All
therefore whatsoever they bid you observe, that observe and do; but
do not do after their works: for they say, and do not. v4 For they
bind heavy burdens…v5…their works they do to be seen of men…
v9…call no man father…

10. A diatribe against teachers: eight woes and a final condemnation:
Matt 23:13 Woe unto you, scribes and Pharisees, hypocrites! v14
Woe…you devour widows houses and for a pretence make long
prayer…v15 Woe…make one proselyte…v 16 Woe…blind…
swear(ing) by the temple…v23 woe…you pay tithe of mint and
aniseed and cumin, and have omitted the weightier matters of the
law…v24 …strain at a gnat and swallow a camel…v25 Woe…make
clean the outside of the cup…within they are full of extortion… v27
Woe… whited sepulchres…full of dead men's bones… v29 woe…

you build the tombs of the prophets... v30 ...we would not have been partakers with them of the blood of the prophets...v31...ye are the children of them which killed the prophets. v32 Fill thee up then the measure of your fathers. v33 Ye serpents, ye generation of vipers, how can ye escape the damnation of hell?

11. Warning of treacherous religious leaders in general:
Matt 24:24 False Christ's, and false prophets, and shall show great signs and wonders; insomuch that, if it were possible, they shall deceive the very elect.

12. Prophetic warning of treacherous religious leaders in the Church:
Matt 24:48 But and if that evil servant shall say in his heart, My Lord delays his coming; v49 And shall begin to smite his fellow servants, and to eat and drink with the [spiritually] drunken; v50 The Lord of that servant shall come in a day when he looketh not for him, and an hour that he is not aware of. v51 And shall cut him asunder, and appoint him his portion with the hypocrites: there shall be weeping and gnashing of teeth.

13. Hardness of heart:
Mark 3:5 And when he had looked round about on them with anger, being grieved for the hardness of their hearts, he said unto the man, stretch forth thine hand...v6...Pharisees... took counsel...how they might destroy him.

14. Hypocrisy:
Mark 7:5...Pharisees...v6...well hath Esaias prophesied of you hypocrites, as it is written, this people honoureth me with their lips, but their heart is far from me. v7 Howbeit in vain do they worship me, teaching as doctrines the commandments of men. v8 For laying aside the commandment of God, ye hold the tradition of men...

15. Appearances:
Mark 12:38 And he said unto them in his doctrine, Beware of the scribes, which love to go in long clothing, and love salutations in the marketplaces. v39 And the chief seats in the synagogues, and the uppermost rooms at feasts...these shall receive greater damnation.

16. Challenge on legalism and the lawful use of the Sabbath:
Luke 6:9…is it lawful on the Sabbath days to do good, or to do evil?
To save life or to destroy it?...v10…stretch forth thine
hand…v11…they [Pharisees] were filled with madness…

17. Futility of outer cleanliness and appearance:
Luke 11:37…he went in, and sat down to meat. v38 And when the
Pharisee saw it he marvelled that he had not first washed before
dinner. v39 And the Lord said unto him, Now do you ye Pharisees
make clean the outside of the cup and the platter; but your inward
part is full of ravening and wickedness…v41…give alms of such
things ye have and behold things are clean unto you…

18. A set of six woes against the religious Leaders and Lawyers:
Luke 11:42 But woe unto you, Pharisees! For ye tithe mint and rue
and all manner of herbs, and pass over judgment and the love of
God…v43 Woe unto you, Pharisees!…you love the uppermost
seats… greetings in the marketplace …v44 Woe unto you…
hypocrites …for ye are as graves which appear not, and the men that
walk over them are not aware of them…v46 Woe unto you also, ye
lawyers. For ye lade men with burdens grievous to be borne, and ye
yourselves touch not the burdens with one of your fingers. v47 Woe
unto you! for ye build the sepulchres of the prophets, and your
fathers killed them…it shall be required of this generation…v52
Woe unto you lawyers! for ye have taken away the key of
knowledge: ye entered not in yourselves, and them were entering in
ye hindered.

19. Hypocrisy of the Pharisees likened to leaven:
Luke 12:1…beware ye of the leaven of the Pharisees, which is
hypocrisy. v2 For there is nothing covered, that shall not be
revealed…

20. The top religious leader rebuked for hypocrisy:
Luke 13:14 And the ruler of the synagogue answered with
indignation, because that Jesus had healed on the Sabbath day…v15
The Lord answered him, and said, thou hypocrite…

21. Covetous Pharisees rebuked:
Luke 16:13…ye cannot serve God and mammon. v14 And the
Pharisees also, who were covetous, heard all these things: and they
derided him. v15 And he said unto them, Ye are they which justify
yourselves before men; but God knoweth your hearts: for that which
is highly esteemed before men is an abomination in the sight of God.

22. Temple corrupted by the Pharisees:
Luke 19:39 And some of the Pharisees….v46 Saying unto them, It is
written, My house is the house of prayer: but you have made it the
den of thieves.

23. Pharisees likened to evil husbandmen:
Luke 20:16 He shall come and destroy these husbandmen, and shall
give the vineyard to others…v19…the chief priests…perceived that
he had spoken this parable against them.

24. Outward show of righteousness condemned:
Luke 20:46 Beware of the scribes, which desire to walk in long
robes, and love greetings in the markets, and the highest seats in the
synagogues, and the chief rooms at the feasts; v47 Which devour
widows' houses, and for a show make long prayers: the same shall
receive greater damnation.

25. Pharisees rebuked for seeking the honour of men:
John 5:42 but I know you, that you have not the love of God in
you…v44 How can you believe, which receive honour one of
another [therefore leaders being addressed] and seek not the honour
that cometh from God only?...There is one that accuses you, even
Moses…v47 But if you believe not his writings, how shall ye
believe my words?

26. Pharisees rebuked for unbelief:
John 8:23 And he said to them, You are from beneath, I am from
above; you are from this world, I am not of this world. I say
therefore unto you, ye shall die in your sins. For if you believe not
that I am he, ye shall die in your sins.

27. Pharisees rebuked for willful disobedience:
John 9:41 Jesus said unto them, If you were blind, ye should have no sin: but now ye say, We see; therefore your sin remaineth.

28. The religious rulers loved the praises of men
John 12:42 nevertheless among the chief rulers also many believed on him; but because of the Pharisees they did not confess him, lest they should be put out of the synagogue: v43 For they loved the praise of men more than the praise of God.

APPENDIX III – further examples of the number 19.

(Relating to the chapter *A 19 year Middle East cycle*)

Events in the cycle are here given in greater detail:

1910 The first Jewish Kibbutz: this was founded in Umm Juni, later renamed Degania Alef. The Israeli leader Moshe Dayan was born there.

1929 The Arab slaughter of Jews: in Jerusalem. On August 14th Jews marched in Tel Aviv chanting "The Wall is ours." Next day hundreds of Jews, some armed with batons, demonstrated at the "wailing" Wall in Jerusalem. Rioting broke out. The lone British policeman in Hebron heard screams down a tunnel-passage and going down it saw an Arab in the act of cutting off a child's head with a sword. Raymond Cafferata recounted, "I shot him low in the groin. Behind him was a Jewish woman smothered in blood with a man I recognised as an Arab police constable….with a dagger in his hand. He saw me and bolted into a room close by…..shouting 'Your honour, I am a policeman'….I got into the room and shot him." During that week fatalities were in the hundreds.

1948 The declaration of a Jewish State: followed immediately by war. Azzam Pasha, Secretary-General of the Arab League, said "This will be a war of extermination and a momentous massacre…." Andre Gromyko of the Soviet Union told the U.N. Security Council; "This is not the first time that the Arab States, which organized the invasion of Palestine, have ignored a decision of the Security Council…." The United States, the Soviet Union, and most other states immediately recognised Israel. The invading forces were fully equipped with the standard weapons of a regular army of the time - artillery, tanks, armored cars and personnel carriers, in addition to machine guns, mortars and the usual small arms in great quantities, and full supplies of ammunition, oil and gasoline.

1967 The Six Day War: after months of tense border incidents, Egypt's closure of the Straits of Tiran to Israeli shipping and the deployment of Egyptian troops to the Sinai, Israel launched a pre-emptive attack against the Egyptian Air Force of 450 Soviet-built planes. The war also involved Syria and Jordan. At its conclusion

Israel controlled the Gaza Strip, the Sinai Peninsula, the West Bank and the Golan Heights.

2005 The withdrawal from Gaza: after thirty-eight years (two nineteen-year cycles) of occupation. By the summer of 2005 the Israelis are completely disengaged from Gaza yet the outcome has been the election of Hamas shortly after the death of Yasser Arafat. This hard-line government has not brought the hoped for post-withdrawal progress. Once the experienced Sharon was replaced by Prime Minister Olmert any provocation from Israel's opponents would inevitably carry an increased risk of conflict. Summer 2006 saw the invasion of the Lebanon in response to the kidnapping of two Israeli soldiers. The outcome was widely regarded as a victory for the Iranian-backed Hezbollah.

It is remarkable that such crucial events in history have been subject to such an exact cycle. The taking of Jerusalem by Saladin in October 1187 after 88 years of Crusader rule is a further case regarding the 19 year cycle.

Saladin, whose real name was Salah al-Din Yusuf, was by nature a chivalrous and merciful man. The Englishman Balian of Ibelin surrendered the city to him after only a brief siege, to avoid unnecessary bloodshed. Negotiations were conducted over the evacuation of the cities inhabitants.

Neighbouring cities such as Tyre resisted the Saracen onslaught but Saladin's ineffectual 1188 campaign, however, established a pattern of halfhearted and indecisive Saracen prosecution of the war. That matters were settled and the die cast for the Holy Land from 1188 was partly due to Balian's clever negotiations with Saladin over the safe passage of women, children and dignitaries out of Jerusalem, a scenario depicted in the 2005 Ridley Scott film *Kingdom of Heaven*.

Peace was substantially achieved in the Holy Land from that time on and a subsequent and final European crusade fizzled out. Muslims held Jerusalem until December 1917, when the Turks surrendered to General Allenby. The 19 year Middle-East-Cycle appears to have run throughout: the period from 1188 to 1910 is 722 years (722 is $2 \times 19 \times 19$). There are several examples of the number 19 in the Bible.

In the book of Revelation, the most symbolic book of all, the number 7 is mentioned 19 times:

Chapter 1: seven Churches, seven Spirits, seven Candlesticks, seven Stars,

Chapter 4: seven Lamps of fire, seven Spirits of God,

Chapter 5: seven Seals, seven Horns, seven Eyes,

Chapter 8: seven Angels, seven Trumpets,

Chapter 10: seven Thunders,

Chapter 11: seven Thousand,

Chapter 12: seven Heads, seven Crowns,

Chapter 15: seven Plagues, seven Golden Vials,

Chapter 17: seven Mountains, seven Kings.

The same pattern also appears in respect of musical instruments, of which there are 19 found in the Bible. These are as follows:

1. KINNOR
2. NEBEL
3. NEBEL ASOR
4. KHALIL
5. SHOPHAR
6. KHATSOTSRAH
7. KEREN
8. UGAB
9. TOPH
10. TSELTS-LIM
11. M'TSIL-TAYIM
12. MENAANEIM
13. SHALISHIM
14. PHA-AMON
15. MASHROKITHA
16. KITHROS
17. SABECA
18. PSANTERIN
19. SUMPONYAH

(From Appendix II *The Music of the Bible*, John Stainer, Novello)

APPENDIX IV - the 107 mentions of the Father.

(Relating to the chapter *John confirms the Grid*)

Occurrences of the actual *mention of the person* of the Father in the gospel of John amount to 107, the Grid total. It is when we include all instances, such as "my Father's", in the count (as in John 14:1 *"In my Father's house are many mansions"*) that we see the total of 107. In this count there will be some statements in which the title 'Father' occurs *twice in a single mention of him,* for example:

> John 10:15 *"As the Father knoweth me, even so I know the Father."*

This is why the 107 count is lower than Bullinger's factual and revealing count of 121 (11^2) for every instance of the word 'Father.' In the above statement from John chapter ten Christ *broaches the subject* of his Father once, (in one breath) but uses the name twice. In this way, through the broaching of the subject, the Father is encoded 107 times in the gospel of John as follows:

1. 1:14 the only begotten of the Father
2. 1:18 from the bosom of the Father
3. 2:16 Make not my Father's house a house of merchandise
4. 3:35 The Father loveth the son and hath given all things into his hand
5. 4:21 ...nor yet at Jerusalem worship the Father. 22 Ye worship ye know not what
6. 4:23 ...true worshippers worship the Father in spirit and in truth: for the Father seeketh such to worship him
7. 5:17 My Father worketh hitherto and I work
8. 5:18 ...but said also that God was his Father, making himself equal with God
9. 5:19 ...the son can do nothing of himself but what he seeth the Father do
10. 5:20 For the Father loveth the son
11. 5:21 For as the Father raised up the dead
12. 5:22 For the Father judgeth no man, but hath committed all judgment unto the son
13. 5:23 ...even as they honour the Father; he that honoureth not the son honoureth not the Father which hath sent him

14. 5:26 For as the Father hath life in himself
15. 5:30 I not mine own will, but the will of the Father which hath sent me
16. 5:36 ...the works that the Father hath given me to finish, the same works that I do bear witness of me, that the Father hath sent me
17. 5:37 ...the Father himself that hath sent me, hath born witness of me
18. 5:43 I am come in my Father's name and ye receive me not
19. 5:45 Do not think that I will accuse you to the Father: there is one that accuses you even Moses
20. 6:27 ...eternal life, which the son of man shall give unto you: for him hath God the Father sealed
21. 6:32...my Father giveth you the true bread from heaven
22. 6:37 All that the Father giveth to me shall come to me
23. 6:39 And this is the Father's will which hath sent me...that I should lose nothing...
24. 6:44 No man can come to me, except the Father which hath sent me draw him
25. 6:45 Every man therefore that hath heard, and hath learned of the Father, cometh unto me.
26. 6:46 Not that any man hath seen the Father, save he which is of God, he hath seen the Father
27. 6:57 As the living Father hath sent me, and I live by the Father
28. 6:65 ...no man can come unto me, except it were given unto him of my Father
29. 8:16 ...for I am not alone but I and the Father that sent me
30. 8:18 ...the Father that sent me beareth witness of me
31. 8:19 Then they said unto him, Where is thy Father?
32. 8:19 ...Jesus answered, ye neither know me nor my Father: if thou had known me, ye should have known my Father also
33. 8:27 They understood not that he spake to them of the Father
34. 8:28 ...as my Father hath taught me I speak these things
35. 8:29 ...the Father hath not left me alone, for I do always those things that please him
36. 8:38 I speak that which I have seen with my Father, and you do that which you have seen with your father [the devil, v44]
37. 8:41...we have one Father, even God
38. 8:42 If God were your Father, you would love me

39. 8:49 I honour my Father and ye do dishonour me
40. 8:54 It is my Father that honoureth me
41. 10:15 As the Father knoweth me, even so I know the Father
42. 10:17 Therefore doeth my Father love, because I lay down my life
43. 10:18 This commandment have I received from my Father
44. 10:25 …the works that I do in my Father's name, they bear witness of me
45. 10:29 My Father that gave them me is greater than all and no man is able to pluck them out of my Father's hand
46. 10:30 I and my Father are one
47. 10:32 Many good works have I shewed you from my Father
48. 10:36 Say ye of him, whom the Father hath sanctified, Thou blasphemest
49. 10:37 If I do not the works of my Father, believe me not
50. 10:38 …the Father is in me, and I in him
51. 11:41 Father I thank thee that thou hast heard me
52. 12:26 …if any man serve me, him will my Father honour
53. 12:27 …Father save me from this hour
54. 12:28 Father, glorify thy name
55. 12:49 …but the Father, which sent me, he gave me a commandment, what I should say, and what I should speak
56. 12:50 …even as the Father said unto me, so I speak
57. 13:1 …his hour was come that he should depart out of this world unto the Father
58. 13:3 …Jesus knowing that the Father had given all things into his hands
59. 14:2 …in my Father's house are many mansions
60. 14:6 …no man cometh unto the Father but by me
61. 14:7 If ye had known me ye should have known my Father also
62. 14:8 And Phillip said, show us the Father
63. 14:9 …he that hath seen me hath seen the Father, and how sayest thou then, show us the Father
64. 14:10 …I am in the Father, and the Father in me
65. 14:10 …I speak not of myself: but the Father that dwelleth in me, he doeth the works
66. 14:11 Believe me that I am in the Father, and the Father in me: or else believe me for the very works sake
67. 14:12 …greater works than these…because I go unto my Father

68. 14:13 And what ever ye shall ask in my name, that will I do, that the Father may be glorified in the son
69. 14:16 And I shall pray the Father, and he shall give you another comforter
70. 14:20 …I am in my Father, and ye in me, and I in you
71. 14:21…he that loveth me shall be loved of my Father
72. 14:23 …if a man love me he shall keep my words: and my Father
will love him, and we will come unto him, and make our abode with him
73. 14:24…the which you hear is not mine, but the Father's which sent me
74. 14:26 But the comforter, which is the holy spirit, whom the Father will send
75. 14:28 …I go unto my Father; for my Father is greater than I
76. 14:31 …I love the Father; and as the Father gave me commandment, even so I do
77. 15:1 I am the true vine and my Father is the husbandman
78. 15:8 Herein is my Father glorified, that ye bear much fruit
79. 15:9 As the Father hath loved me, so have I loved you
80. 15:10…I have kept my Father's commandments
81. 15:15 …all things that I have heard of my Father, I have made known unto you
82. 15:16 …whatsoever ye shall ask of the Father in my name, he may give it you
83. 15:23 …he that hateth me hateth my Father also
84. 15:24 …but now have they both seen and hated both me and my Father
85. 15:26 …from the Father, even the spirit of truth, which proceedeth from the Father
86. 16:3 …because they have not known the Father, nor me
87. 16:10 …because I go to my Father and ye see me no more
88. 16:15 All things that the Father hath are mine
89. 16:16 A little while and ye shall not see me: and again a little while and ye shall see me, because I go to the Father
90. 16:17 Then said some of his disciples…because I go unto the Father
91. 16:23 Whatsoever ye shall ask of the Father in my name, he will give it you

92. 16:25 ...the time cometh when I shall no more speak unto you in proverbs, but I shall show you plainly of the Father
93. 16:26 ...I will pray the Father for you
94. 16:27 For the Father himself loveth you
95. 16:28 I came forth from the Father
96. 16:28 ...I leave the world and go to the Father
97. 16:32 ...I am not alone because the Father is with me
98. 17:1 ...Father, the hour is come; glorify thy son that thy son also may glorify thee
99. 17:5 And now O Father glorify me with thine own self with the glory which I had with thee before the world was
100. 17:11 ...Holy Father, keep through thine own name those whom thou hast given me
101. 17:21 That they may all be one; as thou, Father, art in me
102. 17:24 Father, I will that they also, whom thou hast given me, be with me where I am
103. 17:25 O righteous Father; the world hath not known thee
104. 18:11 ...the cup which my Father hath given me, shall I not drink it?
105. 18:17 ...touch me not, for I am not yet ascended unto my Father
106. 18:17 ...say unto them, I ascend unto my Father, and your Father
107. 18:21 ...peace be unto you: as my Father hath sent me, even so I send you

The one instance in which the Pharisees invoke the name of the Father falls in position 31, the number of deity. This may be indicative of their willful rejection of Him.

Also of importance is Christ's most weighty endorsement of the Ten Commandments in John 15:10 *"I have kept my Father's commandments"* which is seen in position eighty, a number that alludes to Moses (who had received the Ten Commandments) whose work for God began at age 80.

Particularly noteworthy is position 28. At this 'teaching number' position we find the momentous statement: John 6:65 *"...no man can come unto me, except it were given unto him of my Father."* Christ had already stated at position twenty-four: John 6:44 *"No man can come to me, except the Father which hath sent me draw him.*" But at position twenty-eight he re-affirms that 'no man can come to me' by saying that it has to be given, or granted. No other doctrine within this exhaustive list is *reiterated in the manner of John 6:44 and 57*. What is actually the most radical teaching theme of the entire 107 statements concludes at position 28.

An additional Grid verification is seen in the way in which the 107 verses are distributed in successive sets according to the numbers of the Grid and their meaning by column order. The match occurs when the Grid is compared to the 107 verses in order of its columns from left to right: 1 - 44 then 7 - 40 and finally 11 - 4 as follows:

Grid number 1: begettal or its result, birth.

Statement number 1 of 107:- John chapter 1:14 the only begotten (or born. *Gk. gennao: born, begotten*) of the Father

Grid number 44: predominantly begettal. The section begins on the origin of Christ from the intimate bosom of the Father, and ends on what can be understood as the complete security of the 'womb of the Church' for the true follower:

Statement number 2: ch. 1:18 from the bosom of the Father

3. 2:16 Make not my Father's house a house of merchandise
4. 3:35 The Father loveth the son and hath given all things into his hand
5. 4:21 …nor yet at Jerusalem worship the Father. 22 Ye worship ye know not what
6. 4:23 …true worshippers worship the Father in spirit and in truth: for the Father seeketh such to worship him
7. 5:17 My Father worketh hitherto and I work
8. 5:18 …but said also that God was his Father, making himself equal with God
9. 5:19 …the son can do nothing of himself but what he seeth the Father do
10. 5:20 For the Father loveth the son
11. 5:21 For as the Father raised up the dead
12. 5:22 For the Father judgeth no man, but hath committed all judgment unto the son
13. 5:23 …even as they honour the Father; he that honoureth not the

son honoureth not the Father which hath sent him

14. 5:26 For as the Father hath life in himself
15. 5:30 I seek not mine own will, but the will of the Father which hath sent me
16. 5:36 …the works that the Father hath given me to finish, the same works that I do bear witness of me, that the Father hath sent me
17. 5:37 ...the Father himself that hath sent me, hath born witness of me
18. 5:43 I am come in my Father's name and ye receive me not
19. 5:45 Do not think that I will accuse you to the Father: there is one that accuses you even Moses
20. 6:27 …which the son of man shall give unto you: for him hath God the Father sealed
21. 6:32…my Father giveth you the true bread from heaven
22. 6:37 All that the Father giveth to me shall come to me
23. 6:39 And this is the Father's will which hath sent me…that I should lose nothing…
24. 6:44 No man can come to me, except the Father which hath sent me draw him
25. 6:45 Every man…hath heard, and hath learned of the Father, cometh unto me.
26. 6:46 Not that any man hath seen the Father, save he which is of God, he hath seen the Father
27. 6:57 As the living Father hath sent me, and I live by the Father
28. 6:65 …no man can come unto me, except it were given unto him of my Father
29. 8:16 …for I am not alone but I and the Father that sent me
30. 8:18 …the Father that sent me beareth witness of me
31. 8:19 Then they said unto him, Where is thy Father?
32. 8:19 …Jesus answered, ye neither know me nor my Father: if thou had know me, ye should have known my Father also
33. 8:27 They understood not that he spake to them of the Father
34. 8:28 …as my Father hath taught me I speak these things
35. 8:29 …the Father hath not left me alone, for I do always those things that please him
36. 8:38 I speak that which I have seen with my Father, and you do that which you have seen with your father [the devil, v44]
37. 8:41 ...we have one Father, even God

38. 8:42 If God were your Father, you would love me
39. 8:49 I honour my Father and ye do dishonour me
40. 8:54 It is my Father that honoureth me
41. 10:15 As the Father knoweth me, even so I know the Father
42. 10:17 Therefore doeth my Father love, because I lay down my life
43. 10:18 This commandment have I received from my Father
44. 10:25 ...the works that I do in my Father's name, they bear witness of me
45. 10:29 My Father that gave them me is greater than all and no man is able to pluck them out of my Father's hand

Grid number 7: identity of Christ, who begins by identifying himself in the Father and then twice identifies himself in this section by his works.

46. 10:30 I and my Father are one
47. 10:32 Many good works have I shewed you from my Father
48. 10:36 Say ye of him, whom the Father hath sanctified, Thou blasphemest
49. 10:37 If I do not the works of my Father, believe me not
50. 10:38 ...the Father is in me, and I in him
51. 11:41 Father I thank thee that thou hast heard me
52. 12:26 ...if any man serve me, him will my Father honour

Grid number 40: the trials and testing of Christ. This begins with a reference to Christ's biggest trial, "this hour"; then further mention that "his hour had come" i.e. the greatest trial; mention of the "comforter" by which trials may be endured; Christ and his followers are to be hated; the section ends on the trial of no longer seeing Christ and the burden of being spoken to in parables, soon to be ended.

53. 12:27 ...Father save me from this hour
54. 12:28 Father, glorify thy name
55. 12:49 ...but the Father, which sent me, he gave me a commandment, what I should say, and what I should speak
56. 12:50 ...even as the Father said unto me, so I speak
57. 13:1 ...his hour was come that he should depart out of this world unto the Father

58.13:3 ...Jesus knowing that the Father had given all things into his hands
59.14:2 ...in my Father's house are many mansions
60.14:6 ...no man cometh unto the Father but by me
61.14:7 If ye had known me ye should have known my Father also
62.14:8 And Phillip said, show us the Father
63.14:9 ...he that hath seen me hath seen the Father, and how sayest thou then, show us the Father
64.14:10 ...I am in the Father, and the Father in me
65.14:10 ...I speak not of myself: but the Father that dwelleth in me, he doeth the works
66.14:11Believe me that I am in the Father, and the Father in me: or else believe me for the very works sake
67.14:12 ...greater works than these shall he do, because I go unto my Father
68.14:13 And what ever ye shall ask in my name, that will I do, that the Father may be glorified in the son
69.14:16 And I shall pray the Father, and he shall give you another comforter
70.14:20 ...I am in my Father, and ye in me, and I in you
71.14:21 ...he that loveth me shall be loved of my Father
72.14:23 ...if a man love me he shall keep my words: and my Father will love him, and we will come unto him, and make our abode with him
73.14:24 ...the which you hear is not mine, but the Father's which sent me
74.14:26 But the comforter, which is the holy spirit, whom the Father will send
75.14:28 ...I go unto my Father; for my Father is greater than I
76.14:31 ...I love the Father; and as the Father gave me commandment, even so I do
77.15:1 I am the true vine and my Father is the husbandman
78.15:8 ...herein is my Father glorified, that ye bear much fruit
79.15:9 ...as the Father hath loved me, so have I loved you
80.15:10 ...I have kept my Father's commandments
81.15:15 ...all things that I have heard of my Father, I have made known unto you
82.15:16 ...whatsoever ye shall ask of the Father in my name, he may give it you

411

83.15:23 He that hateth me hateth my Father also

84.15:24 ...but now have they both seen and hated both me and my Father

85.15:26 ...from the Father, even the spirit of truth, which proceedeth from the Father

86.16:3 ...because they have not known the Father, nor me

87.16:10 ...because I go to my Father and ye see me no more

88.16:15 All things that the Father hath are mine

89.16:16 A little while and ye shall not see me: and again a little while and ye shall see me, because I go to the Father

90.16:17 Then said some of his disciples......because I go unto the Father

91.16:23 Whatsoever ye shall ask of the Father in my name, he will give it you

92.16:25 ...the time cometh when I shall no more speak unto you in proverbs, but I shall show you plainly of the Father

Grid number 11: the Father's identity. This section of eleven verses contains themes showing from whence Christ came, from the Father. It identifies the Father's true worshippers, concluding with the fact that the world cannot identify the Father.

93.16:26 ...I will pray the Father for you

94.16:27 For the Father himself loveth you

95.16:28 I came forth from the Father

96.16:28 ...I leave the world and go to the Father

97.16:32 ...I am not alone because the Father is with me

98.17:1 ...Father, the hour is come; glorify thy son that thy son also may glorify thee

99.17:5 And now O Father glorify me with thine own self with the glory which I had with thee before the world was

100.17:11 ...Holy Father, keep through thine own name those whom thou hast given me

101.17:21 That they may all be one; as thou, Father, art in me

102.17:24 Father, I will that they also, whom thou hast given me, be with me where I am

103.17:25 O righteous Father; the world hath not known thee

Grid number 4: the Father's power. The power to resurrect and the power to have actual sons (not merely adopted) to do his work. Indeed, an equivalence between Christ's sonship with the Father and (potentially) our own is inferred (item 106) here.

104.18:11 ...the cup which my Father hath given me, shall I not drink it?
105.18:17 ...touch me not, for I am not yet ascended unto my Father
106.18:17 ...I ascend unto my Father, and your Father
107.18:21 ...peace be unto you: as my Father hath sent me, even so I send you.

APPENDIX V - four further Grid verifications.

(Relating to the chapter *Commonplace Patterns*)

The first Grid verification was that the Grid total of 107 is the 28th prime number. Further verifications are shown in this chapter.

In the chapters *The Mystery of 11* and *Prime numbers of the Bible* the first five books (the Torah) were shown to correspond to the 150 Psalms. The Psalms divide into five sets, four of which have prime number totals of chapters as follows:

Genesis	Psalms 1 – 41	compr. 41 chapters	13th prime
Exodus	Psalms 42 – 72	compr. 31 chapters	11th prime
Leviticus	Psalms 73 - 89	compr. 17 chapters	7th prime
Numbers	Psalms 90 – 106	compr. 17 chapters	7th prime
Deuteronomy	Psalms **107** – 150	compr. **44** chapters.	Not a prime

The last set of the Psalms pertains to Deuteronomy and begins with Psalm 107, the number of which is the Grid total. Anyone setting out to investigate Bullinger's discovery of the 7 and 11 pattern would need to incorporate the two symbols, the Sun and the Moon, into their deductions. An investigator would find 44 last of all, so it noteworthy that: *the number 44 leads to the Grid total 107, but the number 107 provides confirmation of 44 as the sixth Grid number.* It is in Deuteronomy that the pattern is seen. The appearance of both numbers in the manner highlighted above therefore constitutes a further Grid verification.

A quite different verification for the Grid is found in the way that the top and bottom rows of the Grid can be harmonised with one another. As neither of the Grid rows treated as the quantities 1,711 and 44,404 are prime numbers they can be factorised:

Firstly, 1,711 reduces to 29 [the 10th prime] × 59 [the 17th prime]

Then: 44,404 simplifies to 4 × 11,101

And further, 11,101 reduces to 17 (7th prime) × 653 (119th prime)

Three of the four prime positional numbers here (7, 10, 17 & 119) are themselves prime numbers; these are: 7, 17 & 119. Of those, the first two produce the third, as in $7 \times 17 = 119$ so it is seen that a simple arithmetic similarity exists between the two dissimilar looking rows of the Grid. The appearance of 7 (*"perfection and completeness"*) and 17 (the 7th prime and *"the perfection of spiritual order"*)[1] is in harmony with the personalities of the Grid. Another approach is add the two prime number factors in each row of the Grid:

> 1,711: prime factors $29 + 59 = 88$
>
> 44,404: of 11,101 prime factors $17 + 653 = 670$, or 10×67
>
> $= 10 \times 19$th prime. Therefore $44,404 = 40 \times 19$th prime

The above is an interesting reversal: in the Grid the total of the top line is 19, the bottom line 88; here the top line offers 88 and the bottom 19.

Another theme of the Grid is that of *duality*, and a list of dualities in the Grid was given in chapter fifteen as follows:

1. Two deities (Father/Son) and their symbols
2. The dual appearance of the number 28 in the primitive Grid
3. Two grid rows (Identity and power/motion)
4. Two columns of 44 (1×44 and 11×4)
5. Two lines of 88 as clues to pi (circle and bottom line)
6. Two forms of the Grid 'perform' in pi, the pair 1711 & 44404 and 171144404
7. Two positions in pi show a pattern, 1st and 2nd
8. Two segments encode a name
9. Two halves of pi match two halves of Genesis one
10. Two numbers 11/44 as in the full Grid string
11. Two commands in Gen1:28, each has a pattern
12. Two sub-segments encode a name
13. Two solid symbols (Sun/Moon) matched by two ephemeral
14. Two mounts (Sinai/Olives) contain Grid numbers

Duality, as either *opposites* or *seconds*, is a Bible theme. Many fundamental concepts are in pairs as seen in the following examples:

Opposites:-
The first Adam: the second Adam (Christ)
The righteous line: the unrighteous line
The old man: the new man

Seconds:-
The Old Testament: the New Testament
The Old Covenant: the New Covenant
Circumcision of the flesh: circumcision of the heart
The present Jerusalem: the new Jerusalem
God the Father and Jesus Christ mentioned at the beginning of
most Pauline Epistles.

There is a fourth way in which the Grid encodes the teaching
number 28. The conventional way is as follows:

Identity: **1**   **7**   **11**

Motion: **44**   **40**   **4**

                        280   28

The reader will recall that the Grid total is 107, which is the 28th
prime number. But there is another way the Grid encodes the
number 28, as follows:

1   **7**   **11**

44   **40**   **4**

Intertwining the numbers to create a pattern not dissimilar to the double helix in DNA produces the sets 1-40-11 and 44-7-4. These also encode the number 28 in the following way:

$$1 \times 40 \times 11 = 440 \qquad \text{and} \qquad 44 \times 7 \times 4 = 1232$$

$$\text{where } 1232 / 440 = 2.8 \qquad \text{which is } 1/10^{\text{th}} \text{ of } 28$$

As explained in the chapter *The Ovum and the Spermatozoon*, the strands of the DNA molecule woven into the structure of human chromosomes contain the reproductive Grid number 44. Each parent provides 23 chromosomes making a total of 46. But there are 44 autosomes used in creating the human frame, plus 2 sex chromosomes to provide the 'embellishment' of sexual identity. Evidence that this indeed may be no more than an embellishment is that there will be "no marrying or giving in marriage" (Matthew 22:30) in the Kingdom of God. *The factors 2 and 44 represent the building blocks of the physical human being*, just as they do in the reproductive message of the Grid, where $2 \times 44 = 88$

The helix shape of DNA is reflected in the Grid. The Grid signals 44 and so does the New Testament word *sperma* as Bullinger attested:

> "(sperma), seed 44 (4 × 11)" page 29, E.W. Bullinger, Number in Scripture

Furthermore, the human sperm and ovum signal 44 and so does the human genome. Thus the number of reproduction is confirmed as 44.

So in four further ways - the Torah/psalms interplay, the numerical relationship between the top and bottom rows of the Grid, duality and the striking 'DNA' pattern - the Grid is further validated.

APPENDIX VI – further patterns in music

(Relating to the chapter *Grid Vibrations*)

Grid numbers, when used as positional notes of the keyboard, are able to produce a pattern. Their frequencies can be inserted into the Grid in the following way:

1	**7**	**11**
A0	D1sharp	G1
27.5	38.891	48.999

44	**40**	**4**
E4	middle C	C1
329.63	261.63	32.703

In this 'vibrating Grid' it can be seen that the sum (addition) of the top line is 115.39 and the sum of the bottom line is 623.963. The product (multiplication) of these two totals is 71999.09 which is very close to 72,000. The Sun and Moon both occupy half a degree in the sky, or $1/720^{th}$ of a circle, which, if simplified gives the number 72. In this additional way the Sun and the Moon appear linked to the world of music via the Grid. Grid numbers are seen in the successive notes of A, the vibrations of which double up perfectly as you move from left to right along the keyboard:

Bottom note	1	27.5	A0
Next A	13	55	A1
" "	25	110	A2
" "	37	220	A3
" "	49	440	A4
" "	61	880	A5
" "	73	1760	A6
"	85	3520	A7
Top note C	88	4186	C8

The above figures remain unchanged from the nineteenth century. Apart from bottom A0 they are all round numbers. Their values are absolutely precise. These eight A's encompass the entire keyboard except for an extra segment at the top. The difference in frequency between top A, the 85th note and top C, the 88th note is:

$$4186 - 3520 = 666$$

If this constitutes a pattern, regardless of whether 666 or really 616 is the number of the beast (see chapter 32), what of the notes from 1 – 85? The frequencies at that end of the keyboard begin with the ultra low 27.5cps for A0 (the bottom note) and end with 3520cps exactly for A7 (note 85). This gives a gap of 3492.5. The whole number simplification (multiply by 2) for this is the number 6985, which breaks down into:

$$5 \times 11 \times 127$$

What might these factors mean? Firstly, multiples of five were common in the building of the (biblical) Temple. The number 11 is the identity number of the Father in the Grid and in the Bible, the chief Deity, but what about the number 127? As a prime number it has no factors. It is the 31st prime number. 31 is the number of deity and *the 11th prime number, pertaining to the Father.* Thus the main range of the keyboard, notes 1 – 85, contains the deity number 31, whereas the remaining segment has the value 666.

The range of A0 to A7 containing 85 notes encompasses 84 intervals. We already know pi's first 31 digits contain segments defining:

1) The Father
2) Jesus the man
3) Christ the divine messenger
4) The Bible

Supposing the 1 - 85 range of keys containing 84 intervals matched the first three segments of pi demonstrated in the chapter *The Deity String.* There is a simple way in which this range of 84 intervals divides into three:

$$3 \times 28 = 84$$

The teaching number in this scheme is 28. If the 1 - 85 range has three segments, it is evident that the remainder has also, because that tiny section of the instrument is comprised of three intervals. Thus there is symmetry in this arrangement. The remainder at the top end of the keyboard is made up of the four notes 85, 86, 87 and 88 (A, B-flat, B and C) and it therefore encompasses three semitone intervals. These three intervals enclose a frequency range of 666 cps. So there is symmetry in two keyboard sections, each with three segments, which encompass the entire keyboard. This invites comparison with the segments of pi already uncovered. A match is obtained when these segments of semitones are compared with the corresponding four segments of the keyboard including the partitioning eights:

3.1415926535897932384626443383279

Intervals:
[...............28.............8........28.......8.....28............8] [3]

$$28 + 8 + 28 + 8 + 28 + 8 + 3 \; = \; 111$$

It has been shown that Grid numbers 7, 11 and 44 appear prominently in the structure of musical scales and the frequencies of the notes with which they are constructed. Numbers that derive from, or are closely similar to, Grid numbers such as 88 (keys) and 111 (the keyboard pattern) also appear as shown above.

One interesting fact is not yet highlighted: the all important middle C represents the other Grid number: it is the 40th note.

A final observation is necessary. The fourth (last) segment of pi was shown to correspond to the number 66, by way of encoding 39 books of the Old Testament and 27 books of the New Testament, in the chapter *The Deity String*. It is the last segment of the keyboard that corresponds to 666, a similar number. This allows for a close correlation between the keyboard of the piano and the four pi segments as shown here:

Segments:	one	two	three	four

3.14159265358979323846264338327 9

28 intervals	*28 intervals*	*28 intervals*	*3 intervals*
(bass notes)	(mid range)	(melodious)	(666cps)

|-----communication range----------|

Father		Jesus	Christ	Bible 66 books

Note that the segments of the instrument most able to communicate via melody correspond to Jesus Christ, the Word, or communicator. Even the Father's Grid number 4 (the power to create) is suggested by the appearance of four segments. Thus all Grid numbers are represented as well as the associated numbers 28, 88 and 111. The numbers of the Grid are woven into the laws governing the construction of the human brain and ear, as well as the behaviour of inanimate copper piano strings.

APPENDIX VII – more observations on the 'three to the power of four law.'

(Relating to the chapter *Numbers: Figment or Fact*)

The author Peter Plichta PhD has shown several hugely significant discoveries in his 1997 book *God's Secret Formula*. A chemist by profession, he had realised that two of the 83 supposedly stable elements did not in reality exist. The true number of such elements was accordingly 81, or 3^4. Moreover, it soon became apparent to him that a law of 'three to the power of four law' applied to the structure of the atom. Atoms contain three basic types of particles: Protons, neutrons and electrons. However, to this 'three-ness' is attached a fourth factor: the energy states of the orbiting electrons. These sit at different distances from the centre in 'shells' the first four of which follow a simple law. The inner shell may only contain one pair of electrons, the next up to four pairs. The maximum number of pairs allowed at successive levels outwards is 1, 4, 9 and 16, after which new rules apply. The numbers 1, 4, 9, and 16 are the squares of 1, 2, 3 and 4.

Another hugely important insight by Plichta is the 'three-ness' and 'four-ness' of matter and space. While an object like a car can be described, or defined volumetrically, by length, breadth and height it is not possible to do this for infinite space. To test this, stand in the corner of a room with both arms extended along the walls, at 90 degrees from one another. If one's torso is infinitely long upwards and downwards, and one's arms were of infinite length, still only ¼ of infinity is proscribed - ¾ is still left out.

However, to 'cover' all of infinity one could start with a window pane and extend it to infinity in all four directions: an area of i^2 is described, or $i \times i$, where i is infinitely long. Of course this shape is not solid. It is completely flat - the glass has no thickness in this illustration. A 'passenger' sliding across this sheet could not move anywhere in infinite space, only along the sheet.

How then can one describe infinite space? The method is to place another identical sheet of glass at 90 degrees to the first one. This sheet also stretches out 'everywhere' and is therefore also i^2. Any point in infinity can now be reached if the sheets can be slid through one another. A static passenger sitting on either sheet can be moved to any place in infinity if the sheets are allowed to pass through one another while remaining at right angles.

Combining the two sheets i^2 and i^2 gives i^4, four dimensions in infinite space. But a solid object can be defined by d^3 where three dimensions, length, breadth and height, describe it. Thus it can be seen that a 3^4 law operates when describing infinite space and the objects within it.

The four forces of nature appear to follow such a pattern, at least in as much as there is a 3+1 aspect. Here gravity is the odd one out, whereas electromagnetism and the strong and weak nuclear forces act over short distances. It would be quite a turn up if the long sought after Grand Unification Theory conformed in some way with the Grid, because it has been said previously that the Grid conforms to this 3^4 law: it possesses three columns each of which has an identity number. The identity numbers 1, 7 and 11 are empowered by the bottom line of the Grid which pertains to power and motion. The bottom line is dominated by the number 4. Each of the three identity numbers is empowered by a number containing the number 4 as a factor: 44, 40 or 4. The bottom line factor is the *determinant* for reproduction, testing and creation.

Plichta also observed that living things broadly follow a 3^4 law. The most fundamental and numerous class, insects, are characterised by the numbers 3 and 6. They possess six legs and three body sections (head, thorax and abdomen). Even their eyes have hexagonal facets. Mammals have three aspects to their structure: head, body and extremities, but they possess four limbs. He proposed the idea that there are two types of life corresponding to two types of space. This idea is not as outlandish as it may initially sound when one considers that insects do not perceive perspective: they see in two dimensions only. This raises a further issue: the ability to see in four dimensions. Mankind ordinarily sees in three dimensions like other mammals, but he, uniquely, may aspire to seeing and perceiving in the fourth dimension. In further fulfillment of the 3^4 law, *The Genesis Grid* is a book that begins in three dimensions but progressively calls on the reader to think in a fourth.

APPENDIX VIII - mathematical findings regarding the Grid, the Golden Ratio and Fractals.

(Also relating to the chapter *Numbers: Figment or Fact*)

A number of observations has been made of the Grid's mathematical nature. Both of the foundational Grid numbers 7 and 11 are prime numbers. As a pair of consecutive primes, 7 and 11 are unique in one respect. They are the only primes in the series that as a fraction produce a recurring double digit pattern of any kind:
7/11 = 0.636363636363r.

The Grid itself is able to show an affinity between pi and the Golden Ratio phi.

The Golden ratio - phi

In the 1997 film *Pi* by Darren Aronofsky, mentioned in the chapter *The Ovum and the Spermatozoon*, the mathematician Maximillian Cohen is accosted by a member of an Orthodox Jewish splinter group.

The young Jew, steeped in numerology, scribbles out two numbers: 144 and 233. Cohen recognizes them immediately – they are adjacent numbers from the Fibonacci numerical sequence. The Jew explains: *"The Garden of Eden, 144 and the tree of knowledge 233; 233 over 144, that's the Golden Ratio."*

What he meant was: 233/144 = 1.61805

Successive numbers in the Fibonacci Numerical Sequence can form fractions, the value of which tends toward the Golden ratio, 1.618…. Like pi, the Golden Ratio cannot be represented by a finite set of digits and is therefore classed as 'irrational.' The name 'the Golden Ratio' has quite recently become abbreviated to *phi*, pronounced "fee." Some mathematicians consider it even more interesting than pi. Pi and phi are certainly two of the most intriguing numbers in the universe.

The Golden Ratio has long been associated with art and beauty. It is derived from the 'Fibonacci numerical series' that begins:

1, 2, 3, 5, 8, 13, 21, 34, 55, 89, 144

In this series each number is obtained by adding together its two predecessors (as in 13 + 21 = 34). A progressively closer approximation to 1.618... is obtained by dividing each successive number of the sequence into its predecessor (as seen above where 233/144 = 1.61805). A link between the 'Golden Ratio' and the Grid can also be demonstrated via music.

Grid numbers are intrinsic to music, as already shown in the chapter *Grid Vibrations,* but so also is the Fibonacci series, from which the Golden Ratio is derived. For example, a major sixth can be obtained by combining the notes A and C vibrating with frequencies of 440 and 264 respectively. Two Fibonacci numbers 5 and 3 appear from their frequencies which have the ratio 5/3. Similarly an augmented fifth is produced by the notes high C down to E. Their ratio is equivalent to Fibonacci numbers eight and five as in 8/5 (from 528/330).

The 11th Fibonacci number of 144, mentioned above in the film *Pi* in connection with the Garden of Eden, is also a number seen in the book of Revelation. In that book it is associated with a future event:

> *Revelation 7:4 "And I heard the number of them that were sealed: and there were sealed an hundred and forty and four thousand..."*

This number 144 is evoked by the left hand column of the Grid, the spirit of the Father/reproduction column, which contains the numbers 1 and 44:

1   7   11

44   40   4

The likeness between these numbers of the spirit of the Father in the Grid and the 11th Fibonacci number 144 (11 being the Grid number for the Father) *points to a relationship* between pi (3.1415926...) and phi (1.6180339...)

Another pointer to a relationship between pi and phi is via the following simple sum involving the number 4: the number of *power* from the bottom line of the Grid. It can be expressed in a concentrated form as 4^2, or 16. The Golden Ratio begins with the digits 1 and 6. Using the number 16 the sum works as follows yielding a value very close to phi:

$$44404 / 1711 / 16 = 1.6220046$$

Another strong indicator of a relationship between pi and phi is the result obtained when a *Deity String* of 31 digits (a pattern demonstrated in pi) is totalled:

$$1+6+1+8+0+3+3+9+8+8+7+4+9+8+9+4+8+4+8+2+0+4+5+8+6+8$$

$$+3+4+3+6+5 = 162$$

What is obtained is a three digit approximation to 1.618, i.e. 1.62. To this result can be added the observation that the initial eight digits in both pi and phi total the same number 31, the number of deity:

Pi 3.1 4 1 5 9 2 6 $3 + 1 + 4 + 1 + 5 + 9 + 2 + 6 = 31$

and:

Phi 1. 6 1 8 0 3 3 9 $1 + 6 + 1 + 8 + 0 + 3 + 3 + 9 = 31$

Much has been made here of pi beginning with the digits 3 and 1, but phi begins with 1 and 6. Of the number 16, Bullinger had found that the only section of scripture that did not produce a multiple of 11 for the occurrences of the Father's name was the book of Luke, which had 16 occurrences. This exception has the consequence of producing a total for the Father of 280, but it is noteworthy in its own right.

The number 16 can be thought of as a concentrated version of 4. In this case the number 4 was found to be the secondary number in the Sun/Father column of the Grid and in Luke the Father was found, uniquely, in multiples of 4. The significance of the number four is shown by:

The four forces of nature (the strong and weak nuclear forces, electromagnetism and gravity)
The four winds
The four Cherubim of Ezekiel moving the throne of God (Ezekiel 1:4-28)
The four prophetic beasts of Revelation (Rev.4:6)
The four rivers of Eden (Genesis 2:10)

The number 4 denotes power and motion, but also the act of creation, specifically in respect to the Earth, as the following examples show:

Four seasons of the year
Four phases of the Moon
Four kinds of flesh: men, beasts, fishes and birds (1 Corinthians 15:39)
Four as the number of the city (*Number in Scripture* page 133)
Fourth Commandment; first that refers to the Earth (Sabbath)
Fourth book; *Numbers*, means wilderness (Earth, *Number in Scripture* p126)
Fourth Psalm; the Book of the Wilderness (*Number in Scripture* p126)

These associations broaden the meaning of the number 4. It is associated not merely with power, but the power to *create*. The bottom row of the Grid 44404 is dominated by the number 4. While the top row *identifies* deities the bottom row concerns their power. The top row of the Grid is readily linked to pi, because the first 11 digits of pi represent the Father. The bottom row of the Grid, dominated by 4's, can be associated not only with power but now with creating and propagating. Accordingly a newly labelled Grid is proposed:

Pi	1	7	11
Phi	44	40	4

(The power to create)

In this arrangement the idea is that, whereas pi was already characterised as a "calling card" (pertaining to identity) in the *Introduction*, phi relates to the process of creating.

There is a well demonstrated match between pi and the Bible in its first 31 digits. It is evident that a *validation number* exists at the beginning of that pi string, namely 3.1 or simply 31. This links the series to God by a sort of label. Similarly phi begins, as will be shown, with a validation number, namely 1.6 or simply 16. Totalling the first 16 digits of phi we find that:

$$1+6+1+8+0+3+3+9+8+8+7+4+9+8+9+4 = 88$$

With this result, matching phi to the *bottom line* of the Grid (above) is confirmed because the bottom line of the Grid totals 88. But if a pattern in phi emerges within its first 16 digits what of the first 16 digits of pi? These are:

$$3+1+4+1+5+9+2+6+5+3+5+8+9+7+9+3 = 80$$

The dominant personality of the Grid emerges from the 88 and 80 results because when the totals are positioned according to their corresponding symbols (pi over phi):

88/80 = 1.1 which is one tenth of 11

This appearance of the primary number of the Grid demonstrates a further similarity between pi and phi. The idea of the number 16 as a concentrated version of 4 is bolstered by the simple observation that phi begins with 1 and 6 and its first 4 digits total 16:

$$1+ 6 + 1 + 8 = 16$$

As the first 4 digits of phi produce the number 16 this is all the more reason to match phi to the *second* line of the Grid. That line is dominated by the number 4. However, it was shown that a string of 31 digits of phi produced the phi approximation of 162, an approximation in the sense that 1/100 of that number is 1.62, very close to 1.618...

Whereas pi is the ratio between a circle and its diameter, phi is derived from a series that has nothing to do with circles. How could so many common points arise between pi and phi? Another rather cryptic pointer is seen in the totals for pi and phi in respect of their first 16 and first 31 digits: these are 80 and 88 and 150 and 162 respectively. So their fractions would be 80/150 for pi and 88/162 for phi. As decimals these are 0.53333333 and 0.54320987654. The difference between the two decimals is an astonishing result:

$$0.00987654321$$

In the light of these findings it is apparent that a previously unknown relationship exists between pi and phi. The means of revealing their similarities has been, in each case, the knowledge of the Grid.

In all of these discoveries it is the Grid that has led the way. The relationship between pi and phi is thus demonstrated beyond all reasonable doubt.

Fractals

It is often said of the next generation: "He's a chip off the old block." There is a way in which fragments from an object can bear a great similarity to their source. These fragments have been named *fractals*.

Some of the characteristics of the Fibonacci numerical sequence have been discussed. In the Fibonacci sequence beginning 1, 2, 3, 5, 8, 13…, where each number is derived by adding its two predecessors, fractals can be seen.

By way of illustration, the fractal quality of a cauliflower is observed when one breaks off a small fragment of the same appearance as the original vegetable. Cauliflowers grow in a manner that demonstrates *self-similarity*. Inanimate objects can demonstrate self-similarity: a rock can resemble a miniature version of the mountain from which it was taken. Fractals are seen when representing the Fibonacci series in the following way:

```
1
10
101
10110
10110101
```

In this pattern a '1' is replaced with a '10' on the next line down and every zero is replaced with a '1' on the next line down. The pattern expands at the same rate as the Fibonacci series:

Positions occupied

1	= 1
10	= 2
101	= 3
10110	= 5
10110101	= 8
1011010110110	= 13
1011010110110110101	= 21
1011010110110101101011011010110110	= 34

Thinking back to the cauliflower, the broken off fragment that resembles the original is equivalent to a section such as 1011010110110. In this way it can be seen that self-similarity can be seen in the expanding strings of 1's and 0's. The positions occupied by chunks like 1011010110110 are proliferating through the same mechanism by which 21 is derived from 13 + 8, as shown below:

1	1	
2	10	
3	101	
5	10110	
8	10110101	8+
13	1011010110110	13
21	1011010110110110101	= 21
34	1011010110110101101011011010110110	
55	etc	
89	etc.	
144	etc.	

It is apparent that each increasing length of digits contains segments of the same *shape* as its predecessor as illustrated:

10110101101101 0110110101 (21 digits)

10110101101101011 0101 *1011010110110* (34 digits)

It can be seen that small repetitive modifications where 'bricks' of 10 and 1 are continually being slotted into a locality lead to runaway growth. The fractal qualities of phi show how different in character it is to pi.

The Fibonacci series of numbers can produce a spiral shape. This is defined as a natural logarithmic spiral, also called the equiangular spiral. It can be constructed using an accumulation of rectangles that represent the successive numbers of the Fibonacci sequence. The resulting spiral shape is seen throughout the universe, from the humble single cell *foraminifera* to a spiral galaxy.

The most impressive display of the *fractal principle* can be seen in the heavens, where galaxies are found in local groups with other galaxies. These galaxies are then found to be in association with similar groups that form a larger cluster, and so on, extending outwards in a self-similar manner.

The equiangular spiral is, like the Fibonacci sequence itself, self-similar. A better understanding of the principle of self-similarity is gained by constructing the spiral from the rectangles that correspond to the series, which again is:

1, 2, 3, 5, 8, 13, 21, 34, 55, 144...

This series can be used to assemble rectangles describing a spiral. Taking the first in the series, number 1, its shape has to be 1×1. The next number is 2 giving a rectangle 1×2. Next is the number 3 giving 2×3 and putting these rectangles together produces the following:

Measuring 3×3

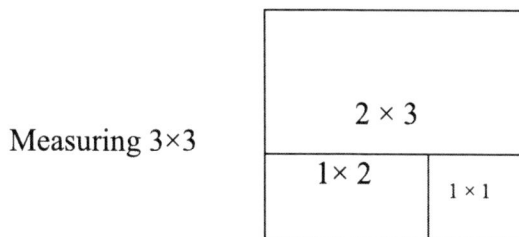

2×3

1×2

1×1

Adding the next two rectangles 3×5 and 5×8 produces a further square:

Measuring 8×8

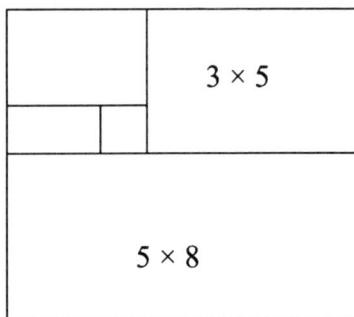

Adding two more rectangles 8×13 and 13×21 produces the next square in the series:

Measuring 21×21

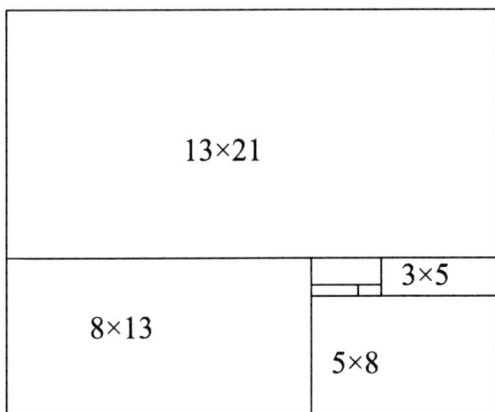

Two steps further the next square is 55 by 55 and two steps after that the next square is 144 by 144. From the first box (1×1) it takes five expansions through boxes of 3×3, 8×8, 21×21, 55×55 to reach 144×144.

The spiral produced by joining successive rectangles is the following shape:

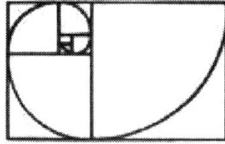

The scientific evidence is mounting that physical objects are constructed according to the ratio within the logarithmic spiral. For example, according to a recent discovery, the microscopic flat terraces of quasi-crystals of aluminium-copper-iron alloy are of two different depths, the ratio of which is 1.618. It is evident that phi is a number associated with the way matter can behave and living things can propagate.

The Grid numbers 7 and 11 are found throughout the Fibonacci series. For example if one takes any ten successive numbers in the series, they contain 7 and 11 as seen in the following example:

$$3 + 5 + 8 + 13 + 21 + 34 + 55 + 89 + 144 + 233 = 605$$

The total 605 divided by the 7th number in the set, 55, produces **11**. This works for any set of ten successive Fibonacci numbers for example:

$$8 + 3 + 21 + 34 + 55 + 89 + 144 + 233 + 377 + 610 = 1584$$

where 1584 divided by the 7th number 144 = **11**

So we can see that the two prime numbers 7 and 11 representing the two creating Deities are woven like a ribbon throughout the underlying substrate of phi, the Fibonacci series.

Prime numbers have been of vital importance, as explained in the chapter *Prime numbers of the Bible*. This is seen on examination of the Father's string of 31415926535. Firstly, the segment of 11 digits pertaining to the Father is comprised of two parts, 31415926 and 535, yielding the totals 31 and 13 respectively. In this way the deity number 31 at the start of pi, appearing as 3.1, is reiterated. Furthermore, 31 and its reflection 13 reappear, as when we arrange the first eight numbers into sets as follows:

433

$$31 \quad 41 \quad 59 \quad 26$$

Only the last of these four pairs does not form a prime number. The prime positional (the place in the series of primes, i.e. 2 is 1st, 3 is 2nd, etc.) values are:

$$31 \quad 41 \quad 59 \quad (26)$$

$11^{th} \quad 13^{th} \quad 17^{th}$ totalling 41, the 13th prime

Although the number 26 is not a prime it contains 13 as a factor. Thus the number 13 features in the 31415926 string three times, as well as appearing in the sub-segment 535 as in $5 + 3 + 5 = 13$. The observation of the reflecting 31/13 arrangement is extended by reversing the above set to produce the following arrangement:

$$62 \quad 95 \quad 14 \quad 13$$

This reversed set contains the following prime factors:

$$31 \quad 19 \quad 7 \quad 13$$

The *deity/apostasy pattern* of 31:13 appears again.

Finding the number 88

There is further observation that pertains to the chapter An Abstract Machine. The number 88 appears in two obvious ways within the Grid: as the sum of the bottom line and the sum of the products of the outer columns, 44 in each case. A third way in which it makes a cryptic appearance is by the rotation of the full Grid string. Delving down a bit more, the number 88 is again seen in a cryptic manner when a *circular* motion is applied to the numbers of the Grid.

$$1 \quad 7 \quad 1 \quad 1 \quad \mathbf{4} \quad 4 \quad 4 \quad 0 \quad 4$$

In this idea the Grid numbers above are rotated around their centre digit 4. If that same number is used as a multiplier for the numbers rotated, then the Grid number then become:

$$(4 + 28 + 4 + 4) + (16 + 16 + 0 + 16) = 88$$

A final observation: the pi Deity String pattern:

31415926535897932384626433**3279**

Of further interest is the fact that the first 4 digits and last 4 digits of the 31digit Deity String, totaled and halved, equals 3210. So that amounts to an average value for 3141 and 3279 of 3210. This number contains the factor 107 (30 × 107). In that way 4 and 4 (or '44') indicates 107, the Grid total.

Yet again the Grid takes us to pi and pi confirms the Grid.

APPENDIX IX – More on the Number 44

(Relating to the chapters *The Ovum and Spermatozoon* & *A Secret Architecture*)

The begettal concept is supported by a pattern seen in the 22nd, 44th and 66th books of the Bible. In this symmetrical distribution there is a progressive process encoded:

1. The 22nd book is the *Song of Solomon*, a unique book picturing Christ and the Church in highly romanticized *sexual prose*.
2. The 44th book is *Acts,* where the process of *begettal* through the spirit began in the new Church age via the spirit, pictured as flames of fire on the heads of the disciples.
3. The 66th book is *Revelation*, vividly describes the final end result of that begettal: the (future) *birth* of sons and daughters of God into the future Kingdom of God.

Thus three phases in chapters 22, 44 and 66 are depicted: romance, begettal and birth. The begettal number is evidently 44. This involves conception rather than birth as it involves seeds, not birth pangs. The gematria (number values) of the names of the twenty kings of Judah, all of the fathers and sons from Rehoboam to Zedekiah, amount to 4,400 (*Number in Scripture*, page 216). The 44 factor is also reflected in *Isaiah*, a major Old Testament book of 66 chapters echoing the Bible's 66 books; it is regarded as a sort of mini-Bible in itself. Like the books of *Acts* (the 44th book) it links 44 with begettal. Its 44th chapter states:

> *"Thus saith the Lord thy redeemer, and he that formed you from the womb, I am the Lord that maketh all things, that stretcheth forth the heavens."*

Nowhere else in Isaiah is there any mention of the womb or that which grows within it. There should be little doubt that the human reproductive system bears the creative fingerprint of God.

APPENDIX X – mysteries of the 'Godhead'

(Relating to the chapter: *And then there were Two*)

One particular issue can be investigated here to shed light on the question of *individual* beings in the Godhead: the question of the *pre-existence of Christ* as born out by the following description of Christ as the Word, the *logos* (or spokesman), of John:

> *John1:1 "In the beginning was the Word, and the Word was with God, and the Word was God...v14 The Word became flesh and dwelt among us."*

This pre-existent personage was the one who had had dealings with Moses and the Israelites. The God of the Old Testament, who gave the Ten Commandments from Mount Sinai, followed the Israelites during their 40 years of wandering in the wilderness:

> *I Corinthians 10:1-4 "...passed through the (Red) sea... baptised unto Moses...they drank of that spiritual Rock that followed them: and that Rock was Christ."*

Christ revealed the future to the prophets of the Old Testament:

> *I Peter 1:10 "Of which salvation the prophets have enquired...v11 Searching what, or what manner of time the spirit of Christ which was in them did signify, when it testified beforehand the sufferings of Christ..."*

Paul discussed Christ's willingness to give up his glorious pre-existence for mankind:

> *Philippians 2:6 "Who, being in the form of God, thought it not robbery to be equal with God: but made himself of no reputation, and took upon him the form of a servant, and was made in the likeness of men v8 And being found in fashion as a man, he humbled himself, and became obedient unto death..."*

The Genesis account of 'creation' (actually a refurbishment of an ancient planet as explained in the chapter *The Great Breach*) testifies to a plurality of Deities:

> *Genesis 1:26 "And God said, Let us make man in our image, after our likeness...v27 So God created man in his own image..."*

437

Some monotheists claim that this was the heavenly host speaking (including angels and cherubs), but man was made in the image of God, not in the image of any other heavenly being. Moses saw the pre-existent Christ. This event is recorded in a detailed account in Exodus chapter thirty-three:

*Exodus 33:18 "And he [Moses] said, I beseech thee, show me thy glory. 19 And he said, I will make all my goodness pass before thee...20 And he said, Thou canst not see my face, for there shall no man see me, and live. 21 And the Lord said, Behold, there is a place by me, and thou shalt stand upon a rock: 22 And it shall come to pass, while my glory passeth by, that I will put thee in the cleft of the rock, and will cover thee with my hand while I pass by: 23 And I will take away mine hand, and thou shall **see my back parts**, but my face shall not be seen."*

The outcome of this unique sighting was the wearing of a veil by Moses:

Exodus 34:35 "And the children of Israel saw the face of Moses, that the skin of Moses' face shone: and Moses put the veil upon his face again..."

The reason for the veil was given by Paul:

2 Corinthians 3:7 "...the children of Israel could not steadfastly behold the face of Moses for the glory of his countenance ...13...Moses put a veil over his face that the children of Israel could not steadfastly look..."

In time the effect apparently faded, as later accounts of Moses' dealings with colleagues record no such problem. But it can be deduced from several passages that the two Deities exist as separate personalities. This has to be true because the New Testament states that no man has seen God, which of necessity (or the Bible would contradict itself) must mean the Father:

John 1:18 "No man has seen God at any time..."

John 5:37 "And the Father himself, which has sent me, has born witness of me. You have neither heard his voice at any time, nor seen his shape."

These two statements in John prove the separateness of Christ from the Father. How fitting that in the gospel of the disciple Christ especially loved, the book that encodes the Grid in its entirety, his personhood is made plain.

A passage in Exodus 24:9-11 appears to confuse matters by stating that not only Moses, but seventy elders also, saw God. Is it possible to see God and yet not see God?

Visions of deities have been recorded, as for example, when John, in the book of Revelation chapter four, saw a vision of God's throne:

"...I was in the spirit: and, behold, a throne..." Revelation 4:2

Ezekiel also saw a vision:

"...I saw visions of God." Ezekiel 1:1

The disciples, Peter, James and John, saw a vision of the glorified Christ (the 'transfiguration') standing with Moses and Elijah. In Matthew Christ charged them:

"...tell the **vision** to no man, until the Son of man be risen from the dead." Matthew 17:9

The vision had a purpose: it was a glimpse of the governmental structure of the future Kingdom of God. So were any of these privileged men wearing veils after the event? There is no evidence that the seventy elders or Ezekiel, Peter, James or John were wearing veils. All the evidence is that they were not, and a lot of explaining would have been needed had their faces supernaturally glowed. Their acquaintances would have known of the account about Moses and his glowing face. Clearly, the disciples' faces did not glow: this was because they only saw a vision, not the real thing.

Some may argue that because the Old and New Testaments are so different from one another that deductions of this type based on single instances from each are too simplistic, particularly in such a contentious area as the nature of God. The two testaments are certainly different as the first describes God dealing with a physical nation and supplying the civil and ceremonial laws required by it; the second describes a spiritual nation. However, it is still the case that scripture is consistent throughout in the application of *principles* because *"I am the Lord, I change not"* (Malachi 3:6) and *"Jesus*

439

Christ, the same yesterday, and today, and for ever. " (Hebrews 13:8). The scriptures are one integrated whole or they are not *the scriptures.* Deductions should certainly be made. Isaiah shows the need for deduction:

> *"For precept must be upon precept...line upon line...v13...that they might go and fall backward, and be broken, and snared, and taken." Isaiah 28:10*

It is logical to deduce that as far as scripture is concerned, if you see a deity your face will glow and a veil is needed; if you see a mere vision you do not need a veil. The plain record of human experience in interacting with God shows this to be true.

Then some who object to a deduction concerning the "back parts" seen by Moses will say that it is a mistake to ascribe a *body* to a deity. They insist that it is foolish to 'limit God' by suggesting he has a body. With an erroneous 'three in one' deity this would admittedly be a problem. But many religious people cling to a concept of God where he is rather like motorway fog. God is everywhere so he cannot have a body, they argue. The 'Holy Spirit' is everywhere as well (and is a personality too, they believe) so 'he' is also amorphous. These ideas are full of internal contradictions: they are nothing more than confusion. But Paul discusses a *body* in respect of the resurrection:

> *"Who shall change our vile body, that it maybe fashioned unto his glorious body..." Philippians 3:21*

That would be the same glorious body seen by Moses, about whom a Deity said would see *"...my back parts..."* The passage referred to above from John 5:37 stating that *"...no man has heard his voice or seen his shape,"* ascribes **shape** to God. This idea is also conveyed by the following:

> *"...it does not yet appear what we shall be: but we know that, when he shall appear, we shall be **like** him; for we shall see him as he is." 1John 3:2*

Can someone be *like* motorway fog? To be like someone you must have a form and so must they. There is no particular translation issue in these passages. The problem is in dislodging error that has been planted, most often throughout childhood, in the minds of millions.

Most people will find it hard to swallow the fact that most theologians believe in a monumental error: a form of monotheism that dictates the existence of only one personality. But God can be correctly understood to be one family.

Paul knew about both deities in this family and gave greetings from them at the start of nearly all of his Epistles, e.g. *"Grace be unto you from God our Father, and from the Lord Jesus Christ"* (Philippians 1:2). To most impartial readers of the Bible this would sound like two people. That there are two is a simple truth reflecting *"the simplicity that is in Christ"* (2 Corinthians 11:3).

APPENDIX XI – the Grid in Revelation.

(Relating to the chapter *Revelation encodes the Grid*)

The chapter structure of Revelation reflects Grid numbers. Consider the following story flow: firstly, Revelation has an introduction which leads into the 7 church eras. These finish at the close of chapter three. The penultimate verse of chapter three refers to *"my Father in his throne"* which provides continuity – a link - into chapter four where there is a vision *of that throne*. This remains the theme to the end of the fourth chapter, with chapter five continuing the account of happenings at the throne. The *Lamb*, Christ, is introduced and at the close of chapter five he is pronounced worthy to unleash punishments on the world.

Chapter six explains those punishments, which are packaged as six seals which must be sequentially broken open by the Lamb. The sixth seal involves great earthquakes and the darkening of the Sun and the Moon. There has been an uninterrupted flow up to this point.

At the end of chapter six something interesting happens. *Chapter seven can be described as an inset chapter*. This is because although its first verse links back to the end of chapter six, (proving that the chapter is in the correct place) the seventh chapter *digresses* from the theme of successive punishments into some completely different material. This makes the seventh chapter odd and special within the story flow, because it is *an interruption to it*. Nevertheless, chapter seven finishes with a reference to the Lamb who, at the start of chapter eight, opens the seventh seal as he is the one deemed worthy to do so. This careful linking at the beginning and end of the chapter proves that its positioning is deliberate. But it is still a *digression* – the first of its kind – from the narrative (of punishments) that *resumes* in chapter eight. At the seventh chapter it amounts to a 'Sabbath rest' from punishments!

This interruption in the story flow with an interlude dealing with *rewards*, not punishments, marks out the seventh chapter as special. The punishments resume immediately at the beginning of chapter eight. Thus the Grid number 7 is encoded in the structure of the book.

What of the Grid number 11? Continuing on through chapter eight and into chapter nine it is apparent that the seventh seal contains six trumpets, each of which announce successive punishments upon the world. These six trumpets encompass chapters eight and nine. Mention of a seventh and final trumpet is made in chapter ten and in chapter eleven it is sounded, at which *"the kingdoms of this world are become the kingdoms of our Lord, and of his Christ; and he shall reign for ever and ever."* The nations are angry (verse 18) but reward is given to the servants of God and verse 19 concludes with a final vision of the heavenly temple.

The section comprising the first 11 chapters of Revelation *could be a book in itself.* From the beginning of chapter twelve the account abruptly reverts about 2000 years to Christ's birth and the establishment of the Church. That small organisation, a 'little flock', is described as fleeing into the wilderness to escape persecution: then an outline history unfolds. That is immediately followed by a different subject in chapter thirteen, that of prophetic beasts that rise, symbolically, out of the sea. In chapter fourteen the subject of the 144,000 is revisited, and so on.

The number 11 is also encoded in the mentioning of *the spirit*, as in the sense of it emanating from God (therefore not including evil spirits, or any other meanings). Other than the set of four mentions of *the seven spirits of God* previously discussed, Revelation chapters two and three contain 7 exhortations to *"listen to what the spirit says to the churches"* (the spirit singular). These are 7 prophetic utterances from God to his congregations. There are 4 further times where the spirit acts an agency of the Father (in one case bringing about a resurrection to a spirit state), or expresses a prophetic utterance of the Father *bringing the total to 11*. Those 4 instances are as follows:

> *"...the spirit of (spirit) life (a resurrection to spirit form) from God entered into them (the two witnesses) and they stood upon their feet...v12...and they ascended up into a cloud..." Rev 11:11*

> *"...blessed are the dead that die in the Lord from henceforth: yea, saith the spirit, that they may rest from their labours; and their works do follow them." Rev 14:13*

"...worship God, for the testimony of Jesus is the spirit of prophecy." Rev 19:10

"And the spirit and the bride say, come. And let him that heareth say, come. And let him that is athirst come..." Rev 22:17

That set of four, in conjunction with the four mentions of the seven spirits of God, suggests 44. Thus it can be shown that in the book of Revelation there are 11 mentions of the Father's spirit acting as his resurrecting and prophesying agent. Thus we can see that the defining numbers of the Grid, 7, 11 and 44 are repeatedly coded in this the last book of the Bible.

APPENDIX XII – the snake on a pole

(Relating to the chapter *The Lady*)

Is it wrong to have symbols of God? The Sun and the Moon are the first symbols mentioned in the Bible. The hypothesis put forward here is that they are secret symbols for the two leading Bible personalities. Some could object to this idea on the grounds that the Bible forbids the worship of physical objects as symbols of God:

> *"Thou shalt have not make unto thee graven image, or any likeness of anything that is in heaven above, or that is in the earth beneath..." Exodus 20:4*

This would appear to prohibit any works of art, except the most abstract. But this is not the meaning of the commandment. Its purpose clarified by the second part:

> *"Thou shalt not bow down thyself to them, or serve them."*

That is the intent of the command. The problem lies in the idolatrous use of the object, not the artifact itself. In Numbers 21:8 Moses told the Israelites in the wilderness to set up a 'fiery serpent' of brass on a pole. Prior to this, the people had rebelled and as a result poisonous serpents had entered the camp. But those who looked upon the brass serpent were healed from their bite.

The serpent Moses had manufactured was a powerful metaphor for Christ:

> *"And as Moses lifted up the serpent in the wilderness, even so must the Son of man be lifted up: that whosoever believeth in him should not perish..." John 3:14-15*

The people only looked upon the brass serpent, as instructed, but did not worship the object at first. The brass serpent was later destroyed by King Hezekiah because the people had begun to worship it as a graven image, breaking the second of the Ten Commandments:

> *"He removed the high places, and brake down the images, and cut down the groves, and brake into pieces the brazen serpent that Moses had made: for unto those days the children of Israel did burn incense to it..." 2 Kings 18:4*

The reason for the commandment is simply that the human mind must be trained to relate to an *invisible* God, because God is a spirit. He must be worshipped *"in spirit and in truth"* (John 4:24). The bowing down to statues and images induces a certain kind of blindness that blocks out understanding. This activity can be witnessed in any Catholic stronghold, such as Notre Dame Cathedral.

The use of the serpent by Moses was biblically legal, as is the *illustration* of the relative powers and attributes of deities by celestial bodies. To illustrate or explain is not the same as to serve, or bow down. So it can be seen that, in biblical terms, a representation of God or Christ by celestial spheres or a grid of numbers is not in itself prohibited.

APPENDIX XIII – more evidence of Middle Eastern migration to the British Isles.

(Relating to the chapter *Samaria on Thames*)

The evidence for a migration of Hebrews from the area of Armenia across Western Europe is considerable. Samuel Lysons' book *Our British Ancestors* was concerned with evidence of Middle East origin for the language of the British Isles. English words and names bear similarity to many Hebrew words, as Lyons demonstrates:

> *"In the name of Ell-barrow, Wiltshire...we recognise the Hebrew* א ל*, the British Heeaul, the Greek Ηλιος, 'the sun'."* page 94.

> *"But to return to the root* א ל*, Al, El. 'In Persia,' says Dr. Prichard, "there is a numerous wandering tribe of fire-worshippers, (ignis solis loco) called Ells."* page 100.

> *"In Horne's Introduction to the critical study of the scriptures we learn that 'Burnt offerings are in Hebrew termed* ע ל ח *, Uley, which signifies to ascend because their offerings...ascended as it were to God...'...In Yorkshire and other northern counties, at Christmas time...there is a custom still for the people to run about in the churches crying Ule, Ule, as a token of rejoicing. it is not a little striking that the Greenlanders to the present day keep a sun feast, or Yul, at the winter solstice, Dec. 22, to rejoice at the return of the sun...A custom in Kent...to encircle the apple and cherry trees...pronouncing...God send us a youl-ing sop...This ceremony is called Youl-ing."* page 110-111

To this piece of evidence one could add that 'Yule logs' have been a part of the pagan celebration of Christmas, together with mistletoe, for millennia in the West.

> *"Belas Knap, otherwise called Hamley Hough, in Charlton Abbots parish, Gloucestershire, carries with its name similar ideas of solar superstition: in Belus [sic], as before observed, we recognise* ב ע ל*, Bel, 'Baal;' and in Hamley.*

> *...* ח ם ל א י *a place of Ham, solar heat, a place of worship. Belas Knap barrow was opened in 1863, and presented all the*

*interesting features of the long tumuli of the Britons..." page
123, ibid.*

Lysons draws a likeness between ancient sites of worship in Britain
and those of the Israelites:

*"The Baal worship of Beth-Aven (or Temple of Aven), in
Canaan, was notorious. The prophet Hosea [Chapter 10]
especially warns God's people against it: 'Come no ye into
Gilgal, neither go ye up into Beth-Aven.' (בית אין)...The
worship at Beth-Aven, in Canaan, and that of Avening in
Gloucestershire, and that of Aven, otherwise Heliopolis, or
Baalbec, above mentioned, were all identical. The stone altars,
the high place, the calves' bones discovered there [see Hosea
10:5], mark the similarity...Rivers, we know, were dedicated to
various deities, and took their name from them: so Avon is the
river of the sun." pages 125-6, ibid.*

Of the name 'Britain' Lysons offers several points gleaned from
various writers:

*"Brit, or Brith, ברית, 'as a noun,' says Parkhurst (page 77) 'is
a purifier, purification, or purification-sacrifice.'...Brithim would
be the worshippers of Baal Berith the sun-god...this idea need not
interfere with Nennius's traditional account of a descent from
Brutus or Brito...kings and chieftains derived their names from
their gods, there is nothing irreconcilable with the in the idea that
Brito, the British founder, was called after Baal, Berith, or Brith.
Mr. Wilford (Asiatic Researches, v3) states that the old Indians
were acquainted with the British Islands, which their books
describe as the sacred Islands of the West, calling one of them
Brit-ash-tan (Stan, as applied to a country, seems to derive from
שת, St, 'to settle, to place...') (ברת אש שת) would be 'the place
of bright burning sacrifices, or fire.' This, if correct, would not
only give us an interesting view of the long-disputed etymon of
the British name, but would illustrate a remarkable manner that
prophecy of Isaiah 14:15 'They shall lift up their voice, they shall
sing for the majesty of the Lord, they shall cry aloud from the sea.
Wherefore glorify thee the Lord in the fires, even the name of the
Lord God of Israel in the isles of the sea. From the uttermost
parts of the earth have we heard songs, even glory to the*

righteous." That Britain was universally known to the ancients as Ultima Thule, 'the uttermost parts of the earth, the extreme boundary of the West,' is shown by hundreds of passages from classical and patristic writers."

Lyons collated a substantial glossary of 120 pages containing over 3000 examples of English words with a phonetic similarity to Hebrew words of a similar meaning. Here is a short selection:

Beryl	ב ו ר ל א	pron. "byrila"	Beryl, a precious stone
Bad	ב ר	pron. "bad"	To separate, good, bad
Castle	ב מ ל	pron. "selac"	Strength, support
Cable	ח ב ל	pron. "chabel"	A cord or rope
Dross	ר ד ש	pron. "drossh"	Trodden down refuse
Duck	ד ה ה	pron. "duchen"	To dive, into water,
Exempt	ע מ ת	pron. "zempt"	Cut off, excluded
Enough	כ א פ	pron. "naph"	To satiate, sufficient
Fulcrum	פ ל ך	pron. "fulc"	A prop, support, a staff
Fresh	פ ר ץ	pron. "fresh"	Bursting forth, anew
Fork	פ ר ך	pron. "phork"	To force, branch off,
Fist	פ ז	pron. "fist or fix"	Solid, compact,
Fizz noise	פ צ ץ	pron. "Fizz"	To burst off with a
Gash	ג ז ז	pron. "gash"	To shear, cut
Gore	ג ר י א	pron. "goria"	A sore wound, a cut

| Harbour | ב ר א | pron. | "arb" | A place of concealment, retreat, lurking place |
| Hubbub | ב ו ב א | pron. | "habub" | Sound of a great noise |

These examples are from pages 366 – 398 of *Our British Ancestors: who and what were they?* Samuels Lysons M.A., F.S.A.

APPENDIX XIV - Global warming.

(relating to the chapter *Two Americans*)

In the year 2003 *eleven* giant solar flares (X-class) erupted in an event numerous scientific websites claimed was the most outstanding solar event in 144 years. Certain sources later claimed 12 flares, the largest of which was ranked as between X28 – X40, but any rate the largest ever recorded. Regardless of whether there were 11 or 12, the crucial point is that it was the Sun's signal. It involved sunspots and therefore the principal number would be 11. This is backed up by the 144 span since the last comparable phenomenon on the surface of the Sun, 144 being the 11th Fibonacci number. The relevance of the Fibonacci series is shown in Appendix VIII.

Today's global-warming science fraternity should wake up to the possibility that increasing temperatures are coming as a punishment on the world. It is a theme in several important prophecies:

> *"...in the day of the great slaughter, when the towers fall.*
> *"Moreover, the light of the moon shall be as the light of the sun, and the light of the sun shall be sevenfold, as the light of seven days, in the day that the Lord bindeth up the breach [sheber: fracture, ruin, crashing destruction] of his people, and healeth the stroke of their wound [makkeh: carnage, pestilence, plague, slaughter]." Isaiah 30:25-26*

The claim that man-made emissions are causing a runaway greenhouse effect remains scientifically contentious. For scientists, the internet age has seen the overthrow of monopolies by which they were able to restrict information. Scientists are coming to be trusted less than politicians. Today the 'experts' are frequently challenged in a manner unheard of a generation ago. Al Gore, master propagandist of the global warming bandwagon, was embarrassed in August 2007 when close ally James Hansen of GISS (NASA's Goddard Institute for Space Studies) was forced to revise temperature figures for the 20th century. It transpired that the 1930's, not the period 1997 - 2007, was the hottest decade. In August 2007 Christopher Booker of *The Sunday Times* observed:

"...the latest satellite figures from the National Oceanographic and Atmospheric Administration [show] that in recent years global temperatures have not continued to rise [as orthodox CO_2 warming theory would suggest] but have flattened out at a level significantly lower than in 1998."

But it still may be true that the weather is changing in the long term. If so, a better explanation for this may be that solar activity is increasing. Eventually a super-heated Sun is to chastise errant mankind:

"And the fourth angel poured out his vial upon the sun; and power was given unto him to scorch men with fire. And men were scorched with great heat, and blasphemed the name of God, which hath power over these plagues: and they repented not to give him glory." Revelation 16:8-9

During a March 2007 BBC investigation *The great global warming swindle* Prof Eigil Friis-Christensen said: *"...the 400 year sunspot chart proves the real source of the temperature rises."* Many experts interviewed spoke out against the orthodox view that man is responsible. When in Matthew chapter twenty-four Jesus Christ predicted, verse 21 *"...great tribulation, such was not since the beginning of the world"* he was well aware of the following prophecies of Isaiah:

"And they shall go into the holes of the rocks, and into the caves of the earth, for fear of the Lord, and for the glory of his majesty, when he ariseth to shake terribly the earth." Isaiah 2:19

Isaiah chapter thirteen continues this theme:

"The burden of Babylon...v4 The noise of a multitude in the mountains, like as of a great people; a tumultuous noise of the kingdoms of the nations gathered together...v5 They come from a far country...to destroy the whole land. v6 Howl ye, the day of the Lord is at hand, it shall come as a destruction from the Almighty. v7 Therefore shall all hands be faint, and every man's heart shall melt: v8 And they shall be afraid: pangs and sorrow shall take hold of them; they shall be in pain as a woman that travaileth: they shall be amazed one at another; their faces shall be as (margin note: faces of the) flames. v9 Behold, the day of

the Lord cometh, cruel both with wrath and fierce anger, to lay the land desolate: and he shall destroy the sinners thereof out of it. v10 For the stars of heaven and the constellations shall not give their light: the sun shall be darkened in his going forth, and the moon shall not cause her light to shine. v11 And I shall punish the world for their evil... v13 Therefore I will shake the heavens, and the earth shall remove out of her place, in the wrath of the Lord of hosts, and in the day of his fierce anger...v19 And Babylon, the glory of kingdoms, the beauty of the Chaldees' excellency, shall be as when God overthrew Sodom and Gomorrah."

A reversal of the Earth's poles is implied by further passages in Isaiah. Sir Ian Niall Rankin discusses the evidence that this has already happened in the past in his book *Doomsday Just Ahead*. He contends that the repeated destruction and flash freezing of Woolley Mammoths with fresh flowers in their stomachs can only be explained by the abrupt tilting of the planet. A further factor cited by him is the presence of iron particles in what was once molten lava pointing to a previous magnetic pole:

Isaiah 24: 1 "Behold the Lord maketh the earth empty, and maketh it waste, and turneth it upside down, and scattereth abroad the inhabitants thereof."

Estimated dates for the step changes in Earth's environment that apparently decimated the Woolley Mammoth could coincide with Joshua's long day, when the Sun stood still in the sky (Joshua 10:13), as well as the recession of the Sun's shadow in Hezekiah's day (2 Kings 20). Such effects could be seen if the sudden tilting of the Earth caused the region of the Middle East to point sunwards for several hours. Such a re-alignment of the Earth's pole of rotation will almost certainly be a feature of the Day of the Lord:

Isaiah 25:19 "The earth shall reel to and fro like a drunkard, and shall be removed like a cottage...v21...in that day the Lord shall punish the host of the high ones...v23 Then the moon shall be confounded and the sun shall not give her light."

A change to the Sun/Earth/Moon system is predicted. Immense volcanic activity would be an inevitable consequence of changes to

453

the earth's crust caused by anything other than a clean 180° degree switch. This would be because the equatorial bulge would reappear in a new position, warping the present shape of the Earth.

It appears from these prophecies that the Earth will move closer to the Sun before settling into a new orbit. If so, this helps explain the predicted increase in the intensity of its heat earthwards. Many ancient records reflect a 360 day year, a point also discussed in *Doomsday just ahead*. This will most likely be restored, as and when the Earth takes a terrifying trip in the direction of the Sun.

One should not be taken in by glib claims that mere human activity can alter weather patterns on Earth. Rotting vegetation produces many times the CO_2 output of humans and the methane (20 times as warming an effect as CO_2) produced by three billion cows exceeds any improvement man can make in reducing global warming. A cow is much more damaging in terms of CO_2 than a four litre Land Rover, according to a survey published by the *Daily Mail* in July 2007.

The world is getting hotter, in part, due to the warming Sun. But belief in lies and fables, including a false claim for scientific unanimity on this and many other subjects, is a common human weakness. As the saying goes, if you want to tell a lie, tell a big one. Massive bias is in evidence among those charged with deliberating over climate change.

According Václav Klaus, President of the Czech Republic, the IPCC climate panel is made up of a group of politicized scientists with one-sided opinions and one-sided assignments:

"It is a political authority without any scientific basis."

Klaus claims that the panel did not include

"neutral scientists, a balanced group of scientists."

Klaus even went on to attack ambitious environmentalism as:

"...the biggest threat to freedom, democracy, the market economy and prosperity."

These are strong words from an acclaimed university lecturer who holds nearly 50 honorary degrees from international universities.

Benny Peiser, a social anthropologist at Liverpool's John Moores University muses:

> *"Global warming on Neptune's moon Triton as well as Jupiter and Pluto, and now Mars has some (scientists) scratching their heads over what could possibly be in common with the warming of all these planets ... Could there be something in common with all the planets in our solar system that might cause them all to warm at the same time?"*

Habibullo Abdussamatov, head of space research at St. Petersburg's Pulkovo Astronomical Observatory in Russia blamed solar fluctuations for Earth's current global warming trend:

> *"Man-made greenhouse warming has (made a) small contribution (to) the warming on Earth in recent years, but (it) cannot compete with the increase in solar irradiance."*

The CATO institute published an advertisement in 2009 quoting a statement by President Obama and a response from one hundred and fifteen named scientists, of whom ninety-three held PhD's:

> *"Few challenges facing America and the world are more urgent than combating climate change. The science is beyond dispute and the facts are clear."*

But the response from the scientists was:

> *"We, the undersigned scientists, maintain that the case for alarm regarding climate change is grossly overstated. Surface temperature changes over the past century have been episodic and modest and there has been no net global warming for over a decade now. After controlling for population growth and property values there has been no increase in damages from severe weather-related events. The computer models forecasting rapid temperature change abjectly fail to explain recent climate behaviour. Mr. President, your characterization of the scientific facts regarding climate change and the degree of certainty informing the scientific debate is simply incorrect."*

APPENDIX XV – concerning the Great White Throne Judgment.

(Relating to the chapter *The 1711171 System*)

Almost at the end of the Bible there is a description of the general resurrection beginning in the 12th verse of Revelation 20:

> *"And I saw the dead, small and great, stand before God; and the books were opened: and another book was opened which is the book of life: and the dead were judged out of those things which were written in the books, according to their works. And the sea gave up the dead which were in it; and death and hell [Gk. hades, the grave] delivered up the dead which were in them: and they were judged every man according to their works."*

This refers to the great general resurrection first described in Ezekiel chapter 37, the valley of dry bones of the famous negro-spiritual: *'Dem bones, dem bones, them dry bones, now hear the word of the Lord.'* But the last thing the masses expect to encounter after death is to have to live their physical lives, all over again, in the same fleshly bodies in a physical resurrection on this earth:

> *Ezekiel 37:1 "...the valley which was of dry bones...v5 Thus saith the Lord God unto these bones; behold I will cause breath to enter into you...v6 And I will lay sinews upon you, and will bring up flesh upon you...v10...and they lived, and stood upon their feet, an exceeding great army. v11 Then he said unto me, Son of man, these bones are the whole house of Israel..."*

But there is a better, or superior, resurrection also described:

> *Hebrews 11:25 "Choosing rather to suffer affliction with the people of God, than to enjoy the pleasures of sin for a season. Esteeming the reproach of Christ greater riches than the treasures in Egypt...v35...others were tortured, not accepting deliverance; that they might obtain a better resurrection."*

This better resurrection is described in I Corinthians:

> *"As we have borne the image of the earthly, we shall also bear the image of the heavenly. v50...flesh and blood cannot inherit the Kingdom of God...v52...for the trumpet shall sound and the dead shall be raised incorruptible..."I Corinthians 15:49*

To be raised incorruptible is to be in possession of a permanent body and mind composed entirely of spirit. The reference to the blowing of a trumpet corresponds with that in Revelation:

> "And the seventh angel sounded; and there were great voices in heaven saying, The kingdoms of this world are become the kingdoms of our Lord and of his Christ..."Rev 11:15

These kingdoms will be controlled by those who appear in the better resurrection:

> "And I saw thrones, and they that sat upon them, and judgment was given unto them: and I saw the souls of them that were beheaded for the witness of Jesus, and for the word of God, and which had not worshipped the beast, neither his image, neither had received his mark upon their foreheads, or in their hands; and they lived and reigned with Christ a thousand years." Rev 20:4

What the next verse says in regard to the dead is highly significant:

> v5 "But the rest of the dead lived not again until the thousand years were finished. This is the first resurrection."

From this it is clear that there is an order for everyone: a first resurrection for those called and trained to rule under Christ on earth for a thousand years, and a second general resurrection for humanity as a whole who will be brought back in the flesh, initially. That is the step referred to in Ezekiel chapter thirty-seven. Revelation also describes this event:

> Revelation 20:11 "And I saw a great white throne, and him that sat on it, from whose face the earth and the heaven fled away...v12 And I saw the dead, small and great, stand before God; and the books were opened; and another book was opened, which is the book of life..."

Those that qualify, after an appropriate period of living again as humans on earth, will be granted eternal life; but the incorrigibly wicked are destroyed:

> Malachi 4:1 "Behold the day cometh, that shall burn as an oven; and all the proud, yea, and all that do wickedly, shall be stubble: and the day that cometh shall burn then up...v3 And ye shall

tread down the wicked; for they shall be ashes under the soles of your feet..."

There are three stages in this plan. People are brought back to life in one of three resurrections:

1. 'First fruits' - the bride of Christ.
2. The general resurrection or White Throne Judgment. A fleshly resurrection to live again on earth and qualify for eternal spirit life, or
3. The second death, which is final. A resurrection to hear judgment pronounced and be destroyed for ever. This will include enemies of God from the present age and those who have rejected God under the administration referred to in item 2 above (see Revelation 20).

The reader will search in vain for a main-stream religion that teaches this truth of the three resurrection phases, as described in the Bible.

APPENDIX XVI – the 888 name pattern

(Relating to the chapter *The key pattern 888*)

Observations about the number 8 are relevant to understanding who Jesus Christ was and what he did. The following excerpts from *Number in Scripture* are useful:

Page 196

In Hebrew the number eight is...Shah'meyn "to make fat," "cover with fat," "to super-abound." As a participle it means "one who abounds in strength" etc. As a noun it is "superabundant fertility," "oil," etc. So as a numeral it is the superabundant number.

Page 198

The Feast of Tabernacles was the only Feast which was kept eight days. The eighth is distinguished from the seventh...

The Lord Jesus was on a mountain eight times (omitting of course the scene in the temptation). Seven times were before the cross, but the eighth time after he rose from the dead.

Page 200

Eight by itself: it is seven plus one. Hence it is the number specially associated with Resurrection and Regeneration, and the beginning of a new era or order. When the whole earth was covered with the flood it was Noah "the eighth person" (2Peter 2:5) who stepped out onto a new earth to commence a new order of things. "Eight souls" (2Peter3:20) passed through it...to the new or regenerated world.

Hence, too, circumcision was to be performed on the eight day (Genesis17:12), because it was the foreshadowing of the true circumcision of the heart, that which was to be "made without hands" even "the putting off of the body of the sins of the flesh by the circumcision of Christ." (Colossians 2:2)

Page 201

Eight is the first cubic number, the cube of two, 2 * 2 * 2. We have seen that three is the symbol of the first plane figure, and *that four is the first square. So here, in the first cube, we see something of transcendent perfection indicated, something, the length and breadth and height of which are equal. This significance of the cube is seen in the fact that the "Holy of holies", both in the Tabernacle and in the Temple, were cubes. In the Tabernacle it was a cube of ten cubits. In the Temple it was a cube of 20 cubits. In Revelation 20 the New Jerusalem is also to be a cube of 12,000 furlongs. Dr. Milo Mahan [The author of Palmoni, mentioned in Bullinger's preface] is inclined to believe that the Ark of Noah, too, had a kind of sacred Shechinah in "the window finished in a cube above" – a cube of one. If so, we have the series of cubes:-*

$$1 = \text{the Ark}$$
$$10 \text{ cubed} = 1{,}000 \text{ the Tabernacle}$$
$$20 \text{ cubed} = 8{,}000 \text{ the Temple}$$
$$12{,}000 \text{ cubed} = 1{,}728{,}000{,}000{,}000 \text{ the New Jerusalem}$$

The miracles of Elijah were eight in number, marking the divine character of his mission...

The miracles of Elisha were double in number, viz. sixteen for his request was, "Let a double portion of thy spirit be upon me." 2 Kings 2:9

Page 203

EIGHT IS THE DOMINICAL NUMBER, for everywhere it has to do with the Lord. It is the number of his name Jesus, Iησους, in Greek:-

I	=	*10*
η	=	*8*
σ	=	*200*
o	=	*70*
υ	=	*400*
ς	=	*200*

		888

This excerpt from *Number in Scripture* culminates in the above well known numerical label, derived from gematria values, of 888 for Jesus the man. That, it bears repeating, is the pattern of his name as a *human being*, whereas the term Christ corresponds to the number 7. (The number 8 on its own corresponds to Noah, as discussed in chapters 32, 39, 40, 47 and 52).

This aspect of his person, the number 7, is his identity as the Messiah. It harks back to his pre-existence as the Deity seen by Moses in Exodus chapter thirty-three and should therefore be preeminent. Moreover, the Greek for Christ has *seven* letters:

χριστος

For this and other reasons given in the chapter *The key pattern 888,* Bullinger's assertion of 8 as the 'dominical' number of Jesus Christ - the Lord of the (seventh day) Sabbath - is erroneous. His principal number as evidenced by the Bible and the Grid discovery is 7.

Anglo-Saxons 10, 19, 228, 242, 247, 249, 259, 281, 284, 288, 293, 295-7, 314, 319, 388

Abu Dhabi 266

Adoption 208

Ahmadi-Nejad, M. 9, 10, 12-3, 195, 197, 215-8

Albedo 27, 30

Aronofsky, Darren 424

Asimov, Isaac 25, 135

Assyria 232-6, 242, 245, 250-1, 253-5, 259, 261-4, 267, 278, 390

Atomic bomb 74, 177, 186, 200, 202, 214

Atomic weight 130-1, 136, 138

Babylon 180, 182-3, 188, 194, 196-8, 204, 236, 300

Barroso, José M. 11

Beast power 187, 191, 196, 327

Beast, number of 169, 195

Berlin Wall 230

Bernanke, Ben 281, 289

Birthright 246, 254, 261-2, 313

Blaine, David 69-74, 142, 209-10, 213-4, 271-3, 299, 303, 318, 381

Blair, Tony 11, 199, 239, 264, 267

British Empire 240-1, 243-4, 250, 254

Brown, Dan 74, 299-300

Brown, Gordon 8, 10, 267, 276, 279, 290-1

Bullinger, E.W. 21-4, 28, 34-9, 55, 64-5, 82, 91, 96, 111, 123, 126-7, 130, 140, 148-50, 153-4, 161, 204, 208, 376

Burning Bush 109-10

Bush, George 10, 176, 219, 251, 289-90

Canopy theory 89

Carter, Jimmy 10, 175, 182, 197, 218

Charlemagne 15, 179, 198

Cherubim 37, 427, 438

Children of Israel 43, 254, 325, 438, 445

China 11, 20, 76, 177, 189, 191, 281, 95-7, 320

Churchill, Winston 15, 175, 187, 190, 202-3, 207, 185-7, 232, 251, 265-6, 215-6, 219, 222, 261, 328, 387

Circles 59-61, 71, 75, 148, 339-40, 428

Clarke, Arthur C. 73, 77

Commandments 81, 88, 90, 100-6, 105, 126, 142-3, 153, 200, 225-6, 323, 326, 330, 345, 347-8, 351-2, 359-61, 369-70, 381, 389, 394, 396, 406-15, 427, 437, 445-6

Conjugal act 141, 380

Constantine 16, 228, 256, 307, 332, 360

Corona 2, 56-7, 65, 101, 131, 142-4, 148, 171, 243, 339-40

Council, Nicaea 16

Council, Laodicea 322, 330

Cubits 43, 155, 158-60, 168, 352-3, 460

Darwinism 115, 134, 141

Dawkins, Richard 133-4

Depression, Great 8, 18, 259, 283-4, 289, 293-4, 297

DNA 37, 56, 137-8, 141, 309-10, 322, 380-1, 417

Donkey 31, 34, 301

Drosnin, Michael 393-4, 386

Duality 81, 108, 153, 380, 414, 417

Eddy, Jack 30

Einstein 7, 37, 45, 55, 163, 355, 373

Elements 119, 129-31, 135-8, 172-3, 191, 199, 367, 379, 381-2, 387, 422

Elijah 39, 272, 335, 337

End-time 180, 193, 197, 201, 233, 237, 255, 257, 263

Ephraim 239-40, 242, 244, 246-7, 249, 253, 255-6, 258-9, 262-3, 267, 278-9

Fast, fasting 36, 69-71, 73-4, 121, 209-10, 213, 267, 271, 273-4, 299-300, 318, 357, 381-2

Feast, Tabernacles 46, 205, 325, 344, 356-7, 359, 362, 374,-5, 386, 459

Fibonacci numbers 206, 424-5, 451

Figure of eight 9, 49, 55, 77, 119, 200, 267, 269-70, 271-72, 302, 342, 382, 388, 391

Fractals 138, 424, 429, 431

Gematria 38, 56, 64, 100, 102, 121, 139-40, 149, 151, 200,

	271, 300, 374, 380, 382, 436, 461	John Paul II	8, 212, 263
Germany	11, 13, 18-20, 27, 135, 184-8, 190-1, 195-9, 202, 223, 229-32, 233-7, 239, 241-2, 244, 250, 253-6, 261- 6, 280, 287, 291, 328, 344	Josephus	234, 248
		Judah	233, 243, 245-6, 253, 259, 278-9, 390, 436
		Keynes, John M.	292, 384
		King of the North	177, 181, 192, 199
		King of the South	177, 180-2, 192
		King, Mervyn	11, 294
Golden Ratio	131, 138, 424-9	Klaus, Václav	454
Gore, Al	216, 451	Levitation	16, 72, 221
Gibbons, Cardinal	202, 329, 347	Lewontin, Richard	133
Gingrich, Newt	10	Livingstone, Ken	210-11
Gorbachev, M.	231, 260	Lunar Eclipse	36, 379
Greek language	38, 56, 64-5, 100, 109, 114, 121, 139, 142, 153, 195, 207-8, 232, 248, 338, 346-7, 374-6, 380, 447, 460-1	Lunation	30
		Magus, Simon	16, 221, 377
		Mahan, M.	21, 30, 39, 97, 152, 271, 460
		Major, John	242, 261
Greenspan, Alan	282-3, 288, 291	Manasseh	240, 244, 246-7, 249, 255, 263, 267, 278-9
Hebrew calendar	8, 45-6, 200-1, 355, 375, 382		
		Mary, worship of	16-7, 223, 225, 228, 245, 363
Hebrew language	22, 100-1, 207, 241		
Henry VIII	228	Maunder Min.	27, 30
Hezekiah	233, 335, 445, 453	Medvedev, Dmitry	17
Hitler, Adolf	9, 13, 19, 31, 95, 175-6, 188, 192, 195, 197-8, 215, 218-9, 223, 231-2, 236, 263-4, 287, 292, 300, 388	Melchizedek	109
		Merkel, Angela	184-5, 225, 293, 315, 352, 419
		Messenger, Angel	109-10, 203
		Millennium	199, 229, 344
Hittite	234-6	Miracles, wonders	70, 192-4, 201, 210, 326-7, 385, 388, 396, 460
Holocaust, Jewish	7, 9, 12, 16, 215, 261, 347		
Holy days	45-6, 48, 104, 120, 156, 174, 205-6, 256, 318, 348-9, 348-9, 355, 361, 363, 378, 382	Monotheism	152, 388, 438, 441
		Moses	28-9, 39, 43, 58, 76, 87, 108-10, 168, 194, 228, 244, 272, 315, 363, 395, 398, 404, 407, 409, 437-40, 445-6, 438, 445, 461
Hussein, Saddam	176, 216-8		
IMF	11, 17, 186, 283, 291		
Independent events	84		
Indulgence	228, 305, 308, 310, 363		
		Napoleon	179, 198, 240
Ingham, Bernard	184	Newton, Isaac	25, 63, 384
Iran	9-13, 45, 176-7, 181-2, 192-9, 204, 215-9, 264, 279, 300, 400-1	NASA	30, 77, 214, 217, 451
		Nazism	186-7, 195-6, 222, 229, 237, 239, 254, 262-3
		Noah	200, 267, 271-2, 301-2, 341, 363, 374, 382, 388, 391
Iraq	9-11, 181-2, 188, 216-9, 258, 264, 275, 277-9, 381, 390, 400-1		
		Northern Rock	7-8, 10, 17, 199, 275, 279, 280-4, 286, 294, 318-9, 328
Israel, State of	9, 12-3, 279		
Jerusalem	45, 82, 157, 174, 180, 193, 205, 233, 242, 245-6, 250, 253, 257-8, 279, 301, 316,		
		Nuclear deterrent	202
		Obama, Barack	8, 18, 189, 216, 276, 292, 455
Jews	45, 142, 188, 215, 224, 233, 241, 246, 253-4, 261, 274, 279, 295, 329-3, 335, 355-60, 390, 394, 400, 424	Ovum	56, 65, 98, 100, 121, 139-141, 147-8, 150, 380, 424
		Palmoni	21, 30, 39, 97, 152, 271, 460

Pascal's triangle	134-5, 138, 379	Solar flares/wind	77, 144, 212-4, 299, 303-4, 381, 451-2, 455
Paulson, Henry	289-91		
Perfect number	158	Spermatozoon	56, 65, 98, 100, 121, 139-142, 147-8, 150, 168, 380, 417, 424, 436
Periodic table	119, 130, 135, 138, 151, 284, 367, 379, 382, 387		
Pharisees	107, 319, 347, 350, 359-60, 394-9, 407	Spiral, logarithmic	431-3
		Stalin	13, 185, 265
Pianos	123-6, 164, 305, 420-1	Stone of Scone	242
Pitch, musical	124-5	Sunday observance	202, 310, 322, 324, 327-9, 346-7
Pius XII	223		
Planets	13, 52-3, 75, 94, 119, 174, 297, 319, 453, 455	Sunspots	74
		Tabernacle	43, 58, 155,-6, 158-60, 168, 193, 369-70, 460
Plichta, Peter	35, 135-8, 422	Thatcher, Margaret	184, 230-1, 260-1
Polar shift	453	Time cycle, 19 yr.	45, 49, 88, 90, 120, 256, 401
Predestination	349		
Prime numbers	33, 35	Tower Bridge	69, 70, 73, 209
Pyramid at Gaza	76	Tradition	28, 37, 47, 242, 248-9, 258, 272, 309, 312, 324, 326, 328, 347-8, 373-4, 394, 396, 448
Queen Elizabeth II	242		
Queen Victoria	244		
Radio dating	89		
Rankin, Ian Niall	453	Trinitarian, Trinity	22-3, 28, 47, 49, 109-10, 147, 226, 152-3, 174, 195, 228, 232, 310-1, 361, 388
Refurbishment	202, 437		
Rock	36, 109, 151, 199, 319, 349, 429, 437-8		
Roman Empire	139, 177-9, 181-3, 187, 191-3,198-9, 201, 204, 206, 221-4, 232-3, 235-7, 239, 256, 261, 273, 303, 310, 321, 328-30, 332-4, 336-7, 346-7, 351, 355, 358-61, 375, 397, 427, 442, 461	U.A.E	266
		U.S.S.R.	190
		Validation	41, 42, 57-8, 75, 80, 171-2, 188, 338, 380, 417, 428
		van Ceulen, L.	6, 76, 377
		Volker, Paul	18, 297
Rumsfeld, Donald	281, 290	Waldenses	224
Sabbath	202, 254, 324, 327, 442	Weimar Republic	293
		Wells, H. G.	222
Samaria	241-2, 245, 253-4, 258, 278, 288, 447, 461,		
Satan	95-6, 193-4, 145, 157, 174, 187, 195, 199-200, 205, 263, 279, 301, 316, 326-7, 330, 343, 351, 357, 385, 389, 391		
Schofield Bible	96		
Scientists	11, 13, 26, 77, 115, 132-3, 145, 214, 384, 451, 454,-5		
Self characteristic	42, 52, 57-8, 75, 79, 82, 84, 118, 380		
Sermon on the Mt.	63, 103-6, 370		
S.E.T.I.	13, 14, 75, 174, 383		
Sex, Sexuality	52, 65, 100, 102, 141, 207, 257, 380, 417, 436		
Shaw, George B.	15, 41, 103		

Lightning Source UK Ltd.
Milton Keynes UK
06 May 2010

153799UK00001B/60/P